T0074162

Lecture Notes in Physics

The Lecture Notes in Physics

The series Lecture Notes in Physics (LNP), founded in 1969, reports new developments in physics research and teaching – quickly and informally, but with a high quality and the explicit aim to summarize and communicate current knowledge in an accessible way. Books published in this series are conceived as bridging material between advanced graduate textbooks and the forefront of research and to serve three purposes:

- to be a compact and modern up-to-date source of reference on a well-defined topic

- to serve as an accessible introduction to the field to postgraduate students and nonspecialist researchers from related areas

- to be a source of advanced teaching material for specialized seminars, courses and schools

Both monographs and multi-author volumes will be considered for publication. Edited volumes should, however, consist of a very limited number of contributions only. Proceedings will not be considered for LNP.

Volumes published in LNP are disseminated both in print and in electronic formats, the electronic archive being available at springerlink.com. The series content is indexed, abstracted and referenced by many abstracting and information services, bibliographic networks, subscription agencies, library networks, and consortia.

Proposals should be sent to a member of the Editorial Board, or directly to the managing editor at Springer:

Christian Caron
Springer Heidelberg
Physics Editorial Department I
Tiergartenstrasse 17
69121 Heidelberg / Germany
christian.caron@springer.com

I. Mann
A.M. Nakamura
T. Mukai (Eds.)

Small Bodies in Planetary Systems

 Springer

Ingrid Mann
Kobe University
Dept. Earth & Planetary Sciences
Rokkodai-cho Kobe
Nada-ku 657-8501
Japan
mann@diamond.kobe-u.ac.jp

Akiko M. Nakamura
Kobe University
Dept. Earth & Planetary Sciences
Rokkodai-cho Kobe
Nada-ku 657-8501
Japan

Tadashi Mukai
Kobe University
Dept. Earth & Planetary Sciences
Rokkodai-cho Kobe
Nada-ku 657-8501
Japan

Mann, I. et al. (Eds.), *Small Bodies in Planetary Systems*, Lect. Notes Phys. 758 (Springer, Berlin Heidelberg 2009), DOI 10.1007/978-3-540-76935-4

ISBN: 978-3-540-76934-7   e-ISBN: 978-3-540-76935-4

DOI 10.1007/978-3-540-76935-4

Lecture Notes in Physics ISSN: 0075-8450

Library of Congress Control Number: 2008929548

Cover design: eStudio Calamar S.L., F. Steinen-Broo, Pau/Girona, Spain

Printed on acid-free paper

9 8 7 6 5 4 3 2 1

springer.com

Preface

The small bodies in planetary systems are indicative of the material evolution, the dynamical evolution, and the presence of planets in a system. Recent astronomical research, space research, laboratory research, and numerical simulations brought a wealth of new and exciting findings on extra-solar planetary systems and on asteroids, comets, meteoroids, dust, and trans-Neptunian objects in the solar system. Progress in astronomical instrumentation led to the discovery and investigation of small bodies in the outer solar system and to observations of cosmic dust in debris disks of extra-solar planetary systems. Space research allowed for close studies of some of the small solar system bodies from spacecraft. This lecture series is intended as an introduction to the latest research results and to the key issues of future research. The chapters are mainly based on lectures given during a recent research school and on research activities within the 21st Century COE Program "Origin and Evolution of Planetary Systems" at Kobe University, Japan.

In Chap. 1, Taku Takeuchi discusses the evolution of gas and dust from protoplanetary disks to planetary disks. Using a simple model, he studies viscous evolution and photoevaporation as possible mechanisms of gas dispersal. He further considers how the dust grows into planetesimals. Motion of dust particles induced by gas drag is described, and then using a simple analytic model, the dust growth timescale is discussed.

Chap. 2 by Mark Wyatt covers the interpretation of observations of small bodies in extrasolar planetary systems. While observations of debris disks trace the distribution of dust in these systems, they can be used to infer the distribution of larger bodies: planetesimals and planets. The chapter describes a theory for the dynamics of dust–planetesimal–planet interactions. Such a theory is essential for a successful interpretation of the observations, and is equally applicable to the study of dust originating in the asteroid and Kuiper belts in the solar system.

The collisional disruption of small bodies is a fundamental process in the formation and evolution of planetary systems. In Chap. 3, Akiko Nakamura and Patrick Michel describe the current knowledge on collision processes of asteroids and relevant laboratory studies. Related recent findings from the Hayabusa mission to the small asteroid 25143 Itokawa are shortly described.

Subsequently, Patrick Michel presents in Chap. 4 the different disruption mechanisms that can affect the physical properties of small bodies in planetary systems over their history. Our current but still very poor understanding of the concept of material strength of those bodies and its role in the action of those mechanisms are also described.

In Chap. 5 Shinsuke Abe discusses meteor observations in connection to properties of the meteoroid parent bodies. Meteors are phenomena that result from interaction of meteoroids entering from interplanetary space with the Earth's upper atmosphere. The derived meteoroid characteristics bear potential information to investigate their parent bodies, which are in majority comets and asteroids.

Except for the meteor observations, the few cases of in-situ detections in space, and the limited studies of cosmic samples in terrestrial laboratories, our information concerning small bodies in planetary systems are based on astronomical observations, especially on the observations of cosmic dust. In Chap. 6, Aigen Li presents the theoretical basis for describing the optical properties of dust and the dust interaction with electromagnetic radiation.

The physical processes of the dust in planetary debris disks are discussed in Chap. 7 by Ingrid Mann. Planetary debris disks are exposed to the brightness of the central star, stellar wind, and energetic particles originating from the system as well as galactic cosmic ray particles. Both stellar radiation and stellar wind give rise to a Poynting–Robertson effect, which limits the lifetime of the dust particles that are in bound orbit about the star (migration-dominated disks). In debris disks with high dust content lifetimes due to mutual collisions are even shorter (collision-dominated disks). Dust collisions are a potential source of second-generation gas in planetary debris disks. The chapter also touches on the role of non-thermal alteration for dust material evolution.

In Chap. 8, Masateru Ishiguro and Munetaka Ueno describe recent developments in observations of interplanetary dust particles. These developments are largely due to the introduction of cooled charge coupled device (CCD) detectors and two-dimensional infrared array detectors used with infrared space telescopes. The new observational data show not only the global structure of the interplanetary dust cloud, e.g., its plane of symmetry, but also faint structures, asteroidal dust bands, and cometary dust trails, seen as brightness enhancements of a few percent above that of the smooth component. Spectrographic observations provide some knowledge about the dynamics and composition of these local components. The observations reveal the connections between interplanetary dust particles and their parent bodies. The chapter ends by describing ongoing and future projects related to the observational study of interplanetary dust.

The past 15 years have seen a renaissance in the study of the outer solar system with the discovery of the Kuiper belt and the unveiling of previously unexpected connections between distinct sub-populations of small bodies throughout the system. In chapter Dave Jewitt focusses on six of the most research-active areas. The chapter includes the analysis of lightcurves for information about structure and binarity, systematics of the densities of small bodies, the color (composition) distribution of Kuiper belt objects, the crystalline state of ice and the nature of two little studied groups, the irregular satellites of the giant planets, and the newly perceived main-belt comets.

In the final and 10th chapter, Yoichi Itoh describes the observational perspectives. In this chapter, various observational methods of searching for extrasolar planets and circumstellar disks are reviewed. These include Doppler shift measurements, transit detection, astrometry, gravitational lensing, spectral energy distribution, direct detection, and coronagraphy.

The synthesis of the presented and many other exciting studies will hopefully converge towards a better understanding of the contents, origin and evolution of our planetary system and of planetary systems in general. We finally would like to thank all those people who helped to write this book: the authors of the different chapters, as well as all lecturers and participants of the Kobe Planetary School 'Small Bodies in Planetary Systems'.

Kobe
December 2007

Ingrid Mann
Akiko M. Nakamura
Tadashi Mukai

Contents

1

From Protoplanetary Disks to Planetary Disks: Gas Dispersal and Dust Growth

T. Takeuchi

Department of Earth and Planetary Sciences, Kobe University, Kobe 657-8501, Japan,
taku@kobe-u.ac.jp

Abstract We discuss the evolution of gas and dust in protoplanetary disks, first considering removal of the gas from the disks. Using a simple model, we argue that it is difficult to remove all the disk gas solely by viscous accretion. We discuss photoevaporation as a plausible mechanism for removing the disk gas, and estimate the timescale of gas removal. We then discuss growth of the dust particles into planetesimals. The dust particles first sediment to the midplane of the disk and then radially migrate toward the central star. We estimate the growth timescale during sedimentation and discuss the growth of dust bodies as they move radially toward the star.

1.1 Introduction

Many young pre-main-sequence stars such as T Tauri stars have circumstellar disks, which are called protoplanetary disks because planetary formation is expected to occur within them. The youngest protoplanetary disks provide information about the initial conditions of planet formation. The mass of the dusty material of the disks around classical T Tauri stars are in the range of $10^{-5} - 10^{-3}$ $M_\odot$ ($3 - 300 M_\oplus$, [11]), or in other words, the disks have a sufficient amount of dust to form rocky planets. Because the disks are optically thick in the optical to millimeter wavelengths, a sufficient mass of small particles is required to produce a large cross-sectional area with the given mass of dusty material observed. It is still difficult to measure the amount of hydrogen molecules, which comprises the main component of the gas in protoplanetary disks. Other molecules such as CO may be depleted, and measuring the amount of such species does not allow us to derive the total gas mass. Thus, we currently have only very crude estimates for the gas masses in the disks, but it is thought that some disks have enough gas to form giant planets [23, 30].

As planet formation proceeds, the protoplanetary disks evolve into planetary disks. The small dust particles grow into much larger bodies, such as planetesimals, comets, and planets. Some of the disk gas forms giant planets,

Takeuchi, T.: *From Protoplanetary Disks to Planetary Disks: Gas Dispersal and Dust Growth*. Lect. Notes Phys. **758**, 1–35 (2009)
DOI 10.1007/978-3-540-76935-4_1

but a large fraction of the gas mass disappears, either by accreting onto the central star or escaping from the disk. For example, loss of the gas component from the solar system is obvious. The total gas mass of all the planets in the solar system is much smaller than the gas mass of the protosolar disk estimated from the total solid mass of the planets multiplied by the interstellar gas-to-dust ratio. Thus, during or after the formation of the planets, most of the gas was removed. How the transition from protoplanetary to planetary disk occurs is not yet clear, because only a few objects have been found that are believed to represent a transitional stage in this process [24]. In contrast to the small number of observed objects in a transitional stage, many candidates for planetary disks have been found.

Some main-sequence stars, such as Vega and β Pictoris, have dust disks and are referred to as Vega-like stars [6, 51]. The disks around Vega-like stars are optically thin even at optical wavelengths, which means that the mass of the small particles that emit observable thermal radiation and the scattered light of the central star is much less than in protoplanetary disks. The small dust particles of protoplanetary disks may have simply disappeared during the transition to planetary disks, but it is thought that most of the solid material actually accumulated into planetesimal-sized ($\sim$ km) or larger bodies. These large bodies are needed to produce the small dust particles observed in Vega-like disks [6]. Small particles in Vega-like disks are easily removed through collisional destruction and subsequent ejection by the radiation pressure of the central star. To reconcile the observed dust mass of Vega-like disks (or more accurately the total cross-sectional area of the dust particles) with the ages of their central stars, the small particles must have been replenished. This resupply probably occurs through collisional destruction of larger bodies such as planetesimals or through evaporation of comets approaching the star. Such resupply of the dust particles actually occurs in the solar system and produces the dust of the zodiacal light. Hence, in Vega-like disks, most of the solid bodies have grown to at least the size of planetesimals or comets ($\sim$km), and most of these disks have very little or no gas [13, 16, 18, 53, 54]. The ratio of the gas-to-solid mass is thus expected to be much smaller than the interstellar value; i.e., a significant fraction of the gas has been removed. In view of the above facts, Vega-like disks are considered to be planetary disks. They contain planetesimals (or asteroids) so that small dust particles are continuously produced, and they consist of very little gas.

In this chapter, we discuss theoretically how protoplanetary disks evolve into planetary disks. We focus here on growth of the dust particles into planetesimal-sized bodies and on removal of the gas. In Sect. 2, we describe a simple protoplanetary disk model that we use as a reference model in the subsequent sections. In Sect. 3, we consider viscous evolution and photoevaporation as mechanisms for removal of the gas from protoplanetary disks. In Sect. 4, we describe the dynamics of dust particles, which are strongly controlled by gas drag. In Sect. 5, we discuss how the dust particles in gas disks grow.

1.2 A Simple Model of Protoplanetary Disks

Many pre-main-sequence stars exhibit protoplanetary disks composed of gas and dust. The main components of the gas are hydrogen molecules and helium atoms with a small fraction of gas-phase molecules of heavy elements, such as CO and N_2. The dust is mainly composed of silicates, organics, and H_2O ices. The dust-to-gas ratio in protoplanetary disks is estimated from the compositional measurements of the interstellar medium, primitive bodies in the solar system, and the solar atmosphere. In the outer part of the disk, the dust-to-gas ratio is 1.4%, and inside the snow line, where H_2O ices evaporate, it is reduced to 0.8% [66]. In this section, we describe a simple model of a gas disk, which is used as a reference model in the subsequent sections.

1.2.1 Equilibrium State of the Gas Disk

We first consider the vertical equilibrium of the gas disk. The vertical component of the gravity of the central star and the gas pressure gradient balance each other. Assuming axisymmetry and using the cylindrical coordinates (r, z), the force balance for $z \ll r$ is

$$-\frac{GM}{r^3}z - \frac{1}{\rho_g}\frac{\partial P}{\partial z} = 0 \ , \tag{1.1}$$

where G is the gravitational constant, M is the mass of the central star, ρ_g is the gas density, $P = \rho_g k_B T/m_g$ is the gas pressure, k_B is the Boltzmann constant, T is the gas temperature, and m_g is the mass of the gas molecules. We assume that the disk gas is isothermal in the vertical direction. The gas density distribution that satisfies (1.1) is

$$\rho_g = \rho_{g,mid} \exp\left(-\frac{z^2}{2h_g^2}\right) \ , \tag{1.2}$$

where $\rho_{g,\ mid}$ is the density at the midplane. The disk scale height is related to the Keplerian angular velocity, $\Omega_K = \sqrt{GM/r^3}$, and the isothermal sound speed, $c_s = \sqrt{k_B T/m_g}$, as

$$h_g = \frac{c_s}{\Omega_K} \ . \tag{1.3}$$

(This definition of h_g differs from the usual definition of the scale height by a factor $\sqrt{2}$.) The surface density is given by

$$\Sigma_g = \int_{-\infty}^{+\infty} \rho_g \ dz = \sqrt{2\pi} h_g \rho_{g,mid}. \tag{1.4}$$

In the radial direction, the balance between gravity, the centrifugal force, and the gas pressure gradient is

$$-\frac{GM}{r^2} - \frac{1}{\rho_g}\frac{\partial P}{\partial r} + r\Omega_g^2 = 0. \tag{1.5}$$

From this equation, the orbital angular speed Ω_g of the gas is given by

$$\Omega_g = \Omega_K(1-\eta)^{1/2} . \tag{1.6}$$

The deviation factor η from the Keplerian speed, which arises due to the gas pressure gradient, is given by

$$\eta = -\frac{1}{r\Omega_K^2\rho_g}\frac{\partial P}{\partial r} \sim \left(\frac{c_s}{v_K}\right)^2 , \tag{1.7}$$

and is on the order of the square of the ratio between the Keplerian velocity, $v_K = r\Omega_K$, and the sound speed, c_s.

1.2.2 Radial Profiles of the Temperature and the Density

The radial temperature distribution of the disk is determined by the energy balance between heating and cooling of the disk material. For the simplest models, as discussed below, the temperature distribution is approximated by a power-law form,

$$T = T_0\left(\frac{r}{r_0}\right)^{-q} , \tag{1.8}$$

where r_0 is a reference radius. We assume that the heating source is either due to disk accretion or stellar irradiation, and that cooling occurs through thermal radiation from the disk surface. For further simplicity, we assume that the disk is vertically isothermal, and we neglect energy transfer along the radial direction inside the disk. Consider first passive disks, in which the heating due to the irradiation from the central star is balanced with the radiative cooling. The luminosity of the star is $L = 4\pi R_*^2\sigma_{SB}T_{eff}^4$, where T_{eff} is the effective temperature, R_* is the radius of the star, and σ_{SB} is the Stefan–Boltzmann constant. If the thickness of the disk is negligible ($h_g \ll R_*$), the incoming energy flux from the star to a unit area of the disk is $L\sin\theta/(4\pi r^2)$, where θ, the mean incident angle of the starlight, is approximated by $0.4R_*/r$. The radiative cooling from a unit area of the disk is $2\sigma_{SB}T^4$ (from the upper and lower surfaces). The balance between heating and cooling gives the temperature profile

$$T = 0.2^{1/4}T_{eff}\left(\frac{r}{R_*}\right)^{-3/4} . \tag{1.9}$$

In the case of flaring passive disks, the disk has a thickness and the ratio of its thickness to the radius h_g/r increases with r. In such disks, the incident angle of the starlight decreases shallower than r^{-1}, and thus the temperature decreases less than $r^{-3/4}$. The disk thickness is related to the gas temperature through (1.3). A self-consistent calculation of T and h_g shows that the power-law index q can be as large as $3/7$ (see, e.g., Sect. 17.3.5 of [75]).

In the case of active disks, disk accretion provides the heating source. The accreting material at large distances from the star ($r \gg R_*$) releases its gravitational energy at a rate $3GM\dot{M}_{acc}/(4\pi r^3)$ per unit area per unit time [67], where $\dot{M}_{acc}$ is the mass accretion rate. This heating is balanced by the radiative cooling $2\sigma_{SB}T^4$. The temperature is

$$ T = T_{eff} \left(\frac{3L_{acc}}{2L} \right)^{1/4} \left(\frac{r}{R_*} \right)^{-3/4} , \qquad (1.10) $$

where $L_{acc} = GM\dot{M}_{acc}/R_*$ is the accretion luminosity. As seen from (1.9) and (1.10), if the accretion luminosity is much smaller than the luminosity of the star, the disk temperature is determined by the irradiation of the star. Taking the solar values for the stellar parameters, the irradiation of the star dominates when $\dot{M}_{acc} \ll 3 \times 10^{-8} M_\odot$ yr^{-1}. If the disk is flared ($q < 3/4$), the disk temperature at large radii is controlled by the irradiation of the star even if the accretion luminosity is comparable to the stellar luminosity. At later stages of disk evolution in which the accretion luminosity is not too large, the disk temperature is well described by the flared passive disk models (e.g., [17, 46]).

The initial density distribution of the disk is determined by the density and angular momentum distributions of the cloud core from which the gas has fallen onto the disk [14, 80], after which it evolves thorough viscous accretion. As discussed in Sect. 1.3.1, the simplest viscous accretion model induces a power-law density profile,

$$ \Sigma_g = \Sigma_{g,0} \left(\frac{r}{r_0} \right)^{-p} , \qquad (1.11) $$

where the power-law index p reflects how the viscosity varies with r. However, it is difficult to theoretically predict the viscosity profile. Our best knowledge about p comes from imaging observations of the disks. Comparison with the observed disk images taken at millimeter wavelengths suggests that the power-law index is $p = 0 - 2$ [5, 47]. The disk masses estimated from observations of the dust and gas are $10^{-3} - 10^{-1}$ $M_\odot$ [4, 12, 22, 30, 64]. The mean free path of the gas molecules at the disk midplane is

$$ l_{free} = \frac{m_g}{\sqrt{2}\rho_{g,mid}\sigma_{mol}} = l_{free,0} \left(\frac{r}{r_0} \right)^{\frac{2p-q+3}{2}} , \qquad (1.12) $$

where the collisional cross section of hydrogen molecules is $\sigma_{mol} = 2 \times 10^{-15}$ cm^2 (p. 228 in [15]).

Fiducial Disk Parameters

When we need numerical values in subsequent discussions, we use the following fiducial model: The central star's mass is $M = 1M_\odot$. The disk temperature and density profiles are $T = 280(r/1\ \mathrm{AU})^{-1/2}$ K and $\Sigma_\mathrm{g} = 3.5 \times 10^2 (r/1\ \mathrm{AU})^{-1}$ g cm^{-2}. In this model, the disk mass inside 100 AU is $M_\mathrm{g} = 2.5 \times 10^{-2} M_\odot$. The midplane gas density is $\rho_\mathrm{g,\ mid} = 2.8 \times 10^{-10}(r/1\ \mathrm{AU})^{-2.25}$ g cm^{-3}, and the mean free path is $l_\mathrm{free} = 4.9(r/1\ \mathrm{AU})^{2.25}$ cm, where the mean molecular mass is $m_\mathrm{g} = 2.34 m_\mathrm{H}$ and m_H is the mass of a hydrogen atom. The dust-to-gas ratio is $f_\mathrm{dust} = 0.01$.

1.3 Disappearance of the Gas Disks

In this section, we consider the possible mechanisms of gas removal from the protoplanetary disks. Viscous accretion of the gas disks is discussed in Sect. 1.3.1. Gas removal due to accretion decelerates as the disk evolves and thus the disk gas can remain for a long time. In Sect. 1.3.2, we consider photoevaporation of the disk gas caused by the extreme ultraviolet photons from the central star. For other mechanisms such as stripping by the stellar wind or tidal torques by embedded protoplanets, see references [39, 78].

1.3.1 Viscous Evolution

A Simple Model of Turbulent Viscosity

In protoplanetary disks, the molecular viscosity, $\nu_\mathrm{mol} = v_\mathrm{T} l_\mathrm{free}/2$, where $v_\mathrm{T} = \sqrt{8 k_\mathrm{B} T/(\pi m_\mathrm{g})}$ is the mean thermal speed, is small, and can be neglected when we consider the global evolution of the disks. The Reynolds number $\mathrm{Re}_\mathrm{mol} = r v_\mathrm{K}/\nu_\mathrm{mol}$ is actually on the order of 10^{14} at 1 AU. Thus, we usually consider turbulent viscosity as a mechanism of disk accretion. We assume that the largest eddy size of turbulence is smaller than the disk scale height, and also that the velocity of the largest eddy is smaller than the sound speed. The largest eddy size and velocity are written as $l_\mathrm{eddy} \sim \alpha_\mathrm{l} h_\mathrm{g}$ and $v_\mathrm{eddy} \sim \alpha_\mathrm{v} c_\mathrm{s}$, respectively. Using the analogy of molecular viscosity, the turbulent viscosity is estimated as

$$\nu_\mathrm{vis} = \alpha h_\mathrm{g} c_\mathrm{s}\ , \qquad (1.13)$$

where $\alpha = \alpha_\mathrm{l} \alpha_\mathrm{v}$. Several mechanisms have been proposed as sources of global disk turbulence, and the most plausible mechanism is the magneto-rotational instability of the disks [7]. Simulations of the magneto-rotational instability suggest that the strength of the viscosity is $\alpha \sim 10^{-2}$ ([37, 69, 76]; however, see [70] for limitations on the use of the parameter α), which is consistent with the value, $\alpha \sim 10^{-3} - 10^{-1}$, obtained from the observed disk accretion rate [33, 35].

If the disk temperature profile is $T = T_0(r/r_0)^{-q}$, then the sound speed is $c_\mathrm{s} = c_{\mathrm{s},0}(r/r_0)^{-q/2}$ and the disk scale height is $h_\mathrm{g} = h_{\mathrm{g},0}(r/r_0)^{(3-q)/2}$. In the simplest so-called "α-viscous model," α is assumed to be constant, and the disk viscosity varies with r as

$$\nu_{\text{vis}} = \nu_{\text{vis},0} \left(\frac{r}{r_0} \right)^{\gamma} , \tag{1.14}$$

where $\nu_{\text{vis},0} = \alpha h_{\text{g},0} c_{\text{s},0}$, and the power law index $\gamma = 3/2 - q$. In the fiducial model ($q = 1/2$), $\gamma = 1$.

Properties of Viscous Evolution

A Keplerian disk has the rotation velocity law ($\Omega_{\text{K}} \propto r^{-3/2}$) in which the inner disk gas rotates faster than the outer gas. Consider two imaginary gas rings inside the disk that contact each other. Through viscous stress at the contact surface, the faster rotating inner ring accelerates the rotation of the outer ring, while the outer ring decelerates the inner ring. The angular momentum is transferred from inside to outside, and consequently, the inner part of the disk shrinks toward the star while the outer part expands. In Keplerian disks, the specific angular momentum increases with the radius as $r^{1/2}$. As the gas at the outer edge travels farther, a smaller mass of gas can carry most of the total angular momentum of the disk. Hence, as viscous evolution proceeds and the disk's outer edge expands, the fraction of the expanding outer part decreases while the inner, infalling portion increases. Even if evolution proceeded until the outer edge expanded to infinity, only an infinitesimally small mass could actually reach infinity, and most of the disk gas would accrete inside an infinitesimally small region around the origin.

Time Evolution of the Disk

We consider the density evolution of the disk due to viscosity. For the simple viscosity model of (1.14), most of the properties of disk evolution can be derived from the estimate using the viscous evolution timescale. (See the textbooks [25, 34] for more detailed discussions.) By dimensional analysis, we see that the disk evolution occurs with the timescale of viscous diffusion,

$$\tau_{\text{vis}} = \frac{r^2}{\nu_{\text{vis}}} = 1.4 \times 10^6 \left(\frac{\alpha}{10^{-2}} \right)^{-1} \left(\frac{r}{100 \text{ AU}} \right)^{2-\gamma} \text{yr} , \tag{1.15}$$

where the numerical value is calculated for the fiducial parameters. The typical timescale is 10^6 yr if we use $\alpha = 10^{-2}$, $r = 100$ AU, and $\gamma = 1$.

Let the outer radius of the disk be initially r_0, and assume that the viscosity does not change with time but varies radially as (1.14). During the initial evolution of the timescale, $\tau_{\text{vis},0} = r_0^2/\nu_{\text{vis}}(r_0)$, the density profile depends on the initial profile. As viscous diffusion dampens initial density fluctuations at a small length scale, the disk "forgets" the initial profile. For times much greater than $\tau_{\text{vis},0}$, the outer radius of the disk is calculated from (1.14) and (1.15) as a function of t,

$$r_{\text{out}} = [\nu_{\text{vis}}(r_{\text{out}})t]^{1/2} = r_0 \left(\frac{t}{\tau_{\text{vis},0}} \right)^{\frac{1}{2-\gamma}} , \tag{1.16}$$

and the density profile smooths out. Consequently, the mass accretion rate of the disk, $\dot{M}_{\mathrm{acc}}(r,t) = 2\pi r \Sigma_{\mathrm{g}} v_{\mathrm{g},r}$, becomes independent of r except for the outermost part of the disk that expands. The accretion velocity is estimated as $v_{\mathrm{g},r} \sim r/\tau_{\mathrm{vis}} \sim \nu_{\mathrm{vis}}/r$. Assuming a power-law density profile, $\Sigma_{\mathrm{g}}(r,t) = \Sigma_{\mathrm{g},0}(t)(r/r_0)^{-p}$, the mass accretion rate is $\dot{M}_{\mathrm{acc}} \sim 2\pi\Sigma_0\nu_{\mathrm{vis},0}(r/r_0)^{\gamma-p}$. Hence, the constant accretion rate with r means $p = \gamma$; that is, the density profile converges to

$$\Sigma_{\mathrm{g}}(r,t) = \Sigma_{\mathrm{g},0}(t)\left(\frac{r}{r_0}\right)^{-\gamma}. \tag{1.17}$$

The total angular momentum of the disk is given by

$$H = \int_{r_{\mathrm{in}}}^{r_{\mathrm{out}}} 2\pi r^3 \Sigma_{\mathrm{g}}\Omega_{\mathrm{K}}\,dr \approx \frac{2\pi r_0^4 \Omega_{\mathrm{K},0}}{5/2 - \gamma}\Sigma_{\mathrm{g},0}(t)\left(\frac{t}{\tau_{\mathrm{vis},0}}\right)^{\frac{5-2\gamma}{4-2\gamma}}, \tag{1.18}$$

where we used (1.16). The integral is evaluated at r_{out} (assuming $\gamma < 5/2$); that is, the disk gas at the outer edge carries most of the angular momentum. From H being constant with time, the time dependence of the density is $\Sigma_{\mathrm{g},0}(t) = \Sigma_{\mathrm{g},0}(\tau_{\mathrm{vis},0})(t/\tau_{\mathrm{vis},0})^{-(5-2\gamma)/(4-2\gamma)}$. The mass accretion rate, which is constant with r, is

$$\dot{M}_{\mathrm{acc}}(t) = \dot{M}_{\mathrm{acc}}(\tau_{\mathrm{vis},0})\left(\frac{t}{\tau_{\mathrm{vis},0}}\right)^{-\frac{5-2\gamma}{4-2\gamma}}, \tag{1.19}$$

where $\dot{M}_{\mathrm{acc}}(\tau_{\mathrm{vis},0}) = 2\pi\Sigma_{\mathrm{g},0}(\tau_{\mathrm{vis},0})\nu_{\mathrm{vis},0}$, and it is $1.7 \times 10^{-8}\ M_\odot\ \mathrm{yr}^{-1}$ for the fiducial model. The disk mass evolves with time as

$$M_{\mathrm{g}} = \int_{r_{\mathrm{in}}}^{r_{\mathrm{out}}} 2\pi r \Sigma_{\mathrm{g}}\,dr \approx M_{\mathrm{g}}(\tau_{\mathrm{vis},0})\left(\frac{t}{\tau_{\mathrm{vis},0}}\right)^{-\frac{1}{4-2\gamma}}, \tag{1.20}$$

where $M_{\mathrm{g}}(\tau_{\mathrm{vis},0}) = 2\pi r_0^2 \Sigma_{\mathrm{g},0}(\tau_{\mathrm{vis},0})/(2-\gamma)$. For the fiducial model ($\gamma = 1$), the disk mass declines as $M_{\mathrm{g}} \propto t^{-1/2}$.

In summary, as viscous evolution proceeds, the timescale of viscous diffusion increases because the outer part of the disk expands. The disk mass decreases only with a power law of the time involved, and it is difficult to disperse all the disk gas via the disk viscosity alone.

1.3.2 Photoevaporation

Extreme ultraviolet (EUV, $h\nu > 13.6$ eV) photons irradiating the disk surface can dissociate and ionize hydrogen molecules. An ionized atmosphere forms above the disk surface with a temperature so high that the gas can escape from the gravity of the central star. This is called photoevaporation of the disk.

The central star or nearby OB stars can be sources of EUV photons. A nearby (<0.3 pc) massive star emits strong EUV (and also far-UV) photons that evaporate the disk gas as observed in the Orion Nebula [8]. The EUV flux from a solar-mass main-sequence star is not high enough to evaporate its disk in 10^7 yr. When the star is young, however, the star has a high chromospheric activity. Analysis of the UV emission lines suggests that classical T Tauri stars emit $10^{41} - 10^{44}$ EUV photons per second, which is $10^3 - 10^6$ times larger than the present solar value [2]. The disk gas accreted onto the stellar surface may also emit EUV photons as it becomes hot through accretion shock. In this section, we consider photoevaporation by EUV photons emitted from the central star.

Mass Loss Rate

The typical temperature of the gas ionized by EUV photons is 10^4 K [63], and its thermal velocity is $v_i \sim 10$ km s^{-1}. The ionized disk gas forms an atmosphere above the disk surface if the ions are gravitationally bound by the central star, i.e., if their thermal velocity is smaller than the escape velocity. However, in the outer part of the disk, the thermal velocity is higher than the escape velocity, and the ionized gas flows away from the disk. The gravitational radius, outside which the ionized gas is unbound, is estimated as

$$r_g = \frac{GM}{v_i^2} = 8.9 \left(\frac{M}{1 M_\odot} \right) \left(\frac{v_i}{10 \text{ km s}^{-1}} \right)^{-2} \text{ AU} . \tag{1.21}$$

The "direct" EUV photons from the central star scarcely penetrate the ionized disk atmosphere but are absorbed by the recombined hydrogen atoms before reaching the disk surface. The disk surface is ionized by the "diffuse" photons emitted from recombinations in the atmosphere (Fig. 1.1a). In the flow region $(r > r_g)$, the number density of the escaping gas (hydrogen ions), which is assumed to be constant with z, is determined by the balance between ionization and recombination, and is approximately given by

$$n_{\text{flow}} = n_g \left(\frac{r}{r_g} \right)^{-\beta} , \tag{1.22}$$

where the power-law index β is numerically estimated as 2.5 by Hollenbach et al. [38]. The density at r_g is roughly estimated by the ionization balance in the Strömgren sphere of radius r_g:

$$n_g = C_{\text{flow,g}} \left(\frac{3\Phi_i}{4\pi\alpha_2 r_g^3} \right)^{1/2} , \tag{1.23}$$

where Φ_i is the number of EUV photons emitted from the star per unit time, $\alpha_2 = 2.6 \times 10^{-13}$ cm^3 s^{-1} is the recombination coefficient to all states except the ground state, and $C_{\text{flow,g}}$ is a coefficient of the order of unity. The total mass loss rate from the disk is

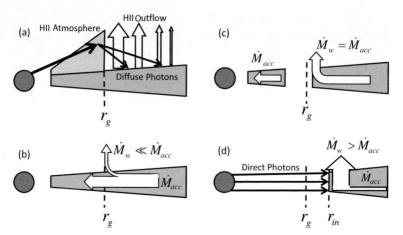

Fig. 1.1. Schematic illustration of disk photoevaporation. (**a**) Most of EUV photons from the central star are absorbed by the ionized atmosphere inside the gravitational radius r_g. "Diffuse" photons emitted from the atmosphere ionize the disk surface. The outflow occurs outside r_g, and its flux decreases as $r^{-5/2}$. (**b**) When the disk is so massive that $\dot{M}_{acc} \gg \dot{M}_w$, the wind barely affects the disk evolution. (**c**) Once $\dot{M}_{acc}$ becomes smaller than $\dot{M}_w$, supply of the accreting gas to the inner disk inside r_g is cut off, and a gap opens. The inner disk still continues to accrete onto the star, resulting in disappearance of the inner disk. (**d**) After clearing of the inner disk, direct photons ionize the inner edge of the residual disk, accelerating evaporation of the whole disk

$$\dot{M}_{\rm w} = 2m_{\rm H}v_i \int_{r_g}^{r_{out}} 2\pi r n_{\rm flow}\ dr \approx 4\pi(\beta - 2)^{-1}m_{\rm H}v_i n_g r_g^2 \ , \qquad (1.24)$$

where $m_{\rm H}$ is the mass of a hydrogen atom, and the factor 2 comes from the upper and lower surfaces of the disk. If $\beta > 2$, most of the wind mass loss originates in the neighborhood of r_g, and thus $\dot{M}_w$ is determined by the density n_g at r_g. Numerically,

$$\dot{M}_{\rm w} = 2.5 \times 10^{-9} C_{\rm flow,g} \left(\frac{\Phi_i}{10^{41}\ {\rm s}^{-1}}\right)^{1/2} \left(\frac{v_i}{10\ {\rm km\ s}^{-1}}\right) \left(\frac{r_g}{10\ {\rm AU}}\right)^{1/2} M_\odot\ {\rm yr}^{-1}.$$
$$(1.25)$$

Gas Dispersal

Here we consider the evolution of a gas disk in which both viscous accretion and photoevaporation play a role. We assume that the EUV photon flux $\Phi_i \sim 10^{41}$ is constant with time. (If the main source of EUV photons was the accreting disk gas onto the star, the flux Φ_i should be weakened as the mass accretion rate decreases, and consequently photoevaporation could not work effectively [57, 68].)

At an early stage, we assume that the disk is as massive as $10^{-2} M_\odot$ and that the viscous accretion rate ($\sim 10^{-8}$ $M_\odot$ yr^{-1}) is larger than the wind mass loss rate ($\sim 10^{-9}$ $M_\odot$ yr^{-1}). During this stage, the wind mass loss barely affects the evolution of the disk (Fig. 1.1b). As viscous accretion proceeds, the disk mass M_g decreases. The accretion rate $\dot{M}_{acc}$ also decreases with time, and in $10^6 - 10^7$ yr it becomes as small as the wind mass loss rate $\dot{M}_w$. After that, most of the accreting gas to r_g does not penetrate inside r_g but escapes as evaporating wind. Consequently, a gap opens at r_g, and the supply of the gas to the inner disk is cut off. However, the inner disk still continues to viscously accrete onto the star, and it disappears at a timescale $\sim r_g^2/\nu_{vis} \sim 10^5 - 10^6$ yr (Fig. 1.1c; some of the gas at the outer part of the inner disk moves outward to r_g and evaporates). After clearing the inner disk (and the ionized atmosphere), EUV photons begin to directly hit the inner edge of the disk (Fig. 1.1d). In this clearing stage, the density of the evaporating gas is estimated by Alexander et al. [3] as

$$n_{flow} = n_{in} \left(\frac{r}{r_{in}} \right)^{-\beta} , \tag{1.26}$$

where the power-law index is numerically calculated as $\beta = 2.42$ and n_{in} is the density at the disk inner edge r_{in}. The direct EUV photons ionize the surface of the inner edge (area $\sim 4\pi r_{in} h_g$). This ionization is balanced by recombination in the ionized gas torus of the volume $\sim 4\pi r_{in} h_g^2$ (radius $\sim 2\pi r_{in}$, height $\sim 2h_g$, and width $\sim h_g$). The density of the ionized gas at r_{in} is estimated as

$$n_{in} = C_{flow,in} \left(\frac{\Phi_i}{4\pi \alpha_2 r_{in}^2 h_g} \right)^{1/2} , \tag{1.27}$$

where the coefficient $C_{flow,in}$ is on the order of unity. The total mass loss rate in this stage is

$$\dot{M}_w = 5.4 \times 10^{-9} C_{flow,in} \left(\frac{\Phi_i}{10^{41} \text{ s}^{-1}} \right)^{1/2} \left(\frac{v_i}{10 \text{ km s}^{-1}} \right) \left(\frac{r_{in}}{10 \text{ AU}} \right)^{1/2}$$

$$\times \left(\frac{h_g/r_{in}}{0.1} \right)^{-1/2} M_\odot \text{ yr}^{-1} . \tag{1.28}$$

The dispersal timescale of the disk of radius r is estimated as

$$\tau_w = \frac{2\pi r^2 \Sigma_g}{\dot{M}_w}$$

$$= 4.6 \times 10^5 C_{flow,in}^{-1} \left(\frac{\Sigma_g(10 \text{ AU})}{35 \text{ g cm}^{-2}} \right) \left(\frac{\Phi_i}{10^{41} \text{ s}^{-1}} \right)^{-1/2} \left(\frac{v_i}{10 \text{ km s}^{-1}} \right)^{-1}$$

$$\times \left(\frac{r}{10 \text{ AU}} \right)^{3/2-p} \left(\frac{h_g/r}{0.1} \right)^{1/2} \text{ yr} . \tag{1.29}$$

As discussed above, a major fraction of the initial disk mass is likely to have accreted onto the star when the inner disk clearing starts. Even if the disk is as massive as the fiducial model (which has 0.02 $M_\odot$ inside 100 AU), the disk inside 100 AU evaporates only in 10^6 yr. Therefore, once photoevaporation begins to clear the disk (when the accretion rate becomes smaller than the wind mass loss rate), the disk clearing proceeds rapidly (probably within 10^6 yr).

1.4 Dust Motion in the Gas Disk

Growth of the dust particles in protoplanetary disks occurs under the gas environment, and the motion of the dust particles is greatly affected by gas drag. In particular, the particle collisional velocities are determined by the vertical or radial motions of the particles induced by gas drag. In this section, we calculate the particle velocities induced by gas drag.

1.4.1 Gas Drag Force

Consider a spherical dust particle of radius s and bulk density ρ_p residing in a gas of density ρ_g and a mean free path of the molecules l_{free}. If the particle radius is much smaller than the mean free path; that is, if $s \ll l_{\text{free}}$, the gas can be treated as free molecules. The gas drag force on a particle that moves with relative velocity Δv to the gas is calculated by the Epstein drag law:

$$F_{\text{drag}} = \frac{4}{3}\pi \rho_g s^2 v_T \Delta v , \tag{1.30}$$

where $v_T = \sqrt{8k_B T/(\pi m_g)}$ is the mean thermal speed. If $s \gg l_{\text{free}}$, however, the gas behaves as a fluid. Under this regime, the drag law varies with the Reynolds number,

$$\text{Re} = \frac{2s\Delta v}{\nu_{\text{mol}}} . \tag{1.31}$$

If $\text{Re} \ll 1$, the drag force is expressed by the Stokes law and is proportional to the radius s, and if $\text{Re} \gg 1$, it is proportional to s^2 [52]. The above drag laws are expressed in a combined form,

$$F_{\text{drag}} = \frac{C_D}{2}\pi \rho_g s^2 \Delta v^2 , \tag{1.32}$$

where the drag coefficient C_D is approximated as follows [82]. When $s < 9l_{\text{free}}/4$ (the Epstein regime), $C_D = 8v_T/(3\Delta v)$. When $s > 9l_{\text{free}}/4$ (the Stokes regime),

$$C_D = \begin{cases} 24\text{Re}^{-1} & \text{Re} \leq 1 \\ 24\text{Re}^{-0.6} & 1 < \text{Re} \leq 800 \\ 0.44 & 800 < \text{Re} \end{cases} . \tag{1.33}$$

The stopping time, $\tau_s = m_p \Delta v/F_{\text{drag}}$, is the time in which a dust particle of the relative speed Δv decelerates by gas drag, where $m_p = (4/3)\pi s^3 \rho_p$ is

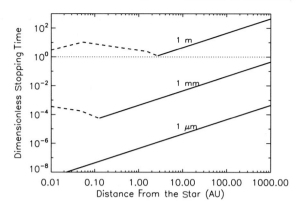

Fig. 1.2. The dimensionless stopping time T_s for dust particles of the sizes $1\,\mu m$, 1 mm, and 1 m. The values are calculated for the fiducial disk model described in Sect. 1.2.2. For large distances in which the gas density is low, the Epstein drag law is applied (shown by the *solid lines*). At high gas density regions, the Stokes drag law is applied (shown by the *dashed lines*). Under the Stokes regime, the stopping time depends on the relative velocity from the gas. In this figure, we assume $\Delta v = \eta v_K/2$. The bends of the lines under the Stokes regime come from the different dependencies of the drag coefficient on the Reynolds number in (1.33)

the particle mass. We define the dimensionless stopping time normalized to the Keplerian time as

$$T_s = \frac{m_p \Delta v}{F_{\text{drag}}} \Omega_K \ . \tag{1.34}$$

Figure 1.2 shows T_s for particles of various sizes in the fiducial disk. Under the Epstein regime, $F_{\text{drag}} \propto \Delta v$ and T_s is independent of Δv, i.e.,

$$T_s = \sqrt{\frac{\pi}{8}} \frac{\rho_p s}{\rho_g h_g} \quad \text{for the Epstein regime ,} \tag{1.35}$$

while under the Stokes regime, T_s depends on Δv. For the Stokes regimes in Fig. 1.2, we simply set $\Delta v = \eta v_K/2$, which is the difference between the gas velocity and the circular Keplerian velocity. This assumed value is appropriate if the particle has a circular Keplerian orbit. As discussed below, if T_s of the particle is much larger than unity, the orbit is nearly Keplerian (but may not be circular). From Fig. 1.2, we see that T_s of 1 m rocks under the Stokes regime is larger (but not much larger) than unity. Thus, the assumption, $\Delta v = \eta v_K/2$, is (marginally) valid for circularly orbiting 1 m rocks. For 1 mm particles, however, the actual stopping time and the relative velocity must be calculated self-consistently. Figure 1.3 shows the dimensional values of the stopping time $\tau_s = T_s \Omega_K^{-1}$ and also shows $T_s^{-1} \Omega_K^{-1}$, which represents the sedimentation timescales (for $T_s \ll 1$) discussed in Sect. 1.4.2.

Behaviors of the dust particles under gas drag are categorized according to the value of T_s. When $T_s \ll 1$, the particle motion is strongly controlled

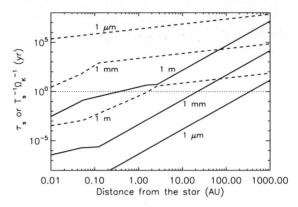

Fig. 1.3. The dimensional values of the stopping time $\tau_s = T_s \Omega_K^{-1}$ (*solid lines*) and $T_s^{-1}\Omega_K^{-1}$ (*dashed lines*) are plotted for dust particles of the sizes 1 μm, 1 mm, and 1 m. The values are calculated for the fiducial disk model described in Sect. 1.2.2

by the gas. In the zeroth order of T_s, the particle has the same trajectory as the gas. The particle orbits circularly on a plane parallel to the midplane of the disk, and its velocity is $v_\theta = (1 - \eta)^{1/2} v_K$. Gas drag induces evolution of the orbit through the first-order perturbation of T_s, and the orbital radius r and distance from the midplane z varies with the time. When $T_s \gg 1$, however, the particle motion is hardly affected by the gas. In the zeroth order of T_s^{-1}, the particle moves on a Keplerian orbit, which may be elliptical and may cross the disk midplane with an inclination angle i. The orbital elements, such as the semimajor axis a, the eccentricity e, and the inclination i evolve through the first-order perturbation of T_s^{-1}. In the intermediate cases, that is, when $T_s \sim 1$, the particle orbit is neither in corotation with the gas nor Keplerian. As seen below, evolution of the particle orbit due to gas drag works most effectively in such cases.

1.4.2 Particle Sedimentation

Dust particles sediment toward the midplane and consequently form a thin dust layer. If no mechanism exists to inhibit sedimentation, the dust layer becomes thinner until it spontaneously fragments by its self-gravity. We first consider dust sedimentation in a laminar disk, and then discuss how gas turbulence inhibits the sedimentation.

Sedimentation Timescale

The timescale of dust sedimentation in a laminar disk is calculated in the following manner.

1. $T_s \ll 1$. The particle orbits circularly on a plane that is parallel to the disk midplane, and this orbital plane (at a height z) moves toward the

midplane. The vertical component of the central star's gravity, $F_{g,z} = \Omega_K^2 m_p z$, accelerates the particle to the midplane. Experiencing gas drag, the particle quickly (in the time $\tau_s = T_s \Omega_K^{-1}$) reaches the terminal velocity v_z. From the balance between the gas drag force, $F_{\text{drag}} = m_p v_z \Omega_K T_s^{-1}$, and the gravity, $F_{g,z}$, we obtain the terminal velocity

$$v_z = T_s z \Omega_K . \tag{1.36}$$

The sedimentation timescale is

$$\tau_{\text{sed}} = \frac{z}{v_z} = T_s^{-1} \Omega_K^{-1}, \tag{1.37}$$

and its value for the fiducial disk is plotted as the dashed lines in Fig. 1.3. The typical timescales of sedimentation at 1 AU is 10^6 yr for 1 μm particles and 10^3 yr for 1 mm particles.

2. $T_s \gg 1$. The particle moves on a Keplerian orbit that has an orbital inclination i to the disk midplane. The orbital inclination gradually dampens
due to gas drag. The damping rate is simply estimated by the stopping
time,

$$\tau_{\text{sed}} = \frac{i}{|di/dt|} \sim T_s \Omega_K^{-1}. \tag{1.38}$$

(This estimate is valid only when $1 \gg i \gg e, \eta$. See [1] for more detailed
calculation. Note also that if the relative velocity is determined by the
inclination, that is, if $i \gg \eta$, the assumption for plotting Fig. 1.3 under
the Stokes regime, $\Delta v = \eta v_K / 2$, cannot be used.)

3. $T_s \sim 1$. As expected from the substitution of $T_s = 1$ into (1.37) or (1.38),
the particle quickly sediments onto the midplane in a Kepler time Ω_K^{-1}
(see [27] for more detailed discussion).
It takes 10^6 yr for the particles of size ~ 1 μm to sediment out if they
do not grow larger. However, if particle growth is taken into account,
sedimentation is considerably accelerated. The particles of initial sizes
~ 1 μm at 1 AU grow to ~ 1 cm when they arrive at the midplane, as
discussed in Sect. 1.5.1, and the time taken for the sedimentation shortens
to 10^3 yr.

Gravitational Instability of the Dust Layer

If no mechanism exists to raise the dust particles, the dust layer becomes thinner as sedimentation continues and its density increases. When the dust density reaches the critical value, when the self-gravity exceeds the tidal force of the central star, the dust layer begins to fragment to form planetesimals through gravitational instability [28]. The critical density is on the order of the Roche density, $\rho_{\text{Roche}} = 3.5M/r^3$. (Linear stability analysis by Sekiya [71] shows that gravitational instability occurs when the dust density exceeds 0.17 ρ_{Roche}.) The thickness of the dust layer for which the gravitational instability occurs is estimated as $h_d/r = r^2\Sigma_d/(3.5M)$. The amount of solid material of the disk, $r^2\Sigma_d$, is estimated as $\sim 1M_\oplus \sim 3 \times 10^{-6} M_\odot$ at 1 AU. Thus,

$$h_d \sim 10^{-6} \times 1 \text{ AU} \sim 100 \text{ km}. \tag{1.39}$$

The thickness of the gas disk at 1 AU is $h_g \sim 3 \times 10^{-2}$ AU. For gravitational instability to occur, the dust layer must be 10^{-4} times thinner than the gas disk and the dust density must be as large as $\rho_d/\rho_g \sim 100$.

If some mechanisms such as gas turbulence or convection lift up the dust, the density of the dust layer may not reach the critical value for gravitational instability. A possible cause of gas turbulence is the Kelvin–Helmholtz instability at the boundary of the dust layer. When the density of the dust layer exceeds the gas density, the gas inside the dust layer begins to be dragged by the dust orbiting at the Keplerian velocity. Because the gas above the dust layer still orbits at a sub-Keplerian velocity, a velocity difference appears at the boundary of the dust layer, leading to turbulence due to the Kelvin–Helmholtz instability [40, 41, 44, 73, 74].

Vertical Diffusion of the Dust by Turbulence

Turbulent motion of the gas stirs the dust particles and diffuses them into the upper layer of the gas disk. Using a simple analogy of molecular diffusion in laminar gas, we discuss the turbulent diffusion of the particles. Following Cuzzi et al. [20], we assume that the dust flux by diffusion is proportional to the gradient of the dust concentration ρ_d/ρ_g, i.e.,

$$\boldsymbol{F}_{\text{dif}} = -D_d\rho_g\nabla\left(\frac{\rho_d}{\rho_g}\right) , \tag{1.40}$$

where the diffusion coefficient,

$$D_d = \frac{\nu_{\text{vis}}}{\text{Sc}} , \tag{1.41}$$

is proportional to the turbulent viscosity. The nondimensional Schmidt number Sc describes the velocity dispersion of the particles induced by turbulence. When the dust particles are small enough and are treated as passive particles, Sc is on the order of unity; if the particles are large and decoupled from the turbulent gas motion, Sc becomes infinity. Numerical simulations by Johansen & Klahr [42] show that typical values of the Schmidt number are ~ 1 for small particles ($T_s < 10^{-2}$).

Consider a steady dust layer in which the particle lift-up by turbulence is balanced by sedimentation. The thickness of such a dust layer, h_d, is estimated as follows. The timescale of particle diffusion in the dust layer is

$$\tau_{\text{dif}} = \text{Sc}\frac{h_d^2}{\nu_{\text{vis}}} . \tag{1.42}$$

In a steady state, this timescale becomes comparable to the sedimentation timescale. Using (1.3), (1.13), (1.37), and (1.42), $\tau_{\text{sed}} = \tau_{\text{dif}}$ results in the estimate of the layer thickness

$$\frac{h_d}{h_g} = \left(\frac{\alpha}{\text{Sc}T_s}\right)^{1/2} . \tag{1.43}$$

For small particles ($T_s \ll \alpha/\text{Sc}$), (1.43) may give $h_d/h_g > 1$. This means that such small particles hardly sediment. If the whole disk is turbulent due to, for example, the magneto-rotational instability, and the viscosity is as strong as $\alpha = 10^{-2}$, only the particles that have $T_s > \alpha/\text{Sc} \sim 10^{-2}$ can sediment. This value of T_s corresponds to the size ~ 1 cm at 1 AU.

We now consider how low the strength of turbulence must be for the gravitational instability of the dust layer to occur. The density of the dust layer is

$$\rho_d \sim \frac{\Sigma_d}{h_d} \sim \frac{\Sigma_d}{h_g} \left(\frac{\alpha}{\text{Sc}T_s}\right)^{-1/2} . \tag{1.44}$$

The value of α must be quite small for the dust density to exceed the critical density, ρ_{Roche}, for the gravitational instability. Suppose that the dust layer is mainly composed of 1 cm particles ($T_s = 10^{-2}$), which is the expected size at 1 AU after the particle growth during sedimentation (see Sect. 1.5.1). Then, the viscosity must be as small as $\alpha < 10^{-11}\text{Sc}$. Turbulence induced by the magneto-rotational instability generates largest eddy sizes similar to the gas disk thickness [36, 69, 76]. This is much larger than the thickness of the dust layer required for the gravitational instability ($h_d/h_g \sim 10^{-4}$), leading to α as large as 10^{-2}. Such a strong turbulence inhibits formation of a sufficiently thin dust layer for gravitational instability. Turbulence induced by the Kelvin–Helmholtz instability develops when the midplane dust density exceeds the gas density, and eddy sizes are similar to the thickness of the dust layer ($h_d/h_g \sim 10^{-2}$; [44]). Turbulence from the Kelvin–Helmholtz instability inhibits the dust layer from becoming thinner and keeps the midplane dust density similar to the gas density, which is much smaller than the critical density for gravitational instability. Therefore, turbulence seems to suppress gravitational instability. However, Sekiya [72] argued that if the initial dust-to-gas ratio of the disk is as large as ~ 0.1, turbulence due to Kelvin–Helmholtz instability cannot raise the dust sufficiently because of the increased inertia of the dust layer, and then the midplane dust density reaches the critical density. In addition, recent numerical simulations have shown that although turbulence suppresses the increase in the average density of the dust layer, it also forms dust clumps whose densities are much higher (by a factor of ~ 100) than the average [26, 42, 44]. Further studies are needed to determine if such dense dust clumps collapse gravitationally to form planetesimals.

1.4.3 Radial Infall Toward the Star

The disk gas usually has an outward pressure gradient and orbits with a sub-Keplerian velocity, as seen in (1.6). The dust particles do not experience the pressure gradient force and they tend to orbit with the Keplerian velocity, i.e., faster than the gas. Consequently, due to gas drag in the azimuthal direction, the particles lose their angular momentum and fall toward the central star. The radial velocity of the particles is easily estimated for two limiting cases of $T_s \ll 1$ and $T_s \gg 1$ [82].

1. $T_s \ll 1$. Due to strong gas drag, the particle orbits with almost the same azimuthal velocity as the gas, $v_\theta = (1 - \eta)^{1/2}v_K$. For such a particle, the gravity is stronger than the centrifugal force and the residual gravity, $F_r = \eta m_p r \Omega_K^2$, accelerates the particle toward the star. This acceleration is balanced by the gas drag force in the r-direction, $F_{drag} = m_p v_r \Omega_K T_s^{-1}$, when the particle reaches the terminal velocity,

$$v_r = T_s \eta v_K \ . \tag{1.45}$$

2. $T_s \gg 1$. The particle orbits with the Keplerian velocity, and the velocity difference from the gas is $v_K - v_{g,\theta} \approx \eta v_K/2$ (for a circular orbit). The angular momentum loss of the particle due to gas drag is $dH/dt = \eta m_p r v_K \Omega_K T_s^{-1}/2$. The particle orbit gradually shrinks with the radial velocity

$$v_r = \frac{2}{m_p v_K} \frac{dH}{dt} = T_s^{-1}\eta v_K \ . \tag{1.46}$$

3. $T_s \sim 1$. As expected from (1.45) and (1.46), $v_r \sim \eta v_K$ when $T_s \sim 1$. Detailed calculations show that $v_r = \eta v_K/2$ for $T_s = 1$ [1, 58, 77].
   The timescale of the radial infall is

$$\tau_{infall} = \frac{r}{v_r} = \begin{cases} T_s^{-1}\eta^{-1}\Omega_K^{-1} & \text{for } T_s \ll 1 \\ T_s\eta^{-1}\Omega_K^{-1} & \text{for } T_s \gg 1 \end{cases} \ . \tag{1.47}$$

As seen from the comparison to (1.37) and (1.38), the radial infall timescale is $\eta^{-1} \sim 10^3$ times longer than the sedimentation timescale. Hence, the particles settle on the midplane first, and then move toward the star. Let us derive the altitude at which the radial velocity of the particle becomes larger than the vertical velocity. For a particle of $T_s < 1$, the vertical velocity is $v_z = T_s z \Omega_K$, and the radial velocity is $T_s \eta v_K$. Thus, $v_r = v_z$ occurs when the particle arrives at

$$z_r = \eta r \ . \tag{1.48}$$

For particles of $T_s > 1$, a similar conclusion can be obtained.

Radial Infall During Gas Dispersal

The particles have the maximum infall speed when their stopping time is comparable to the Kepler time ($T_s = 1$). The timescale of the fastest infall is $2\eta^{-1}\Omega_K^{-1}$ and is only 10^3 yr at 1 AU. In the fiducial disk, meter-sized particles have the maximum infall speed. Thus, the dust distribution in the disk is expected to quickly change compared to the disk lifetime ($10^6 - 10^7$ yr), unless most of the dust mass resides either in the size ranges much smaller or much larger than 1 m. As the disk gas dissipates, gas drag becomes weaker, and when the gas density becomes low enough, the orbital evolution of the dust due to gas drag becomes negligible. We estimate how low the gas density must be for that to occur. In the gas disk, which is more tenuous than the fiducial disk, the gas drag is described by the Epstein law. From (1.4) and (1.35), the stopping time follows as $T_s = \pi\rho_p s/(2\Sigma_g)$. From $T_s = 1$, the size of the particles that have the maximum infall speed is $s = 2\Sigma_g/(\pi\rho_p)$. When the disk is as massive as the fiducial model, the surface density of the gas disk at 1 AU is $\Sigma_g = 3 \times 10^2$ g cm^{-2}, and meter-sized bodies have the maximum infall velocity. As the gas dissipates, this size decreases, and when the gas density becomes 10^{-6} times smaller, $1\,\mu$m particles have the maximum velocity. In such tenuous gas disks, gas drag does not change the orbits of particles much larger than $1\,\mu$m, but micrometer-sized particles are still rapidly infalling. Note that the maximum infall velocity, $\eta v_K/2$, is independent of the gas density. Thus, even in disks 10^{-6} times more tenuous than the fiducial disk, that is, in disks of the masses $10^{-8}M_\odot \sim 10^{-2}M_\oplus$, the gas drag still has some effects on the dust evolution.

1.5 Dust Growth

As discussed in the previous section, the particles first settle on the midplane, and then migrate radially. We consider particle growth during sedimentation, and show that the particles grow to pebble sizes (~ 1 cm) by the time they arrive at the midplane. Particle growth after that is not well understood. Two possible paths may be involved in forming planetesimal-sized bodies. One possibility is that planetesimals form through gravitational instability in the thin dust layer, as discussed in Sect. 1.4.2. The other possibility is that collisional growth continues and planetesimals form before they migrate radially for large distances by gas drag. In Sect. 1.5.3, we discuss how large the dust bodies can grow during radial infall.

1.5.1 Growth During Sedimentation

Consider a dust particle that collides with other particles and sticks to them. Let the mean collisional velocity be v_{col}, and the collisional cross section be πs^2. (This expression is for cases in which the particle collides mainly with particles of much smaller sizes. If the collision is with a similar-sized particle, the cross-section is $\pi(2s)^2$, but we neglect the factor 4 difference for simplicity.) The equation for the time evolution of the particle mass is

$$\frac{dm_p}{dt} = C_{stk}\pi s^2 \rho_d v_{col} , \qquad (1.49)$$

where C_{stk} is the probability of sticking. We further assume that the dust particle is compact and the bulk density ρ_p does not change during its growth. Then, (1.49) becomes

$$\frac{ds}{dt} = \frac{C_{stk}\rho_d v_{col}}{4\rho_p} . \qquad (1.50)$$

During sedimentation, the mean collision velocity is determined by the difference in the sedimentation velocity or by the Brownian motion. Due to Brownian motion, the collision velocity of a particle of mass m_p colliding with the smallest particle (of the assumed size $s_0 = 0.1$ μm and the mass $m_0 = 4 \times 10^{-15}$ g) is

$$v_B = \sqrt{\frac{8k_B T(m_p + m_0)}{\pi m_p m_0}} \approx v_T\sqrt{\frac{m_g}{m_0}} \approx 3 \times 10^{-5} v_T , \qquad (1.51)$$

where the velocity is determined by the smallest mass m_0. The collision velocity due to sedimentation is approximately the vertical speed of the larger particle, $v_z = \sqrt{\pi/8}T_s(z/h_g)v_T$. Thus, for particles of $T_s > 3 \times 10^{-5}$, unless the particle is close to the neighborhood of the midplane ($z \ll h_g$), the collision velocity is determined by the sedimentation velocity. In the subsequent discussion, we neglect Brownian motion. Substitution of $v_{col} = |v_z| = |dz/dt|$ into (1.50) gives

$$\frac{ds}{dz} = -\frac{C_{stk}\rho_d}{4\rho_p} , \qquad (1.52)$$

for $z \geq 0$. This equation is easily integrated. Suppose that a particle of the initial size s_0 starts sedimentation and growth from a high altitude. When the particle arrives at the midplane, its size becomes

$$s_1 = s_0 + \frac{C_{stk}\Sigma_d}{8\rho_p} \approx 1.25 C_{stk} \left(\frac{\Sigma_d}{10 \text{ g cm}^{-2}}\right)\left(\frac{\rho_p}{1 \text{ g cm}^{-3}}\right)^{-1} \text{cm} . \qquad (1.53)$$

At 1 AU in the fiducial disk, particles of $\rho_p = 1$ g cm^{-3} and $C_{stk} = 1$ grow to 1 cm at the midplane. This final size after sedimentation does not depend on the drag law, as long as the collision velocity is given by the sedimentation velocity. The timescale of growth, however, does depend on the drag law.

Next, we estimate the growth timescale. When the particle is small enough and the drag law is under the Epstein regime, the growth timescale is calculated from (1.3), (1.35), (1.36), and (1.50) as

$$\tau_{\text{grow,E}} \equiv \frac{s}{ds/dt} = \frac{8\sqrt{2}f_{\text{dust}}^{-1}}{\sqrt{\pi}C_{\text{stk}}}\left(\frac{z}{h_g}\right)^{-1}\Omega_{\text{K}}^{-1}, \tag{1.54}$$

where the dust-to-gas ratio $f_{\text{dust}} = \rho_d/\rho_g$ is the ratio of the local densities of dust and gas. The value of f_{dust} can increase from the interstellar value $\sim 10^{-2}$ due to sedimentation. Note that the growth timescale is independent of the particle size s. It is only ~ 100 orbital time for the particles of $C_{\text{stk}} = 1$ and $f_{\text{dust}} = 10^{-2}$ at a high altitude ($z \sim h_g$). Thus, the dust growth is a rapid process, unless the sticking probability is too small. When the particle size has become larger than the mean free path of the gas molecules, gas drag obeys the Stokes law. Under the Stokes regime (and if Re < 1), the gas drag force increases only linearly with the size s, while under the Epstein regime it increases as $F_{\text{drag}} \propto s^2$. This means that when the particle enters the Stokes regime, the terminal sedimentation velocity increases more rapidly with size. As the particle grows, the sedimentation is accelerated, and consequently the growth timescale (for Re < 1) is also reduced as

$$\tau_{\text{grow,S}} = \left(\frac{s}{9l_{\text{free}}/4}\right)^{-1}\tau_{\text{grow,E}}. \tag{1.55}$$

Using the above timescales, we estimate the time required for a particle to grow from an initial size s_0 to s_1. If the final size s_1 is smaller than the mean free path of the gas, growth occurs only under the Epstein regime. Approximating, $z \approx h_g$ and ignoring increase in f_{dust} due to sedimentation, the growth time is, from (1.54),

$$\tau_{\text{grow,sed}} = \tau_{\text{grow,E}}\log\frac{s_1}{s_0}. \tag{1.56}$$

(For the exact calculation, taking into account, the time dependence of z, see [58].) If the final size s_1 is larger than $9l_{\text{free}}/4$, then the growth time is (again with the approximation $z \approx h_g$),

$$\tau_{\text{grow,sed}} = \tau_{\text{grow,E}}\left[\log\frac{9l_{\text{free}}/4}{s_0} + \left(1 - \frac{9l_{\text{free}}/4}{s_1}\right)\right], \tag{1.57}$$

where the first term is the time for the growth to $s = 9l_{\text{free}}/4$ (under the Epstein regime) and the second term is the time for further growth to s_1 (under the Stokes regime). Because in the Stokes regime growth accelerates as the size increases, the growth time under this regime is at most $\tau_{\text{grow,E}}$ and is a factor $\log(4s_0/9l_{\text{free}})$ times smaller than the growth time under the Epstein regime. Summarizing the above estimates, the growth time during sedimentation is

$$\tau_{\text{grow,sed}} \approx \tau_{\text{grow,E}}\log\frac{\min(s_1, 9l_{\text{free}}/4)}{s_0}. \tag{1.58}$$

At 1 AU in the fiducial disk, where $f_{\text{dust}} = 10^{-2}$ and $s_1 \sim l_{\text{free}} \sim 1$ cm, a particle of 0.1 µm grows to 1 cm in $\tau_{\text{grow,sed}} \sim C_{\text{stk}}^{-1}1.2 \times 10^3$ yr.

1.5.2 Growth of Dust Aggregates

We have so far considered compact dust particles, whose bulk densities do not change during growth. In realistic collisions, however, the resulting products are probably not compact particles, but aggregates in which small particles combine weakly to form a particle cluster with a large porosity. When collisions occur between small aggregates at slow velocities, they result in sticking of the aggregates without restructuring or destroying their internal structures. When the colliding aggregates are large or the collision velocity is high, the structure of the resulting aggregates are modified and probably compressed [21]. After a series of such compressions, the aggregates may become compact.

An aggregate is modeled as a cluster of monomers of mass m_0 and size s_0. The total mass of the aggregate is expressed by

$$m_{\mathrm{p}} = m_0 \left(\frac{s}{s_0} \right)^D , \qquad (1.59)$$

where s is the effective size of the aggregate (for the definition of s see [45]). The fractal dimensions D have been measured by numerical and experimental simulations of aggregate formation. If the aggregate forms through collisions between a cluster and a much smaller particle, that is, through particle-cluster aggregation (PCA), the fractal dimension is close to $D = 3$ [62]. In contrast, a cluster–cluster aggregation (CCA), which is a collision between two similar-sized aggregates, produces much smaller fractal dimensions of $D = 1.4$–1.8 [45, 50, 65].

Consider sedimentation of aggregates of the fractal dimensions $D < 2$ (CCA-like clusters). First, suppose that the aggregate is small ($T_{\mathrm{s}} \ll 1$) and it is under the Epstein regime. The cross-sectional area of the aggregate is nearly proportional to the mass (or to the number of monomers). This is because for $D < 2$, only a few constituent monomers are hidden by others [45]. Under the Epstein regime, the drag force is proportional to the cross-sectional area, i.e., to the mass for $D < 2$. Thus, the stopping time, $T_{\mathrm{s}} \propto m_{\mathrm{p}}/F_{\mathrm{drag}}$, does not vary with the aggregate mass. The sedimentation velocity, given by $v_z = T_{\mathrm{s}} z \Omega_{\mathrm{K}}$, is also independent of the mass; that is, growth of the aggregate does not accelerate sedimentation. The aggregate keeps the slow sedimentation velocity that it had as a monomer. Therefore, the aggregate barely settles on the midplane. When collisional compression or PCA-like growth increases its fractal dimension greater than 2, some monomers inside the aggregate begin to be hidden by other monomers. The cross-sectional area increases more slowly than the mass as the aggregate grows, and consequently, the sedimentation velocity accelerates. If the fractal dimension continues to be smaller than 2, the aggregate probably stays at a high altitude until it grows as large as the mean free path of the gas molecules, l_{free}. When the aggregate reaches a size corresponding to the Stokes regime, the drag force weakens compared to the Epstein drag (and is proportional to the size for Re < 1). Thereafter the sedimentation velocity accelerates as the aggregate grows. Therefore, rapid sedimentation of aggregates is delayed until the aggregates have either been compressed sufficiently or have become as large as l_{free}. During this initial growth of the aggregates without settling, the bulk density ρ_p decreases ($\rho_p \propto s^{D-3}$). Thus, the final radii of the aggregates when they arrive at the midplane, given by (1.53), can be much larger than the estimates for the compact particles. To calculate the final size and growth timescale, we need to develop a coagulation model that accounts for the collisional compression of aggregates as well as for the evolution of the bulk density [61].

1.5.3 Growth During Radial Infall

Growth to $T_s = 1$

After a particle has settled on the midplane and it has become a body as large as a pebble (~ 1 cm), it moves inward toward the central star. When the body arrives at the altitude z_r, given by (1.48), its radial velocity exceeds its vertical velocity. At this point, the body has probably not grown large enough for its stopping time T_s to become larger than unity, unless its bulk density ρ_p is extremely low (see (1.53)). As the body falls toward the star, it collides with other smaller particles because of the difference in their radial velocities. The collision velocity is approximated by the radial velocity of the body, i.e., $v_{\text{col}} = |v_r| = T_s \eta v_K$. As the body grows, the stopping time T_s increases (toward unity) and the radial velocity also accelerates. Acceleration of radial infall continues until the stopping time exceeds unity, or until the body enters the dust-dominant layer where the dust-to-gas ratio is larger than unity and the gas drag is weakened. After that, the infall begins to decelerate and finally stops. For falling bodies to grow to planetesimal sizes, they must exceed the size for which $T_s = 1$ or they have to enter the dust-dominant layer before they fall onto the star.

We consider first whether the bodies can grow to the size for which $T_s = 1$. Suppose that a body of the initial size s_1 starts infall from the distance r_1. Substituting $v_{\text{col}} = |v_r|$ into (1.50), integration with r gives the body size at r_2 as

$$s_2 = s_1 + \frac{C_{\text{stk}}}{4\rho_p} \int_{r_2}^{r_1} \rho_d \, dr \; . \tag{1.60}$$

This integration is evaluated at r_2, if ρ_d decreases more steeply than r^{-1}. We approximate the integration by $\rho_d r_2$, and then

$$s_2 \sim \frac{C_{\text{stk}}\rho_d r_2}{4\rho_p} \sim 4C_{\text{stk}} \left(\frac{\rho_d}{10^{-12} \text{ g cm}^{-3}}\right) \left(\frac{\rho_p}{1 \text{ g cm}^{-3}}\right)^{-1} \left(\frac{r_2}{1 \text{ AU}}\right) \text{ cm} \; . \tag{1.61}$$

The bodies are expected to grow to several centimeters at 1 AU if we take the fiducial value ($\rho_d = 10^{-12}$ g cm^{-3}) for the dust density, that is, if we do not take into account density enhancements due to dust sedimentation or due to other mechanisms. For larger distances from the star, the body size is smaller, if ρ_d decreases steeper than r^{-1}.

Sedimentation enhances the dust density ρ_d. If the density enhancement is large enough, then the body can grow larger than the size for which $T_s = 1$ and the infall stops. Consider the body that falls to r_2 and its size becomes s_2 given by (1.61). Then derive the condition for its T_s to be larger than unity in the cases when the Epstein drag law is relevant (i.e., at a sufficiently low gas density; $r \geq$ several AU). From (1.35) and (1.61), the stopping time of the body of the size s_2 is given by

$$T_s = \frac{\sqrt{\pi}C_{\text{stk}}\rho_d r_2}{8\sqrt{2}\rho_g h_g} . \tag{1.62}$$

The condition for $T_s > 1$ is satisfied if the dust-to-gas ratio increases to

$$\frac{\rho_d}{\rho_g} > 8\sqrt{\frac{2}{\pi}}C_{\text{stk}}^{-1}\frac{h_g}{r} . \tag{1.63}$$

Because $h_g/r \sim 0.1$, the dust-to-gas ratio has to be close to unity even for perfect sticking ($C_{\text{stk}} = 1$). This means, however, that the body must be inside a dust-dominant region. In such a place, the gas is dragged by the Keplerian rotating dust, and the gas drag force per dust particle is weakened.

We consider next the other possibility that the infall stops in the dust-dominant layer. When the dust layer becomes so thin that the dust density exceeds the gas density, the gas is dragged by the dust and orbits with a nearly Keplerian velocity. Consequently, gas drag cannot effectively induce radial infall of the dust. Calculation of Nakagawa et al. [58] showed that in the dust-dominant layer, the dust bodies hardly migrate in the radial direction, but continue settling until they induce gravitational fragmentation of the dust layer. We estimate how long the bodies migrate in the radial direction before entering the dust-dominant layer. The half-thickness of the dust-dominant layer must be smaller than $z_d = f_{\text{dust}}h_g$, where $f_{\text{dust}} \sim 0.01$ is the initial dust-to-gas ratio. Consider a body that starts its settling and infall from (r_0, z_0). We assume that the settling body enters the dust-dominant layer when it arrives at z_d. This means that by the time the typical body arrives at z_d, most of the dust mass also has sedimented below z_d [59]. From (1.36) and (1.45), the trajectory of the body before arriving at z_d is described by

$$\frac{dr}{dz} = \frac{\eta r}{z} = \eta_0 \left(\frac{r}{r_0}\right)^a \frac{r}{z} , \tag{1.64}$$

where we substitute $\eta = \eta_0(r/r_0)^a$ and $a = 1 - q$. Integration of this equation shows that when the body sediments from z_0 to z_d, it migrates in the radial direction for a distance

$$\Delta r \approx \eta_0 \log\left(\frac{z_d}{z_0}\right) r_0 . \tag{1.65}$$

Radial migration before entering the dust-dominant layer is as short as $\Delta r \sim 10^{-2} r_0$. Thus, the body hardly migrates in the radial direction, although its size is only ~ 10 cm at 1 AU when it enters the dust-dominant layer [58, 59].

When the thin dust layer begins to orbit with Keplerian speed, the gas outside the layer, which rotates with a sub-Keplerian velocity, may exert a negative torque on the layer, leading to inward migration of the layer [29, 84]. If this inward motion of the layer is too fast, we still have the problem that the dust bodies inside the layer cannot survive. Calculations of the torque exerted on the dust layer by the upper layer of the gas have been performed by Youdin & Chiang [88] and Weidenschilling [86], assuming that all the particles in the dust layer are the same size. They showed that when the Kelvin–Helmholtz instability at the layer boundary sets in, turbulence effectively transfers angular momentum from the dust layer to the upper gas. Consequently, the inward velocity of the dust layer approaches the value that an individual particle of the same size in the sub-Keplerian (unaffected by the dust) gas flow has. This means that even a dense dust layer cannot protect bodies inside from the infall. Their calculations assume single-sized particles, however, and it is still unclear if larger bodies in the layer can avoid rapid infall and settle on the midplane as proposed by Nakagawa et al. [58]. In the following discussion, we assume that some bodies manage to grow to the size for which $T_s = 1$.

Growth After $T_s > 1$

When T_s of the body exceeds unity, the infall speed decelerates as the body grows. Numerical simulation by Weidenschilling [85] suggests that further growth proceeds by catching smaller particles that are still moving inward. The maximum collision velocity between the body and the swallowed small particles is $\eta v_K / 2$. If the body acquires mainly small particles of $T_s < 1$, the collision velocity arises from the velocity difference in the azimuthal direction, because the body ($T_s > 1$) orbits with the Keplerian velocity, while the small particles ($T_s < 1$) co-orbit with the gas (i.e., with a sub-Keplerian velocity). The gas velocity at the midplane may be a nearly Keplerian velocity if the midplane dust density is so enhanced that it exceeds the gas density. Thus, the collision velocity is almost $\eta v_K / 2$. On the other hand, if the body absorbs mainly particles of $T_s > 1$, the collision velocity results from the velocity difference in the radial direction. In this case, the maximum collision velocity is also $\eta v_K / 2$. Let us estimate the timescale of forming planetesimal-sized objects. We assume that the supply of small particles continues and the collision velocity is $v_{col} = \eta v_K / 2$. The timescale for forming kilometer-sized objects is

$$\tau_{grow} \sim 9 \times 10^4 C_{stk}^{-1} \left(\frac{\eta}{10^{-3}} \right)^{-1} \left(\frac{\rho_d}{10^{-10} \text{ g cm}^{-3}} \right)^{-1} \left(\frac{\rho_p}{1 \text{ g cm}^{-3}} \right)$$
$$\times \left(\frac{s}{1 \text{ km}} \right) \left(\frac{r}{1 \text{ AU}} \right)^{1/2} \text{ yr},$$

$$(1.66)$$

where we take a reference value of the dust density $\rho_d = 10^{-10}$g cm^{-3}, because dust sedimentation increases the dust density. Note that this timescale should be considered as a lower limit because we assume the maximum collisional velocity $\eta v_K/2$.

1.5.4 Growth in Turbulent Disks

Turbulent gas motion in the disk induces random velocities of the dust particles, and may accelerate particle growth. The mean relative velocities of two particles in isotropic turbulence with the Kolmogorov energy spectrum have been calculated in several studies [19, 56, 60, 81, 83]. Consider two particles in the turbulence. The larger and smaller particles have the dimensionless stopping times of $T_{s,1}$ and $T_{s,2}$, respectively. Let the dimensionless turnover times of the largest eddy and smallest eddy be T_L and T_i, respectively. The approximate expression of the relative velocity is

$$v_{rel} = \begin{cases} v_g \dfrac{T_{s,1} - T_{s,2}}{\sqrt{T_i T_L}} & \text{for } T_{s,1} < T_i \\ 1.7 v_g \sqrt{\dfrac{T_{s,1}}{T_L}} & \text{for } T_i < T_{s,1} < T_L \text{ and } T_{s,2} \ll T_{s,1} \\ v_g \left(\dfrac{1}{1 + T_{s,1}/T_L} + \dfrac{1}{1 + T_{s,2}/T_L} \right)^{1/2} & \text{for } T_L < T_{s,1} \end{cases}$$

(1.67)

where v_g is the velocity dispersion of the gas. (Note that the second expression is valid only for $T_{s,2} \ll T_{s,1}$, see [60] for details.) If the relative velocity due to turbulence is larger than that due to sedimentation or radial drift, turbulence can accelerate particle growth. For particles of $T_i < T_{s,1} < T_L$, the relative velocity v_{rel} due to turbulence decreases with decreasing $T_{s,1}$ as $v_{rel} \propto T_{s,1}^{1/2}$, while the relative velocity due to sedimentation or radial drift decreases more steeply ($v_{rel} \propto T_{s,1}$). Thus, collision velocities between small particles are probably determined by turbulent motion. For particles of $T_{s,1} \sim 1$, the contribution of sedimentation or radial drift becomes great. The maximum relative velocity due to turbulence is $v_{rel} = v_g$, which is for collisions between a large particle ($T_{s,1} > T_L$) and a small particle ($T_{s,2} < T_L$). However, when $T_{s,1} = 1$, the radial velocity reaches its maximum, $v_{r,max} = \eta v_K/2 \sim (h_g/r)c_s$. (The sedimentation velocity is smaller than v_r for particles that have settled onto the midplane, i.e., $z < z_r$.) Thus, for particles of $T_{s,1} = 1$, the radial velocity may possibly determine the collision velocity. For the turbulent motion of $T_s \sim 1$ particles to accelerate their growth, v_g must be larger than $v_{r,max}$. (We assume $T_L \le 1$.) Because $\eta \sim (h_g/r)^2$, this means that $v_g > (h_g/r)c_s \sim 0.1c_s$. However, the turbulent velocity is usually much smaller than the sound speed. Hence, the acceleration of particle growth for $T_s \sim 1$ is probably not important.

Turbulence may accelerate particle growth by trapping particles and producing high dust densities. Barge & Sommeria [9] and Tanga et al. [79] showed that particles accumulate in long-lived vortices. In an anticyclonic vortex, the Coriolis force is directed toward the vortex center, and the dust particles, which are not supported by the gas pressure, move toward the vortex center. Klahr & Henning [48] also proposed that in large eddies, particles concentrate at the point where the centrifugal force due to the eddy motion balances with the vertical component of the central star's gravity. Recent numerical simulations have shown that dust clumps form in turbulent gas disks and that in some cases, the dust densities increase by a factor ~ 100 [26, 42, 44]. In such dust concentrations, it is expected that particle growth is accelerated.

Several other mechanisms that enhance the particle density have also been proposed. The particles migrate radially in the opposite direction of the gas pressure gradient. Thus, if a place exists in the gas disk where the pressure has a local maximum (with regard to r), the particles would accumulate there [31, 32]. Even though no pressure maximum exists, the inwardly migrating particles may cause "traffic jams" if the radial flux $|2\pi r \Sigma_d v_r|$ increases with r [88, 91]. Once the particle density starts to increase at some locations, the increased back reaction of the dust on the gas modifies the gas orbital speed and decelerates the inward particle drift. Consequently, these dense parts catch faster drifting particles from larger radii and become denser, leading to a positive feedback for density enhancement [29]. Recently, it was also shown that the relative motion of the dust and the gas cause streaming instability that can trigger the formation of dust clumps [43, 89, 90]. Such dust accumulation mechanisms may play an important role in planetesimal formation.

1.6 Conclusions

We discussed dispersal of the disk gas due to viscous accretion and photoevaporation. Photoevaporation seems to be a plausible mechanism for gas removal, provided that the central star emits enough ionizing photons. It is not well known, however, whether the pre-main-sequence stars continue to emit enough photons during the gas dispersal. Studies on the origin and magnitude of the ionizing photons are essential, and the timing of gas removal may determine the mass of gas giant planets. Numerical simulations showed that gas giant planets continue to swallow the disk gas even after the planets open gaps around their orbits in the gas disk [10, 55]. Thus, the masses of giant planets are thought to be determined by the time the disk gas disappears [49]. In the theory of planet formation, the lifetime of the gas is an important unknown parameter, and we need further theoretical and observational studies on gas removal from protoplanetary disks.

How the planetesimals form is another major question in planet formation theory. We do not know through which path the planetesimals form, whether via gravitational instability or collisional growth. For the initial growth of the dust during sedimentation, a large uncertainty exists about the internal structure and porosity of the dust aggregates. If dust growth occurs through cluster–cluster aggregation, the bulk densities of the dust aggregates decrease with their growth. Consequently, the sizes of the aggregates when they have settled on the midplane can be much larger than those in the case of compact particles [(1.53)]. The settled dust bodies form a thin dust layer at the midplane. The dynamics of the dust bodies in the dust layer are also not well understood. When the density of the dust layer exceeds that of the gas, the dust layer begins to orbit faster than the gas above and beneath it, and turbulence due to Kelvin–Helmholtz instability develops. We do not understand well, how the dust bodies respond to the turbulent motion of the gas. For example, it is not clear whether the dust particles are completely stirred and mixed in the dust layer, or whether some large particles can continue to sediment toward the midplane. Dust concentration is an important topic because an increase in the dust density accelerates collisional growth and may also trigger gravitational instability. Several mechanisms underlying the formation of dust concentration, such as traffic jams, anticyclonic vortices, and streaming instability, have been proposed, and it is important to investigate the density of the concentrations induced by these mechanisms. Gravitational instability and Kelvin–Helmholtz instability of the dust layer require further investigation, taking into account different responses of the dust to the gas as a function of the particle size. In this chapter, we treat the sticking parameter C_{stk} as an unknown parameter. It is important to know C_{stk} as a function of the particle size and of the collisional velocity. For particles of $T_{\mathrm{s}} = 1$, the collisional velocity due to the difference in the radial velocity can be as large as $100 \mathrm{~m~s}^{-1}$ (at 1 AU). It is not known whether such high velocity collisions result in sticking or destruction (see [87] for collision experiments). These questions must be addressed to elucidate how the planetesimals form.

Acknowledgments

I thank Ingrid Mann and Yoshitsugu Nakagawa for valuable discussions, and Sei-ichiro Watanabe, Hidekazu Tanaka, and Mark C. Wyatt for helpful comments that improved the manuscript. This work was supported in part by the Grant-in-Aid for Scientific Research, Nos 17740107 and 17039009, of the Ministry of Education, Culture, Sports, Science, and Technology of Japan.

Appendix: List of Symbols

C_{D}	drag coefficient
$C_{\mathrm{flow,g}}$, $C_{\mathrm{flow,in}}$	coefficients of the escaping gas densities at r_{g} and r_{in}
C_{stk}	sticking probability
c_{s}	sound speed
D	fractal dimension
D_{d}	diffusion coefficient
e	eccentricity

F_{dif}	mass flux due to diffusion
F_{drag}	gas drag force
$F_{\mathrm{g},z}$	vertical component of the gravity
f_{dust}	dust-to-gas ratio
G	gravitational constant
h_{d}	thickness of the dust layer
h_{g}	scale height of the gas disk
H	angular momentum
i	inclination
k_{B}	Boltzmann constant
L	stellar luminosity
L_{acc}	accretion luminosity
l_{free}	mean free path of gas molecules
M	stellar mass
$\dot{M}_{\mathrm{acc}}$	mass accretion rate
$\dot{M}_{\mathrm{w}}$	wind mass loss rate
M_{g}	disk mass inside 100 AU
m_0	monomer mass
m_{g}	mass of a gas molecule
m_{H}	mass of a hydrogen atom
m_{p}	particle mass
n_{flow}	number density of the escaping gas
$n_{\mathrm{g}}, n_{\mathrm{in}}$	number densities of the escaping gas at r_{g} and r_{in}
P	gas pressure
p	power-law index of the gas density profile
q	power-law index of the gas temperature profile
R_*	stellar radius
Re	Reynolds number
r_{g}	gravitational radius
$r_{\mathrm{in}}, r_{\mathrm{out}}$	disk inner and outer radii
Sc	Schmidt number
s	particle radius
s_0	monomer radius
s_1	particle radius when sedimentation have finished
s_2	particle radius during radial drift
T	gas temperature
T_{eff}	effective temperature of the star
$T_{\mathrm{i}}, T_{\mathrm{L}}$	sizes of the smallest and largest eddies
T_{s}	dimensionless stopping time
v_{B}	relative velocity due to Brownian motion
v_{col}	collision velocity
v_{g}	velocity dispersion of the gas
$v_{g,r}, v_{g,\theta}$	velocity of the gas
v_{i}	thermal velocity of the ionized gas
v_{K}	Keplerian velocity

v_r, v_θ, v_z	particle velocity
v_{rel}	relative velocity
v_T	mean thermal velocity
z_d	thickness of the dust layer when $\rho_d = \rho_g$
z_r	height where $v_r = v_z$
α	viscosity parameter
α_2	recombination coefficient
β	power-law index of the escaping gas density profile
γ	power-law index of the viscosity profile
Δv	relative velocity between the dust and the gas
η	deviation factor from the Keplerian speed
ν_{mol}	molecular kinematic viscosity
ν_{vis}	turbulent kinematic viscosity
ρ_d	dust density
ρ_g	gas density
$\rho_{g,mid}$	gas density at the midplane
ρ_p	particle bulk density
ρ_{Roche}	Roche density
Σ_d	dust surface density
Σ_g	gas surface density
σ_{mol}	collisional cross section of hydrogen molecules
σ_{SB}	Stefan–Boltzmann constant
τ_{dif}	timescale of viscous diffusion in the vertical direction
τ_{grow}	growth timescale
$\tau_{grow,E}$	growth timescale under the Epstein regime
$\tau_{grow,S}$	growth timescale under the Stokes regime
$\tau_{grow,sed}$	growth timescale during sedimentation
τ_{infall}	infall timescale
τ_s	stopping time
τ_{sed}	sedimentation timescale
τ_{vis}	timescale of viscous diffusion in the radial direction
τ_w	wind mass loss timescale
Φ_i	number of EUV photons emitted from the star per unit time
Ω_g	angular velocity of the gas
Ω_K	Keplerian angular velocity
subscript 0	evaluated at the reference distance r_0 (except for s_0 and m_0)

References

1. I. Adachi, C. Hayashi, and K. Nakazawa: *The gas drag effect on the elliptical motion of a solid body in the primordial solar nebula.*, Prog. Theor. Phys. **56**, 1756 (1976)
2. R. D. Alexander, C. J. Clarke, and J. E. Pringle: *Constraints on the ionizing flux emitted by T Tauri stars*, Mon. Not. Royal Astron. Soc. **358**, 283 (2005)

3. R. D. Alexander, C. J. Clarke, and J. E. Pringle: *Photoevaporation of proto-planetary discs – I. Hydrodynamic models*, Mon. Not. Royal Astron. Soc. **369**, 216 (2006)
4. S. M. Andrews and J. P. Williams: *Circumstellar dust disks in Taurus-Auriga: the submillimeter perspective*, Astrophys. J. **631**, 1134 (2005)
5. S. M. Andrews and J. P. Williams: *High-resolution submillimeter constraints on circumstellar disk structure*, Astrophys. J. **659**, 705 (2007)
6. P. Artymowicz: *Beta pictoris: an early solar system?*, Annu. Rev. Earth Planet. Sci. **25**, 175 (1997)
7. S. A. Balbus and J. F. Hawley: *Instability, turbulence, and enhanced transport in accretion disks*, Rev. Mod. Phys. **70**, 1 (1998)
8. J. Bally, C. R. O'Dell, and M. J. McCaughrean: *Disks, microjets, windblown bubbles, and outflows in the Orion Nebula*, Astron. J. **119**, 2919 (2000)
9. P. Barge and J. Sommeria: *Did planet formation begin inside persistent gaseous vortices?*, Astron. Astrophys. **295**, L1 (1995)
10. M. R. Bate, S. H. Lubow, G. I. Ogilvie, and K. A. Miller: *Three-dimensional calculations of high- and low-mass planets embedded in protoplanetary discs*, Mon. Not. Royal Astron. Soc. **341**, 213 (2003)
11. S. V. W. Beckwith and A. I. Sargent: *Circumstellar disks and the search for neighbouring planetary systems*, Nature **383**, 139 (1996)
12. S. V. W. Beckwith, A. I. Sargent, R. S. Chini, and R. Guesten: *A survey for circumstellar disks around young stellar objects*, Astron. J. **99**, 924 (1990)
13. A. Brandeker, R. Liseau, G. Olofsson, and M. Fridlund: *The spatial structure of the β Pictoris gas disk*, Astron. Astrophys. **413**, 681 (2004)
14. P. Cassen and A. Moosman: *On the formation of protostellar disks*, Icarus **48**, 353 (1981)
15. S. Chapman and T. G. Cowling: *The Mathematical Theory of Non-Uniform Gases: An Account of the Kinetic Theory of Viscosity, Thermal Combustion and Diffusion in Gases (Cambridge Mathematical Library)*, (Cambridge Univ. Press, Cambridge, 1991)
16. C. H. Chen and I. Kamp: *Are giant planets forming around HR 4796A?*, Astrophys. J. **602**, 985 (2004)
17. E. I. Chiang and P. Goldreich: *Spectral energy distributions of T Tauri stars with passive circumstellar disks*, Astrophys. J. **490**, 368 (1997)
18. I. M. Coulson, W. R. F. Dent, and J. S. Greaves: *The absence of CO from the dust peak around ε Eri*, Mon. Not. Royal Astron. Soc. **348**, L39 (2004)
19. J. N. Cuzzi and R. C. Hogan: *Blowing in the wind I. Velocities of chondrule-sized particles in a turbulent protoplanetary nebula*, Icarus **164**, 127 (2003)
20. J. N. Cuzzi, A. R. Dobrovolskis, and J. M. Champney: *Particle–gas dynamics in the midplane of a protoplanetary nebula*, Icarus **106**, 102 (1993)
21. C. Dominik and A. G. G. M. Tielens: *The physics of dust coagulation and the structure of dust aggregates in space*, Astrophys. J. **480**, 647 (1997)
22. A. Dutrey, S. Guilloteau, G. Duvert, L. Prato, M. Simon, K. Schuster, and F. Menard: *Dust and gas distribution around T Tauri stars in Taurus-Auriga. I. Interferometric 2.7mm continuum and ^{13}CO J=1-0 observations* , Astron. Astrophys. **309**, 493 (1996)
23. A. Dutrey, S. Guilloteau, and M. Guelin: *Chemistry of protosolar-like nebulae: the molecular content of the DM Tau and GG Tau disks*, Astron. Astrophys. **317**, L55 (1997)

24. A. Dutrey, A. Lecavelier Des Etangs, and J.-C. Augereau: *The observation of circumstellar disks: dust and gas components*, in Comets II, M. C. Festou, H. U. Keller, and H. A. Weaver (Eds.), (Univ. of Arizona Press, Tucson, 2004) 81

25. J. Frank, A. King, and D. Raine: *Accretion Power in Astrophysics*, (Cambridge Univ. Press, Cambridge, 1992)

26. S. Fromang and R. P. Nelson: *On the accumulation of solid bodies in global turbulent protoplanetary disc models* Mon. Not. Royal Astron. Soc. **364**, L81 (2005)

27. P. Garaud, L. Barriere-Fouchet, and D. N. C. Lin: *Individual and average behavior of particles in a protoplanetary nebula*, Astrophys. J. **603**, 292 (2004)

28. P. Goldreich and W. R. Ward: *The formation of planetesimals*, Astrophys. J. **183**, 1051 (1973)

29. J. Goodman and B. Pindor: *Secular instability and planetesimal formation in the dust layer*, Icarus **148**, 537 (2000).

30. J. S. Greaves: *Dense gas discs around T Tauri stars*, Mon. Not. Royal Astron. Soc. **351**, L99 (2004)

31. N. Haghighipour and A. P. Boss: *On pressure gradients and rapid migration of solids in a nonuniform solar nebula*, Astrophys. J. **583**, 996 (2003a)

32. N. Haghighipour and A. P. Boss: *On gas drag-induced rapid migration of solids in a nonuniform solar nebula*, Astrophys. J. **598**, 1301 (2003b)

33. P. Hartigan, S. Edwards, and L. Ghandour: *Disk accretion and mass loss from young stars*, Astrophys. J. **452**, 736 (1995)

34. L. Hartmann: *Accretion Processes in Star Formation*, (Cambridge Univ. Press, Cambridge, 1998)

35. L. Hartmann, N. Calvet, E. Gullbring, and P. D'Alessio: *Accretion and the evolutiion of T Tauri disks*, Astrophys. J. **495**, 385 (1998)

36. J. F. Hawley, C. F. Gammie, and S. A. Balbus: *Local three-dimensional magnetohydrodynamic simulations of accretion disks*, Astrophys. J. **440**, 742 (1995)

37. J. F. Hawley, C. F. Gammie, and S. A. Balbus: *Local three-dimensional simulations of an accretion disk hydromagnetic dynamo*, Astrophys. J. **464**, 690 (1996)

38. D. Hollenbach, D. Johnstone, S. Lizano, and F. Shu: *Photoevaporation of disks around massive stars and application to ultracompact H II regions*, Astrophys. J. **428**, 654 (1994)

39. D. J. Hollenbach, H. W. Yorke, and D. Johnstone: *Disk dispersal around young stars*, in Protostars and Planets IV, V. Mannings, A. P. Boss, and S. S. Russell (Eds.), (Univ. of Arizona Press, Tucson , 2000), 401

40. N. Ishitsu and M. Sekiya: *Shear instabilities in the dust layer of the solar nebula III. Effects of the Coriolis force*, Earth Planets Space **54**, 917 (2002)

41. N. Ishitsu and M. Sekiya: *The effects of the tidal force on shear instabilities in the dust layer of the solar nebula*, Icarus **165**, 181 (2003)

42. A. Johansen and H. Klahr: *Dust diffusion in protoplanetary disks by magnetorotational turbulence* Astrophys. J. **634**, 1353 (2005)

43. A. Johansen and A. Youdin: *Protoplanetary disk turbulence driven by the streaming instability: nonlinear saturation and particle concentration*, Astrophys. J. **662**, 627 (2007)

44. A. Johansen, T. Henning, and H. Klahr: *Dust sedimentation and self-sustained Kelvin–Helmholtz turbulence in protoplanetary disk midplanes*, Astrophys. J. **643**, 1219 (2006)
45. S. Kempf, S. Pfalzner, and T. K. Henning: *N-Particle-simulations of dust growth. I. Growth driven by Brownian motion*, Icarus **141**, 388 (1999)
46. S. J. Kenyon and L. Hartmann: *Spectral energy distributions of T Tauri stars – Disk flaring and limits on accretion*, Astrophys. J. **323**, 714 (1987)
47. Y. Kitamura, M. Momose, S. Yokogawa, R. Kawabe, M. Tamura, and S. Ida: *Investigation of the physical properties of protoplanetary disks around T Tauri stars by a 1 arcsecond imaging survey: evolution and diversity of the disks in their accretion stage*, Astrophys. J. **581**, 357 (2002)
48. H. H. Klahr and T. Henning: *Particle-trapping eddies in protoplanetary accretion disks*, Icarus **128**, 213 (1997)
49. E. Kokubo and S. Ida: *Formation of protoplanet systems and diversity of planetary systems*, Astrophys. J. **581**, 666 (2002)
50. M. Krause and J. Blum: *Growth and form of planetary seedlings: results from a sounding rocket microgravity aggregation experiment*, Phys. Rev. Lett. **93**, 021103 (2004)
51. A.-M. Lagrange, D. E. Backman, and P. Artymowicz: *Planetary material around main-sequence stars*, in Protostars and Planets IV, V. Mannings, A. P. Boss, and S. S. Russell (Eds.), (Univ. of Arizona Press, Tucson: 2000), 639
52. L. D. Landau and E. M. Lifshitz: *Fluid Mechanics (Course of Theoretical Physics)* , Butterworth-Heinemann (1987)
53. A. Lecavelier des Etangs, et al.: *Deficiency of molecular hydrogen in the disk of β Pictoris*, Nature **412**, 706 (2001).
54. R. Liseau: *Molecular line observations of southern main-sequence stars with dust disks: α Ps A, β Pic, ε Eri and HR 4796 A. Does the low gas content of the β Pic and ε Eri disks hint of planets?*, Astron. Astrophys. **348**, 133 (1999)
55. S. H. Lubow, M. Seibert, and P. Artymowicz: *Disk accretion onto high-mass planets*, Astrophys. J. **526**, 1001 (1999)
56. W. J. Markiewicz, H. Mizuno, and H. J. Völk: *Turbulence induced relative velocity between two grains*, Astron. Astrophys. **242**, 286 (1991)
57. I. Matsuyama, D. Johnstone, and L. Hartmann: *Viscous diffusion and photoevaporation of stellar disks*, Astrophys. J. **582**, 893 (2003)
58. Y. Nakagawa, M. Sekiya, and C. Hayashi: *Settling and growth of dust particles in a laminar phase of a low-mass solar nebula*, Icarus **67**, 375 (1986)
59. H. Nomura and Y. Nakagawa: *Dust size growth and settling in a protoplanetary disk*, Astrophys. J. **640**, 1099 (2006)
60. C. W. Ormel and J. N. Cuzzi: *Closed-form expressions for particle relative velocities induced by turbulence*, Astron. Astrophys. **466**, 413 (2007)
61. C. W. Ormel, M. Spaans, A. G. G. M. Tielens: *Dust coagulation in protoplanetary disks: porosity matters* Astron. Astrophys. **461**, 215 (2007)
62. V. Ossenkopf: *Dust coagulation in dense molecular clouds: the formation of fluffy aggregates*, Astron. Astrophys. **280**, 617 (1993)
63. D. E. Osterbrock: *Astrophysics of Gaseous Nebulae and Active Galactic Nuclei*, (Univ. Science Books, Mill Valley, 1989)
64. M. Osterloh and S. V. W. Beckwith: *Millimeter-wave continuum measurements of young stars*, Astrophys. J. **439**, 288 (1995)
65. D. Paszun and C. Dominik: *The influence of grain rotation on the structure of dust aggregates*, Icarus **182**, 274 (2006)

66. J. B. Pollack, D. Hollenbach, S. Beckwith, D. P. Simonelli, T. Roush, and W. Fong: *Composition and radiative properties of grains in molecular clouds and accretion disks*, Astrophys. J. **421**, 615 (1994)

67. J. E. Pringle: *Accretion discs in astrophysics*, Annu. Rev. Astron. Astrophys., **19**, 137 (1981)

68. S. P. Ruden: *Evolution of photoevaporating protoplanetary disks*, Astrophys. J. **605**, 880 (2004)

69. T. Sano and J. M. Stone: *The effect of the Hall term on the nonlinear evolution of the magnetorotational instability. II. Saturation level and critical magnetic Reynolds number*, Astrophys. J. **577**, 534 (2002)

70. T. Sano, S. Inutsuka, N. J. Turner, and J. M. Stone: *Angular momentum transport by magnetohydrodynamic turbulence in accretion disks: gas pressure dependence of the saturation level of the magnetorotational instability*, Astrophys. J. **605**, 321 (2004)

71. M. Sekiya: *Gravitational instabilities in a dust–gas layer and formation of planetesimals in the solar nebula*, Prog. Theor. Phys. **69**, 1116 (1983)

72. M. Sekiya: *Quasi-equilibrium density distributions of small dust aggregations in the solar nebula*, Icarus **133**, 298 (1998)

73. M. Sekiya and N. Ishitsu: *Shear instabilities in the dust layer of the solar nebula I. The linear analysis of a non-gravitating one-fluid model without the Coriolis and the solar tidal forces*, Earth Planets Space **52**, 517 (2000)

74. M. Sekiya and N. Ishitsu: *Shear instabilities in the dust layer of the solar nebula II. Different unperturbed states*, Earth Planets Space **53**, 761 (2001)

75. S. W. Stahler and F. Palla: *The Formation of Stars*, (Wiley-VCH Verlag GmbH & Co.KGaA, Berlin, 2004)

76. J. M. Stone, J. F. Hawley, C. F. Gammie, and S. A. Balbus: *Three-dimensional magnetohydrodynamical simulations of vertically stratified accretion disks*, Astrophys, J. **463**, 656 (1996)

77. T. Takeuchi and D. N. C. Lin: *Radial flow of dust particles in accretion disks*, Astrophys. J. **581**, 1344 (2002)

78. T. Takeuchi, S. M. Miyama, and D. N. C. Lin: *Gap formation in protoplanetary disks*, Astrophys. J. **460**, 832 (1996)

79. P. Tanga, A. Babiano, B. Dubrulle, and A. Provenzale: *Forming planetesimals in vortices*, Icarus **121**, 158 (1996)

80. S. Terebey, F. H. Shu, and P. Cassen: *The collapse of the cores of slowly rotating isothermal clouds*, Astrophys. J. **286**, 529 (1984)

81. H. J. Völk, F. C. Jones, G. E. Morfill, and S. Roeser: *Collisions between grains in a turbulent gas*, Astron. Astrophys. **85**, 316 (1980)

82. S. J. Weidenschilling: *Aerodynamics of solid bodies in the solar nebula*, Mon. Not. Royal Astron. Soc. **180**, 57 (1977)

83. S. J. Weidenschilling: *Evolution of grains in a turbulent solar nebula*, Icarus **60**, 553 (1984)

84. S. J. Weidenschilling: *Radial drift of particles in the solar nebula: implications for planetesimal formation*, Icarus **165**, 438 (2003)

85. S. J. Weidenschilling: *From icy grains to comets* in Comets II, M. C. Festou, H. U. Keller, and H. A. Weaver (Eds.), (Univ. of Arizona Press, Tucson, 2004), 97

86. S. J. Weidenschilling: *Models of particle layers in the midplane of the solar nebula*, Icarus **181**, 572 (2006)

87. G. Wurm, G. Paraskov, and O. Krauss: *Growth of planetesimals by impacts at ∼25 m/s*, Icarus **178**, 253 (2005)
88. A. N. Youdin and E. I. Chiang: *Particle pileups and planetesimal formation*, Astrophys. J. **601**, 1109 (2004)
89. A. N. Youdin and J. Goodman: *Streaming instabilities in protoplanetary disks*, Astrophys. J. **620**, 459 (2005)
90. A. Youdin and A. Johansen: *Protoplanetary disk turbulence driven by the streaming instability: linear evolution and numerical methods*, Astrophys. J. **662**, 613 (2007)
91. A. N. Youdin and F. H. Shu: *Planetesimal formation by gravitational instability*, Astrophys. J. **580**, 494 (2002)

2

Dynamics of Small Bodies
in Planetary Systems

M.C. Wyatt

Institute of Astronomy, University of Cambridge, Cambridge CB3 0HA, UK,
wyatt@ast.cam.ac.uk

Abstract The number of stars that are known to have debris disks is greater than that of stars known to harbor planets. These disks are detected because dust is created in the destruction of planetesimals in the disks much in the same way that dust is produced in the asteroid belt and Kuiper belt in the solar system. For the nearest stars, the structure of their debris disks can be directly imaged, showing a wide variety of both axisymmetric and asymmetric structures. A successful interpretation of these images requires a knowledge of the dynamics of small bodies in planetary systems, since this allows the observed dust distribution to be deconvolved to provide information on the distribution of larger objects, such as planetesimals and planets. This chapter reviews the structures seen in debris disks, and describes a disk-dynamical theory which can be used to interpret those observations. In this way much of the observed structures, both axisymmetric and asymmetric, can be explained by a model in which the dust is produced in a planetesimal belt which is perturbed by a nearby, as yet unseen, planet. While the planet predictions still require confirmation, it is clear that debris disks have the potential to provide unique information about the structure of extrasolar planetary systems, since they can tell us about planets analogous to Neptune and even the Earth. Significant failings of the model at present are its inability to predict the quantity of small grains in a system, and to explain the origin of the transient dust seen in some systems. Given the complexity of planetary system dynamics and how that is readily reflected in the structure of a debris disk, it seems inevitable that the study of debris disks will play a vital role in our understanding of extrasolar planetary systems.

2.1 Introduction

Planetary systems are not just made up of *planets*, but are also composed of numerous small bodies ranging from asteroids and comets as large as 1000 km down to sub-micrometer-sized dust grains. In the solar system, the asteroids and comets are confined to relatively narrow rings known as the asteroid belt and the Kuiper belt (see chapters by Nakamura and Jewitt). These belts are the source of the majority of the smaller objects seen in the solar

Wyatt, M.C.: *Dynamics of Small Bodies in Planetary Systems.* Lect. Notes Phys. **758**, 37–70 (2009)
DOI 10.1007/978-3-540-76935-4_2

system, since such objects are inevitably created in collisions between objects within the belts (see chapter by Michel). Sublimation of comets as they are heated on approach to the Sun is another source of dust in the solar system.

It is known that extrasolar systems also host belts of planetesimals (a generic name for comets and asteroids) that are similar to our own asteroid belt and Kuiper belt. These were first discovered using far-IR observations of nearby stars, which showed excess emission above that expected to come from the stellar photosphere [5]. This emission comes from dust that is heated by the star and which reradiates that energy in the thermal infrared, at temperatures between 40 and 200 K, depending on the distance of the dust from the star. The lifetime of the dust is inferred to be short compared with the age of the star, and so it is concluded that the dust cannot be a remnant of the protoplanetary disk that formed with the star (see chapter by Takeuchi), rather it must originate in planetesimal belts much in the same way that dust is created in the solar system [7].

Over 300 main-sequence stars are now known with this type of excess emission [10, 50, 74], and such objects are known either as Vega-like (after the first star discovered to have this excess) or as debris disks. Statistical studies have shown that $\sim 15\%$ of normal main-sequence stars have debris disks, although it should be stressed that the disks which can be detected with current technology have greater quantities of dust than is currently present in the solar system by a factor of at least 10 [25]. Nevertheless, this indicates that debris disks are common, more common in fact that extrasolar planets which are found around $\sim 6\%$ of stars [18]. Studying these disks provides a unique insight into the structure of the planetary systems of other stars. Indeed, the nearest and brightest debris disks can be imaged, and such studies have provided the first images of nearby planetary systems. These images reveal the distribution of dust in the systems, which can in turn be used to infer the distribution of parent planetesimals, and also the architecture of the underlying planetary system. However, to do so requires an understanding of both the mechanism by which dust is produced in planetesimal belts and its consequent dynamical evolution, as well as of the dynamical interaction between planets and planetesimals and between planets and dust.

This chapter reviews our knowledge of debris disks from observations (Sect. 2.2) and describes a simple model for planetesimal belt evolution which explains what we see (Sect. 2.3), as well as how the detailed interaction between planets and planetesimals imposes structure on that planetesimal belt (Sect. 2.4), and how those perturbations translate into structures seen in the dust distribution (Sect. 2.5). Conclusions, including what has been learned about the planetary systems of nearby stars from studying these disks, are given in Sect. 2.6.

2.2 Observed Debris Disk Structures

The debris disks with well-resolved structure are summarized in Table 2.1.[1] There are two types of debris disk structure: axisymmetric structure (i.e., dust or planetesimal surface density as a function of distance from the star) and asymmetric structure (i.e., how that surface density varies as a function of azimuth). I will deal with each of these in turn.

Table 2.1. Summary of observed properties of debris disks the structure of which has been significantly resolved at one wavelength or more. Asymmetries are identified as: W:Warp, C:Clump, S:Spiral, B:Brightness asymmetry, O:Offset, H:Hot dust component, N:No discernible asymmetry

Name	Sp type	Age, Myr	r, AU	i, °	$f = L_{ir}/L_\star$	Asymm.	Ref.
HD141569	B9.5e	5	34–1200	35	84×10^{-4}	S	[12, 17]
HR4796	A0V	8	60–80	17	50×10^{-4}	B	[67, 76]
β Pictoris	A5V	12	10–1835	~3	26×10^{-4}	WC	[28, 29, 77]
HD15115	F2V	12	31–554	~0	5×10^{-4}	B	[37]
HD181327	K2V	12	68–104	58	25×10^{-4}	N	[69]
AU Mic	M1Ve	12	12–210	~0	6×10^{-4}	WC	[48]
HD32297	A0V	<30	40–1680	10	27×10^{-4}	B	[33]
HD107146	G2V	100	80–185	65	12×10^{-4}	N	[2]
HD92945	K1V	100	45–175	29	8×10^{-4}	N	[21]
Fomalhaut	A3V	200	133–158	24	0.8×10^{-4}	(CB)O	[30, 35, 71]
HD139664	F5V	300	83–109	<5	0.9×10^{-4}	N	[36]
Vega	A0V	350	90–800	~90	0.2×10^{-4}	C	[29, 38, 51, 73]
ε Eridani	K2V	850	40–105	65	0.8×10^{-4}	CO	[26]
η Corvi	F2V	1000	1.5, 150	45	5.3×10^{-4}	CH	[93]
HD53143	K1V	1000	55–110	45	2.5×10^{-4}	N	[36]
τ Ceti	G2V	10000	~55	60–90	0.3×10^{-4}	N	[25]

2.2.1 Axisymmetric Structure

The most basic information about the structure of a debris disk that we can obtain is the distance of the dust from the star. This can be deduced without resolving the dust location by looking at the Spectral Energy Distribution (SED), since this indicates the temperature of the dust, which by thermal balance with the stellar luminosity tells us its distance from the star. For black body dust

---

[1] Resolved disks have also been reported for the following stars: HD92945, HD61005, HD10647, HD202917, and HD207129 (see http://astro.berkeley.edu/~kalas/lyot2007/agenda.html), and HD15745 (Kalas et al., ApJ, submitted). I have excluded these images from the discussion, since they have yet to appear in the literature at the time of writing.

$$T_{\rm bb} = 278.3 L_\star^{0.25}/\sqrt{r}, \tag{2.1}$$

where $L_\star$ is in $L_\odot$ and r is distance from star in AU. Thus dust location, r, can be estimated as long as the level of dust emission has been measured at two or more wavelengths from which its temperature can be estimated.

However, such estimates suffer large uncertainties, since the exact temperature of the dust depends on its size and composition (see chapter by Li). Assuming black body emission for the grains can underestimate (or overestimate) the distance of the dust from the star by a factor of 3 or more if the dust is small [69], since small grains emit inefficiently at long wavelengths and so attain equilibrium temperatures that are significantly higher than black body [91]. Likewise, an SED which can be fitted by a black body emission spectrum does not necessarily indicate that all of the dust is at a single distance from the star, any more than one that requires multiple temperatures indicates that the disk is broad, since dust at multiple distances can appear to have one temperature, and dust with a range of sizes at the same distance from the star have a range of temperatures [53]. This underlines the fact that the interpretation of SEDs is degenerate, and that in order to determine the radial structure of a disk it needs to be spatially resolved. On the other hand, once the radial location of the dust is known the information in the SED is extremely valuable, since it allows a determination of the emission properties of the grains, and hence of their size and/or composition [45, 90].

Nevertheless, it seems that the majority of the known debris disks have SEDs that are dominated by dust at a single temperature, and are seen in images to be dominated by dust at a distance from the star that is compatible with that temperature. More often than not that distance is > 30 AU from the star, which means that debris disks are analogous to our own Kuiper belt [92]. Naturally, the fact that these disks have inner holes similar in size to the planetary system in our solar system leads to the intriguing possibility that there is an (as yet) unseen planetary system sweeping these regions free of both planetesimals and dust. I will return to the putative planetary system in Sect. 2.4. Here, I simply note that while these inner holes are usually dust free [82], a few systems are known with dust within this hole, such as η Corvi [93] and Vega [1]. The hot dust in these systems is thought to be transiently regenerated [94], and care would need to be taken when interpreting observations of the hot dust within the framework described in this chapter (see Sect. 2.6).

Exactly how broad these disks are, is a matter for debate. Optical imaging suggests that there are two types of disk: narrow and broad [36]. However, detectability may be an issue in some cases, since a disk's outer edge is often difficult to detect, as the fraction of intercepted starlight falls off with radius, much in the same way that it was not known for a long time whether or not the outer edge of the Kuiper belt is abrupt [80]. Some disks clearly are extended though, such as that of β Pictoris, which is seen to extend out to > 1000 AU in optical imaging [70], but which is seen as close as in 10–20 AU in mid-IR

imaging [77]. While disks with a dust distribution as broad as that of β Pictoris are rare, the presence of dust at large distances from the star is becoming
more common-place. It is now known that dust in the archetypal debris disk
Vega is not confined to ∼90 AU as suggested by sub-millimeter images [29].
Rather the dust distribution seen at 24 and 70 μm extends out to 800 AU
[73]. This defies intuition, since if the dust is at a range of distances, the disk
should appear smaller at the shorter wavelengths (since shorter wavelengths
tend to probe hotter dust). This intuitive behavior is indeed seen in the disk
of Fomalhaut [71]. It is thought that the counter-intuitive behavior of Vega's
disk arises because the grain size distribution changes with distance: the dust
seen at large distances is small, of order a few micrometers, and so is heated
above black body and emits very inefficiently in the sub-millimeter, whereas
that seen at ∼90 AU is large, millimeter- to centimeter-sized, and emits efficiently in the sub-millimeter at (relatively low) black body-like temperatures.
A similar change in size distribution with distance is seen in the extended dust
distributions of β Pictoris and AU Mic. The extension of these disks is not
seen in mid- and far-IR images, but in optical and near-IR images of scattered
starlight, and the colors and polarization of the scattered light show that the
dust at large distances in these systems is small, sub-micrometer in size [23].

2.2.2 Asymmetric Structure

While to first-order debris disks are rings of material, even if the location and
breadth of the rings is wavelength dependent, on closer inspection those that
have been imaged with sufficient clarity also exhibit significant asymmetries.
Different types of asymmetries have been identified which can be grouped into
the following categories: warps, spirals, offsets, brightness asymmetries, and
clumps. The observed structures are summarized in Fig. 2.1 and are discussed
in more detail below.

Warps

A warp arises when the plane of symmetry of a disk varies with distance from
the star. It is only edge-on extended debris disks for which a warp can be seen,
since this orientation allows the plane of symmetry at any given distance to be
readily identified with the location of the maximum surface brightness there.
Both of the edge-on disks with significant extension, β Pictoris and AU Mic,
are warped [28, 48], as is the structure of the zodiacal cloud in the solar system
[91]. Recent observations of β Pictoris suggest that its warp may in fact not
be continuous, and that there are two separate disks with different planes
of symmetry [20]. Since the images which show warps in debris disks have
been made in scattered light, it is important to point out that care must be
(and has been) taken when interpreting these observations, since asymmetric
scattering (i.e., the effect that causes back scattering to be stronger than
forward scattering) can introduce perceived asymmetries into observations of

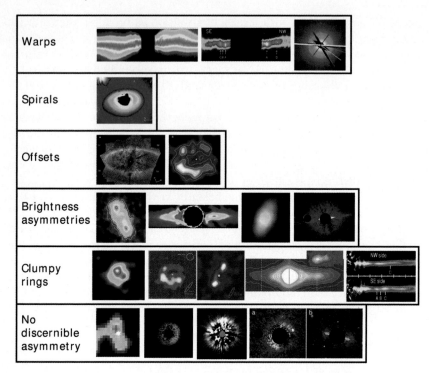

Fig. 2.1. Summary of asymmetries seen in the structures of debris disks. References for the images are from left to right: warps (β Pictoris [28], AU Mic reprinted with permission from AAAS [48], TW Hydra [65]); spirals (HD141569 [12]); offsets (Fomalhaut reprinted by permission from Macmillan Publishers Ltd: Nature [35] copyright 2005, ε Eridani [26]); brightness asymmetries (HR4796 [76], HD32297 [68], Fomalhaut [71], HD15115 [37]); clumpy rings (Vega [29], ε Eridani [24], Fomalhaut [30], β Pictoris [77], AU Mic [39]); no discernible asymmetry (τ Ceti [25], HD107146 [2], HD181327 [69], HD53143 [36], HD139664 [36])

an otherwise axisymmetric disk [34]. A warp has also been identified in the face-on disk of TW Hydra [65] from analysis of the emission spectrum which is affected by the fact that the warp prevents light from the star reaching the outer portions of the disk. This method of detecting a warp was possible for TW Hydra which has a classical T Tauri disk, but is not possible for face-on debris disks which are optically thin.

Spirals

The disk of HD141569 is seen to be significantly extended with dust out to 1200 AU where there are two M stars of similar age which are likely to be weakly bound to the star [81]. The radial distribution of dust is peaked at 150 and 250 AU. Optical coronagraphic imaging shows that both of these rings

is a tightly wound spiral [12]. The diffuse emission from 300 to 1200 AU also forms a more open spiral structure, with one, possibly two arms. Two armed open spiral structure is also reported in younger transition disks, such as AB Aur [46]. Recent observations of Vega suggest that its extended sub-millimeter emission is condensed into two spirals (W.S. Holland, priv.comm.).

Offsets

The star is not always at the center of the rings. This effect was first predicted [91], but then later dramatically seen in optical images of the Fomalhaut disk [35]. The Fomalhaut disk is narrow, and its proximity of 7.8 pc allowed the radius to be measured with great accuracy as a function of azimuth. With a mean disk radius of 133 AU, an offset of 15 AU was measured with significant confidence. The center of the ε Eridani disk is also seen to be offset [26], however the lower resolution of the sub-millimeter observations and its more complicated clumpy structure make the interpretation of this measurement less clear. Nevertheless, it is interesting to note that for the cases where such measurements can be made (nearby bright disks), an offset is seen.

Brightness Asymmetries

The offset effect was first predicted from observations of the HR4796 disk [76]. This edge-on disk was seen to be $\sim 5\%$ brighter on the NE than the SW side, an asymmetry which was attributed to an offset. However, there are other interpretations of brightness asymmetries, since all of the spiral, offset, and clump structures could appear as brightness asymmetries when seen edge-on. In other words, this class is likely another manifestation of one of the other types of structure. Indeed, the β Pictoris structure now attributed to a clump (see below) was originally seen as an asymmetry [43]. Likewise, the brightness asymmetry seen in mid- to far-IR images of the Fomalhaut disk [71], and which gets stronger at shorter wavelengths, can likely be attributed to the offset seen in optical images [35]. Other disks with brightness asymmetries include HD32297 [33] and HD15115 [37], for which the asymmetries are particularly pronounced. The latter is an example of a *needle* disk, which is seen to extend to significantly larger distances on one side of the star than the other. It is not clear if this is a brightness asymmetry (and the shorter side extends out to the same distance but at a level below the detection limit) or whether the two sides really are truncated at different outer radii.

Clumps

The most common type of asymmetry seen in debris disks is a change in brightness with azimuth around the ring, with much of the emission concentrated in one or more clumps. The clearest example of this phenomenon is the ε Eridani disk which is a narrow ring at 60 AU with a well-resolved inner hole [24]. The sub-millimeter images show four clumps of varying brightness within this ring. The interpretation of this structure has been confounded by the ubiquity of background galaxies which appear randomly across sub-millimeter images. However, the rapid proper motion of this star, which is at 3.6 pc, has allowed non-moving background objects to be identified, with three of the clumps confirmed as real using imaging covering a time-span of ∼5 years [26]. While the inner hole of the Vega disk is seen less clearly in 850 μm imaging, the emission in this disk, which is being seen close to face-on, is concentrated in two clumps that are equidistant from the star, but asymmetric in brightness [29]. The clumps are confirmed in millimeter-wavelength interferometry [38, 83], but appear at different locations in 350 μm imaging [51], and not at all in far-IR images [73], although that may be because of the low resolution of these observations. Other disks with clumps include Fomalhaut [30], although this may be a manifestation of the offset, β Pictoris [77], for which a brightness asymmetry appears to be originate in a clump with a sharp inner edge, and AU Mic [48], for which clumps are seen at a range of offsets from the star (although note that given the interpretation of the axisymmetric disk structure [72], all of these clumps are likely to be at the same distance from the star, just seen in projection).

No Detectable Asymmetry

Some of the resolved disks from Table 2.1 exhibit no discernible asymmetry in their structure. These are τ Ceti [25], HD107146 [2], HD181327 [69], HD53143 and HD139664 [36], and HD92945 [21]. However, this does not necessarily mean that the disks are symmetrical, since some of these images do not have the resolution and/or sensitivity to detect even large-scale asymmetries.

2.3 Debris Disk Models

The observed radial distribution of dust in debris disks can be explained as a consequence of planetesimal belt dynamics. Here, I build up a disk dynamical theory which explains how dust is created in a planetesimal belt, and how the combination of gravity, collisional processes, and radiation forces conspire to make the radial distribution of dust vary as a function of grain size.

2.3.1 The Planetesimal Belt

First, it is assumed that the outcome of planet formation was to create a ring of planetesimals at a radius r and of width dr. The dominant force acting on these planetesimals is the gravity of the star, and all material within the belt orbits the star. These orbits are defined by their semimajor axis, a, eccentricity, e, and orbital inclination, I, along with three angles defining the orientation of the orbit (longitude of pericentre, ϖ, and longitude of ascending node, Ω), and the position within it (e.g., mean longitude, λ, or true anomaly, f). There is a distribution of orbital elements which is assumed to be independent of size for the largest planetesimals. This is not the case during planet formation, wherein larger objects grow rapidly specifically because they have lower eccentricities and inclinations than smaller objects.

The size of planetesimals ranges from some maximum diameter $D_{\max}$ down to dust of size $D_{\min}$, and the size distribution is defined by the amount of cross-sectional area $\sigma(D)\mathrm{d}D$ in each size bin of width $\mathrm{d}D$; cross-sectional area is defined such that a spherical particle has an area of $\sigma = \pi(D/2)^2$. Taking the size distribution to be described by a power law,

$$\sigma(D) \propto D^{2-3q}, \tag{2.2}$$

it follows that, as long as the index q is in the range of $5/3$ to 2, the total amount of cross-sectional area in the belt, σ_{tot}, is dominated by the smallest objects within it, whereas its mass is dominated by the largest objects.

2.3.2 Collisions

While eccentricities and inclinations of planetesimals are assumed to be small, the resulting relative velocities are large enough that collisions are destructive. This is necessary if dust is to be produced in collisions rather than lost in growth to larger sizes [15].

Within the planetesimal belt collisions between planetesimals of different sizes are continually occurring. The result of such collisions is that the planetesimals are broken up into fragments with a range of sizes. If the outcome of collisions is self-similar (i.e., the size distribution of the fragments is scale invariant), and the range of sizes in the distribution is infinite, then the resulting size distribution has an exponent with $q = 11/6$ [75]. In this situation, the planetesimal belt forms what is known as a collisional cascade, and the size distribution remains constant, with mass flowing from large planetesimals to small grains.

The outcome of a collision depends on the specific incident kinetic energy, Q. Catastrophic collisions are defined as collisions in which the largest fragment produced in the collision has less than half the mass of the original object. In general, particles are destroyed in collisions with similar-sized particles. In the strength regime, $D < 150$ m, the outcome of a collision is determined by the strength of a planetesimal and the specific incident kinetic energy required to destroy it, $Q_{\mathrm{D}}^{\star}$, decreases with size. In the gravity regime, $D > 150$ m, the fragments created in the collision tend to reassemble under the action of their own gravity, so that a larger input energy is needed to catastrophically destroy a planetesimal, and in that regime $Q_{\mathrm{D}}^{\star}$ increases with size.

The mean time between collisions for dust in the size range which contributes the majority of the total cross-sectional area in the collisional cascade can be approximated by [91]:

$$t_{\rm col} = t_{\rm per}/4\pi\tau_{\rm eff}, \tag{2.3}$$

in years, where $t_{\rm per} = a^{1.5}M_\star^{-0.5}$ is the orbital period and $\tau_{\rm eff} = \sigma_{\rm tot}/(2\pi r\ dr)$ is the effective optical depth of the belt, a (wavelength independent) geometrical quantity that could also be called the surface density of cross-sectional area.

Equation (2.3) usually applies to the smallest dust grains in the cascade. Larger objects have much longer collisional lifetimes, since there is a lower cross-sectional area in the cascade with sufficient incident energy to induce catastrophic destruction. Their lifetime scales $\propto D^{5-3q}$ (i.e., $\propto D^{-0.5}$ for a canonical collisional cascade size distribution). For details of how the planetesimal strength, $Q_{\rm D}^\star$, and orbital eccentricity e affect the collision lifetime the reader is referred to [90, 94].

2.3.3 Radiation Forces

The orbits of small grains are affected by the interaction of the grains with stellar radiation [9]. This is caused by the fact that grains remove energy from the radiation field by absorption and scattering, and then re-radiate that energy moving with the particle's velocity. The resulting radiation force has two components: a radial force, known as radiation pressure, and a tangential force, known as Poynting–Robertson drag (P–R drag). The parameter β is the ratio of the radiation force to that of stellar gravity and is mostly a function of particle size (since both forces fall off $\propto r^{-2}$)

$$\beta = F_{\rm rad}/F_{\rm grav} = 7.65 \times 10^{-4}(\sigma/m)\langle Q_{\rm pr}\rangle_{T_\star} L_\star/M_\star, \tag{2.4}$$

where σ/m is the ratio of a particle's cross-sectional area to its mass (in m^2 kg^{-1}), $Q_{\rm pr}$ depends on the optical properties of the particle, and $L_\star$ and $M_\star$ are in solar units. For large spherical particles $\beta = (1150/\rho D)L_\star/M_\star$, where ρ is the particle density in kg m^{-3}, and D is in μm. For smaller particles, β tends to a value which is independent of size (see chapter by Li).

Radiation pressure

Radiation pressure essentially causes a particle to *see* a smaller mass star by a factor $(1 - \beta)$. It is immediately clear that particles with $\beta > 1$ are not bound and leave the system on hyperbolic trajectories. However, the effect of radiation pressure is also seen at lower values of β, since it means that particles orbiting at the same semimajor axis have different orbital periods, since $t_{\rm per} = [a^3/M_\star(1 - \beta)]^{0.5}$.

Most importantly, though, particles created in the destruction of a parent planetesimal have a range of sizes and so β. All particles start with the same position and velocity as the parent, but have different orbital elements because they move in different potentials. For a parent with an orbit defined by a and e broken up at a true anomaly f, the new orbital elements are

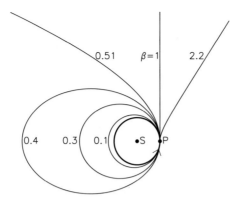

Fig. 2.2. Orbits of particles of different size (and so different β) created in the destruction of a planetesimal originally on a circular orbit [91]. The collision event occurs at point P. Particles with $\beta > 0.5$ are on unbound orbits

$$a_{\mathrm{new}} = a(1 - \beta) \left[1 - 2\beta \left[1 + e \cos f\right] \left[1 - e^2\right]^{-1}\right]^{-1}, \tag{2.5}$$

$$e_{\mathrm{new}} = \left[e^2 + 2\beta e \cos f + \beta^2\right]^{0.5} / (1 - \beta) \tag{2.6}$$

(see Fig. 2.2). This means that, with a small dependence on where around the orbit the collision occurs, it is particles with $\beta > 0.5$ that are unbound and leave the system on hyperbolic trajectories. Since particles just above the radiation pressure blow-out limit survive much longer than orbital timescales, this rapid loss causes a truncation in the collisional cascade for small sizes below which $\beta > 0.5$.

P–R Drag

P–R drag causes dust grains to spiral into the star while at the same time circularizing their orbits (with no effect on the orbital plane). For an initially circular orbit, this means that particles migrate in from a_1 to a_2 on a timescale

$$t_{\mathrm{pr}} = 400 M_\star^{-1} \left(a_1^2 - a_2^2\right) / \beta \tag{2.7}$$

in years. On their way in particles can be destroyed in collisions with other particles, become trapped in resonance with planets [14], pass through secular resonances, be scattered out of the system by those planets [56], or be accreted onto the planets. If none of these occurs, the particle sublimates close to the star once its temperature reaches above ~ 1500 K. This drag force is thus another potential loss mechanism for dust from the collisional cascade.

It is evident that, since $t_{\mathrm{pr}} \propto D$ and $t_{\mathrm{col}} \propto D^{0.5}$, P–R drag can only be relevant for small particles. Assuming that particles affected by P–R drag contribute little to the total cross-sectional area, the particle size at which P–R drag becomes important can be estimated from (2.3) and (2.7),

$$\beta > \beta_{\rm pr} = 5000\tau_{\rm eff}(r/M_\star)^{0.5}. \qquad (2.8)$$

Since the smallest grains that may be influenced by P–R drag are those with $\beta \approx 0.5$ it follows that P–R drag does not affect the evolution of any grains in the disk if $\tau_{\rm eff} > 10^{-4}(r/M_\star)^{0.5}$, as in this case all bound grains have collisional lifetimes that are shorter than their P–R drag lifetimes.

This back-of-the-envelope calculation was demonstrated more quantitatively in [87] which considered the ideal case of a planetesimal belt which produces grains all of the same size. The spatial distribution of such grains as they evolve due to collisions and P–R drag is given by

$$\tau_{\rm eff}(r) = \tau_{\rm eff}(r_0) / \left[1 + 4\eta_0(1 - \sqrt{r/r_0})\right], \qquad (2.9)$$

where $\eta_0 = t_{\rm pr}/t_{\rm col} = 5000\tau_{\rm eff}(r_0)\sqrt{r_0/M_\star}\beta^{-1}$ and r_0 is the radius of the planetesimal belt (see Fig. 2.3). For $\eta_0 \ll 1$ the majority of the grains make it to the star without suffering a collision, whereas for $\eta_0 \gg 1$ the grain population is significantly depleted before the grains make it to the star and so are confined to the vicinity of the planetesimal belt. This model also illustrates how it is not possible to invoke P–R drag to create a large dust population close to the star, since the maximum possible surface density of grains that reach the star in this model is $5 \times 10^{-5}\beta M_\star^{0.5} r^{-0.5}$.

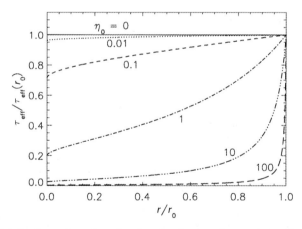

Fig. 2.3. Distribution of surface density for dust grains evolving from a point of origin in a planetesimal belt at r_0 inwards due to P–R drag while also being depleted due to mutual collisions [87]

2.3.4 Disk Particle Categories

The preceding discussion motivates the division of a debris disk into distinct particle categories which is summarized in Fig. 2.4:

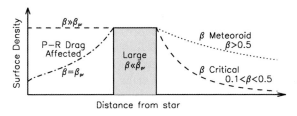

Fig. 2.4. Surface density distribution of particles created in a planetesimal belt [85]. Particles of different sizes have different β and so have different radial distributions. The main categories are: large grains, which have the same distribution as the planetesimals; β critical and β meteoroid grains, which extend much further from the star; and P–R drag affected grains, which extend inwards toward the star

- **Large grains** ($\beta \gg \beta_{\rm pr}$): these are unaffected by radiation forces and have the same spatial distribution as the planetesimals;
- **P–R drag affected grains** ($\beta \approx \beta_{\rm pr}$): these are depleted by collisions before reaching the star;
- **P–R drag affected grains** ($\beta_{\rm pr} < \beta < 0.5$): these are largely unaffected by collisions and evaporate on reaching the star;
- β **critical grains** ($0.1 < \beta < 0.5$): these are on bound orbits and while the inner edge of their distribution follows that of the planetesimals, the outer edge extends out to much larger distances;
- β **meteoroid grains** ($\beta > 0.5$): these are blown out on hyperbolic orbits as soon as they are created.

The presence of different categories in a disk depends on the density of the planetesimal belt. Broadly speaking, the large, β critical and β meteoroid categories are always present (even if the quantities of the latter two relative to the large grain population are not well known). Thus there are two main types of disk: dense disks that are dominated by collisions which have few P–R drag affected grains, and tenuous disks that are dominated by P–R drag in which P–R drag affected grains are present.

Collision Versus P–R Drag-Dominated Disks

The majority of the debris disks that can be detected at present are squarely in the collision-dominated regime, since $\eta_0 \gg 1$ [87]. In the absence of P–R drag, the spatial distribution of material becomes very simple. It is even possible to make the simplifying assumption that the β meteoroid population is negligible, because such grains are lost on timescales that are short compared with even the shortest lifetimes in the large grain population. Further ignoring complications due to the eccentricities of the β critical grains, the disk can be modeled as material entirely constrained to the planetesimal belt with a size distribution that extends in a single power law (2.2) down to the blow-out limit [90]. While clearly a simplification, this model of the disk is far better than one in which it is comprised of grains all of the same size, since it acknowledges that the dust we see has to originate somewhere. Numerical simulations have also been performed to determine the size and spatial distribution in the collision-dominated limit in more detail [79, 78], and further analytical quantification of the distributions in this limit is also possible [72].

The high sensitivity of Spitzer means that more recently relatively low density disks have been detected for which η_0 is as low as 1 meaning that P–R drag is expected to sculpt the inner edges of these disks [95]. The effect of P–R drag also needs to be accounted for when studying dust in the solar system, since $\eta_0 \approx 2 \times 10^{-3}$ in the asteroid belt. It is important to emphasize this point, since it means that the dynamics of dust in the zodiacal cloud is fundamentally different to that of extrasolar systems, albeit in an understandable way. It is also becoming clear that, while stellar wind forces are relatively weak in the solar system providing a drag force $\sim 1/3$ that of P–R drag (see chapter by Mann), such forces may be important for other stars. While the mass loss rates, dM_{wind}/dt, of main-sequence stars are poorly known, it is thought that the low luminosity of M stars means that this force may be responsible for the dearth of disks found around late type stars [61]. Since stellar wind forces act in a similar manner to P–R drag, they can be accounted for in the models by multiplying η_0 by a factor of $[1 + (dM_{wind}/dt)c^2/L_\star]^{-1}$ [32, 53].

Thus, while it is usually the case that the collision-dominated approximation is most appropriate, models which describe the distribution of material evolving under the action of collisions and drag forces continue to be of interest. While a study which takes into account the full range of sizes in the disk has yet to be undertaken, it is possible to see that since grains are typically destroyed in collisions with similar-sized objects, the outcome of such a model will be similar to assuming that grains of different sizes have spatial distributions that can be characterized by different η_0, with large grains having high η_0 and small grains having low η_0. This means that the size distribution would be expected to vary significantly with distance from the star.

2.3.5 Comparison with Observations

This model has had considerable success at explaining the observed radial structure of debris disks. For example, using the collision-dominated assumption with the dust confined to the planetesimal belt provides an adequate fit to the emission spectrum of disks like that of Fomalhaut [90] for which the radius of its planetesimal belt is well known [30]. It can also explain the structures of the disks which are seen to be considerably extended and to exhibit a gradient in grain size throughout the disk (AU Mic and β Pictoris). These observations are explained as β critical dust being created in planetesimal belts which are closer to the star [3, 4, 72]. Further, the emission spectrum of the TWA7 disk is consistent with the distribution of dust expected from inward migration from the planetesimal belt by stellar wind drag [53].

Thus these studies show that the observed dust distributions can be successfully explained within the framework of a realistic physical model. Such a model is an absolute requirement if any asymmetries seen in the disk structure are to be interpreted correctly, since even the axisymmetric dust distribution is different to that of the planetesimals, which hints that its asymmetric distribution may also differ. On a more basic level, this shows that the location of the dust in a debris disk does not necessarily directly pinpoint the location of the planetesimals.

Despite the successes of the disk dynamical theory it is important to point out that it is not yet a predictive theory. There are too many uncertainties regarding the expected size distribution at very small sizes (e.g., because it depends on the size distribution created in collisions), and regarding the optical properties of those grains and the magnitude of stellar wind drag, to predict how bright a disk known from far-IR measurements (of its large grain population) will appear in scattered light images (which are sensitive to the β critical and β meteoroid grains).

2.4 Interaction Between Planets and Planetesimal Belt

Consider now one modification to the planetesimal belt model described in Sect. 2.3, which is that there is a planet orbiting in this system. The gravitational perturbations of that planet will affect the orbits of both the planetesimals and the dust. It turns out that these perturbations are predicted to cause exactly the same set of features as observed in debris disks (Fig. 2.1).

The planet's perturbations can be broken down, both mathematically and conceptually, into three types: secular, resonant, and short-period perturbations. For a detailed description of this dynamics, the reader is referred to Murray & Dermott (1999) [59]. Its secular perturbations are the long-term consequence of having the planet in the disk, and these perturbations are equivalent to the perturbations from the wire that would be obtained by spreading the mass of the planet around its orbit with a density in accordance with its velocity at each point; they affect all material in the disk to some extent. Its resonant perturbations are the forces which act at specific radial locations in the disk where planetesimals would be orbiting the star with a period that is a ratio of two integers times that of the planet. At such locations the planetesimal receives periodic kicks from the planet which can make such locations either extremely stable or extremely unstable. All other perturbations are short period and can be assumed to average out on long enough timescales, although they are responsible for important processes such as scattering of planetesimals.

2.4.1 Secular Perturbations

To first order a planet's secular perturbations can be separated into two components: one arising from the eccentricity of its orbit and the other from its inclination. For high eccentricities and inclinations, these two components are linked, and here I only consider the low eccentricity and inclination case.

Planet Eccentricity: Spirals and Offsets

The consequence of the planet's eccentricity is to impose an eccentricity onto the orbits of all planetesimals in the disk. It does this in such a way that a planetesimal's eccentricity vector, defined by $z = e \times \exp i\varpi$, precesses around a circle centred on the forced eccentricity vector, z_f; i.e.,

$$z(t) = z_f + z_p(t), \tag{2.10}$$

where the forced eccentricity is set by a combination of the planet's eccentricity and the ratio of the planetesimal and planet semimajor axes

$$z_f = [b_{3/2}^2(\alpha_{pl})/b_{3/2}^1(\alpha_{pl})]z_{pl}, \tag{2.11}$$

and the proper eccentricity precesses around a circle the radius of which is determined by the initial conditions

$$z_p(t) = e_p \times \exp i(At + \beta_0), \tag{2.12}$$

at a fixed rate given by

$$A = 0.25n(M_{pl}/M_\star)\alpha_{pl}\bar{\alpha}_{pl}b_{3/2}^1(\alpha_{pl}). \tag{2.13}$$

In the above equations, $\alpha_{pl} = a_{pl}/a$ and $\bar{\alpha}_{pl} = a/a_{pl}$ for $a_{pl} < a$ and $\alpha_{pl} = \bar{\alpha}_{pl} = a/a_{pl}$ for $a_{pl} > a$, and $b_{3/2}^s(\alpha_{pl})$ are the Laplace coefficients. These equations have been given for the case of a system with one planet. However, the same decomposition into forced and proper elements is also true in a system with multiple planets, except that the equations for the forced eccentricity and precession rate A involve sums over all planet properties [91].

The evolution of a planetesimal's eccentricity vector is shown in Fig. 2.5, which shows how the orbits of planetesimals at 1.4, 1.45, and 1.5 times semimajor axis of the planet evolve if they start on initially circular orbits. This is equivalent to a situation which might arise following the formation of a planet on an eccentric orbit, since the planetesimals would have formed on roughly circular orbits. As well as a small change in forced eccentricity for planetesimals at different distances from the planet, their precession rates are substantially different. This means that the dynamical structure of an extended planetesimal disk evolves following the formation of the planet: planetesimals close to the planet, which have completed several precessions can be considered to have eccentricity vectors evenly spread around circles centred on the forced eccentricity, while those further away have pericentre orientations and eccentricities that change with distance from the star.

The resulting dynamical structure can be readily translated into a spatial distribution by creating a model in which planetesimals are distributed randomly in longitude, λ, and with other orbital elements taken from appropriate

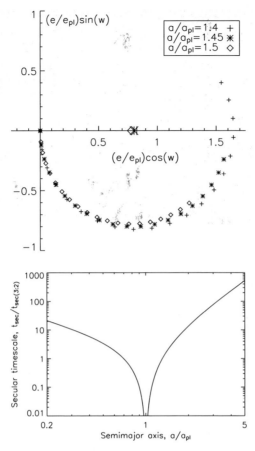

Fig. 2.5. Effect of the secular perturbations of an eccentric planet on planetesimal orbits [88]. *Top* Evolution of the eccentricity vectors of planetesimals at 1.4, 1.45, and $1.5a_{\rm pl}$. All vectors start at the origin (*circular orbits*) and precess around the forced eccentricity imposed on them by the planet. The symbols are plotted at equal timesteps. *Bottom* Precession rate for planetesimals at different distances from the planet. These are given as the timescale to complete one precession $(2\pi/A)$ relative to that timescale for planetesimals at $a = 1.31a_{\rm pl}$ which is given by $0.651a_{\rm pl}^{1.5}M_{\star}^{0.5}/M_{\rm pl}$

distributions for the given time. This is shown in Fig. 2.6 for the planetesimals outside the planet's orbit. It is seen that the planetesimals exhibit spiral structure which propagates away from the planet. A similar spiral structure is formed in the planetesimal disk interior to the planet, again propagating away from the planet with time.

It is possible that the effect seen in Fig. 2.6 may be the explanation for the tightly wound spiral structure seen at 325 and 200 AU in the HD141569 disk [12]. Since the rate at which the spiral propagates away from the planet is determined by the planet's mass, this means that the observed structure

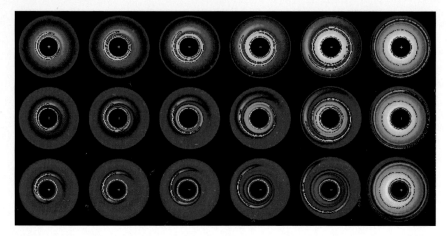

Fig. 2.6. Spatial distribution of planetesimals affected by the secular perturbations of an eccentric planet: spiral structure propagating outward through an extended planetesimal belt outside a planet [88]. The subpanels show, from left to right, snapshots of the disk at times $0.1, 0.3, 1, 3, 10, 100$ times the secular precession timescale at $1.31a_{\mathrm{pl}}$ since the perturbing planet was introduced into the disk. From top to bottom, the panel shows the impact of planets with an eccentricity of 0.05, 0.1, and 0.15

allows the planet's mass to be estimated, assuming the time since the planet formed can also be estimated. For HD141569, this results in the putative planet having a mass greater than that of Saturn, given the 5 Myr age of the star. A tightly wound spiral structure is also seen in Saturn's rings [11], which is explained by a similar model in which the secular perturbations arise from the oblateness of the planet (rather than from an eccentric perturber), and the ring material is assumed to have formed in a relatively recent event (rather than with the planet).

At late times, when the material at the same semimajor axis has eccentricity vectors that are evenly distributed around circles, the resulting disk no longer exhibits spiral structure, but it does exhibit an offset. This is illustrated in Fig. 2.7, although it is also apparent in Fig. 2.6 at late times. The offset is proportional to the forced eccentricity imposed on the planetesimals (and so proportional to the planet's eccentricity), with material on the side of the forced pericentre being closer to the star than that on the side of the forced apocentre.

This offset was originally predicted from a brightness asymmetry in the HR4796 disk [76, 91], and was called *pericentre glow* because the asymmetry was thought to arise from the material on the pericentre side being hotter than that on the apocentre side. It was found that the observed brightness asymmetry could have been caused by a planet with an eccentricity as small as 0.02, demonstrating that even moderate planet eccentricities can have observable signatures. However, little information is available from this structure about

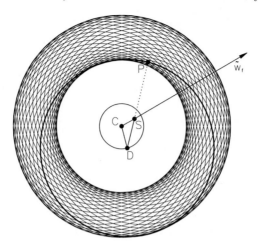

Fig. 2.7. Spatial distribution of planetesimals affected by the secular perturbations of an eccentric planet: offset structure imposed at late times on planetesimals all at the same semimajor axis a [91]. The planetesimals form a uniform torus around the star at S, but one which has its centre at a point D which is offset from the star by a distance ae_f in the direction of the forced apocentre

the mass of the planet, except that it must be sufficiently massive for the pericentres to have been randomized given the age of the star. For HR4796, this means that its putative planet would have to have a mass $> 10 M_\oplus$, although the interpretation of this asymmetry is complicated by the stellar mass binary companion to HR4796A, the orbit of which is unknown at present, but which could also be responsible for an offset of the required magnitude. Nevertheless, an offset has been seen directly in the structure of the Fomalhaut disk [35]. This star also has a common proper motion companion [6], but this is too distant to be responsible for an offset of the observed magnitude.

Planet Inclination: Warps

The consequence of secular perturbations caused by the planet's inclination is directly analogous to the consequence of its eccentricity (Sect. 2.4.1), except that in this case, it is the planetesimal's inclination vector, $y = I \times \exp i\Omega$, which precesses around a forced inclination. The precession rate is also the same, except that it is reversed in sign (i.e., the inclination vector precesses clockwise on a figure analogous to Fig. 2.5). In a system with just one planet the forced inclination vector is simply the orbital plane of the planet ($y_f = I_{pl} \times \exp i\Omega_{pl}$). Since the choice of the zero inclination plane is arbitrary, it can be set to be the planet's orbital plane ($y_f = 0$) making it easy to see that at late times, planetesimals at the same semimajor axis will have orbital planes distributed randomly about the orbital plane of the planet. However at early times, should the initial orbital plane of the planetesimals be different to that of the planet at say y_{init}, then the situation will be that material close to the planet will be distributed randomly about the planet's orbital plane y_{pl}, while that far from the planet will still be on the original orbital plane y_{init}. A smooth transition between the two occurs at a distance from the star which depends on the mass of the planet and the time since the planet formed, much in the same way that spiral structure propagates away from the planet.

This has been proposed as the explanation of the warp in the β Pictoris disk, since other lines of evidence have pointed to a Jupiter mass planet at ~ 10 AU [8, 66], and given the age of the star ~ 12 Myr, it is reasonable to assume that a warp would be seen at ~ 80 AU at the current epoch (if the planet formed very early on). Many observations of the disk, including the warp and the radial distribution (see Sect. 2.2.1), can be explained with such a model [4], although it would be worth revisiting this model in the light of the observations which showed the warp is less of a smooth transition between two orbital planes and looks more like two distinct disks [20].

This mechanism does not just produce a warp in a young disk. As long as there are two or more planets in the system on different orbital planes a warp would also be seen at late times, once all the planetesimals have precessed so that their distribution is symmetrical about the forced inclination plane, since multiple planets would mean that the forced inclination plane varies with distance from the star (e.g., it is aligned with each of the planet's orbital planes at the semimajor axes of those planets). The zodiacal cloud in the solar system is an example of a warped old disk [91].

2.4.2 Resonant Perturbations

Mean motion resonances are locations at which planetesimals orbit the star an integer p times for every integer $p+q$ times that the planet orbits the star. The nominal location of a resonance is at a semimajor axis of

$$a_{(p+q):p} = a_{\text{pl}}(1 + q/p)^{2/3}. \tag{2.14}$$

Planetesimals at a range in semimajor axis about this nominal value may be trapped in resonance, but not all of those in this range are necessarily in the resonance.

Resonant Geometry

The importance of a resonance can be understood purely from geometrical reasons. Figure 2.8 shows the path of planetesimals on resonant orbits in the frame co-rotating with the planet. The pattern repeats itself so that the planetesimal always has a conjunction with the planet (i.e., the two are at the same longitude) at the same point in its orbit for $q = 1$ resonances, or at the same two points in its orbit for $q = 2$ resonances. This is important because the perturbations to the planetesimal's orbit are dominated by those

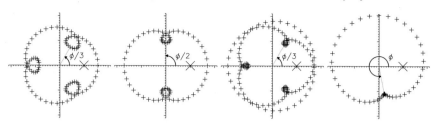

Fig. 2.8. Path of resonant orbits in the frame co-rotating with a planet [86]. On all panels the planet, located at the *cross*, is on a circular orbit, while the planetesimals' orbits have an eccentricity of 0.3. The planetesimals are plotted with a plus at equal timesteps through their orbit, each point separated by 1/24 of the planet's orbital period. The resonances shown are from left to right, with increasing distance from the planet, the 4:3, 3:2, 5:3, and 2:1 resonances

at conjunction which means that the planetesimal receives periodic kicks to its orbit from the planet which are always in the same direction (if the orbit is unchanged by those perturbations). This is not quite true for the $p = 1$ resonances, since the cumulative effect of the perturbations around the orbit are also relevant in this case.

The resonant geometry discussed in the preceding paragraph can be used to infer the loopy patterns on a figure such as Fig. 2.8, but it does not specify the location of the planet with respect to those loops. That is specified by the planetesimal's resonant argument

$$\phi = (p+q)\lambda - p\lambda_{\mathrm{pl}} - q\varpi. \tag{2.15}$$

The resonant argument is important, since the ratio ϕ/p is the relative longitude of the planet when the planetesimal is at pericentre, an angle which is noted in Fig. 2.8; i.e., it determines where along the planetesimal's orbit it receives kicks from the planet's gravity.

The same combination of angles occurs in the planet's disturbing function, and the forces associated with a resonance are those involving the relevant resonant argument [59]. A planetesimal is said to be in resonance if its resonant argument is librating about a mean value (e.g., a sinusoidal oscillation), rather than circulating (e.g., a monotonic increase or decrease). The mean value about which the resonant argument librates is typically 180°, since in this configuration it can be shown that the resonant forces impart no angular momentum to the planetesimal. However, in some instances, asymmetric libration occurs, where $\langle \phi \rangle \neq 180°$, because the equilibrium solution requires resonant forces to impart angular momentum to the planetesimal (see section on resonant trapping and Sect. 2.5.2). Asymmetric libration also occurs for the $p = 1$ resonances (e.g., the 2:1 resonance), because in this configuration angular momentum imparted to a planetesimal at conjunction is balanced by the cumulative effect of the resonant forces around the rest of the orbit [58].

Resonant Trapping

While resonances have non-zero width in semimajor axis, their width is finite and they only cover a narrow region of parameter space. Thus if a planet was introduced into an extended planetesimal belt, while planetesimals at suitable semimajor axes might end up trapped in resonance, such planetesimals would be relatively few. However, resonances can be filled by either the planet or the planetesimals' semimajor axes undergoing a slow migration, since when a planetesimal encounters a planet's resonances the resulting forces can cause the planetesimal to become trapped in the resonance. Resonant forces could then either halt the planetesimal's migration, or make it migrate with the planet, thus ensuring that the planetesimal maintains the resonant configuration. For example, it is thought that Pluto and most of the other Kuiper belt objects that are in resonance with Neptune attained their resonant orbits during an epoch when Neptune's orbit expanded following its formation [49]. There are a number of mechanisms which can be invoked to cause planets to migrate outward, one of which is angular momentum exchange caused by scattering of planetesimals [27], and another is interaction with a massive gas disk [52].

The question of whether a planetesimal becomes trapped once it encounters a planet's resonances is determined by two main factors: the mass of the planet and the rate at which the planet or planetesimals are migrating. For example, the probability of a low eccentricity planetesimal being trapped into any given resonance with a planet migrating on a circular orbit is determined by two parameters, $\mu = M_{\mathrm{pl}}/M_\star$ and $\theta = \dot{a}_{\mathrm{pl}}/\sqrt{M_\star/a}$ [86]. It is expected that the eccentricities of the planet and planetesimal orbits would also affect the trapping probability. Another important factor in determining which resonances are filled by planet migration is the initial distribution of the planetesimals with respect to the planet, since this determines how many planetesimals would encounter a given resonance in the course of the migration, given that some may already have been trapped in another resonance. The dominant resonances in the Kuiper belt are the 4:3, 3:2, 5:3, and 2:1 resonances, which can partly be explained by the fact that first-order resonances (i.e., $q = 1$ resonances) are stronger than second-order resonances (i.e., $q = 2$ resonances), and so on.

Resonances are important not only for a disk's dynamical structure, but also for the spatial distribution of material, since Fig. 2.8 illustrates how planetesimals that are all in the same resonance that have similar resonant arguments would tend to congregate at specific longitudes relative to the planet. That is, while any one planetesimal is on an elliptical orbit, that orbit would take it in and out of regions of high planetesimal density, i.e., clumps. These clumps would appear to be orbiting the star with the planet. The number of clumps formed by resonances is given by p (e.g., Fig. 2.8). An important factor in the formation of resonant clumps is the planetesimals' eccentricities, since the clumps only become pronounced at high eccentricities;

resonant planetesimals on circular orbits have an axisymmetric distribution. Once trapped resonant forces can excite the planetesimals so that they become eccentric.

This mechanism was invoked to explain the clumpy structure of Vega's debris disk (Fig. 2.9; [86]). In that model two clumps form in the planetesimal distribution because of the migration of a Neptune mass planet from 40 to 65 AU over 56 Myr. As suggested by Fig. 2.8, the clumps are the result of trapping of planetesimals into the 3:2 resonance, with an asymmetry caused by planetesimals in the 2:1 resonance. The planet's mass and migration rate are constrained within the model, but not uniquely, since it did not consider the origin of the planet's migration; e.g., the same structure would arise from a three Jupiter mass planet which completed the same migration over 3 Myr. To break this degeneracy models would be required which cause both planet migration and resonant trapping simultaneously, a task which requires significant computing power. Nevertheless, this model shows that observed structures have the potential to tell us not only about the planets in a system, but also about that system's evolutionary history.

The model has made predictions for: (i) the location of the planet (none has been found at the level of $< 3M_{\mathrm{Jupiter}}$, [54]); (ii) the orbital motion of the clumps which should be detectable on decade timescales (these observations will be made in the coming year, and in the meantime a marginal detection of orbital motion has been found in the clumpy structure of the ε Eridani disk, [62]); (iii) lower level structure associated with the 4:3 and 5:3 resonances (which may have been seen in 350 μm observations of the disk [51]).

Resonance Overlap

So far I have discussed the stabilizing properties of resonances. However, resonances can also be destabilizing. As mentioned above, resonances have finite width, and the $q = 1$ resonances are strongest. There is a region nearby the planet where its first-order resonances overlap, and planetesimals in such a region have chaotic orbits and are rapidly ejected [84]. The width of this region is

$$|a - a_{\mathrm{pl}}|/a_{\mathrm{pl}} = 1.3(M_{\mathrm{pl}}/M_\star)^{2/7}. \qquad (2.16)$$

Instabilities of this type have been invoked to explain the cleared inner regions of debris disks [16], as well as to estimate the location of a planet inside an imaged planetesimal belt [64]. The same planet also imposes eccentricities on the planetesimals at the edge of the resonance overlap region, and the magnitude of those eccentricities is dependent on the mass of the planet, with more massive planets imposing larger eccentricities. Since those eccentricities result in a sloping inner edge, it is also possible to use the sharpness of the inner edge of a dust ring to set constraints on the mass of the planet. In this way the sharp inner edge of the Fomalhaut ring was used to determine that its planet must be less massive than Saturn [64].

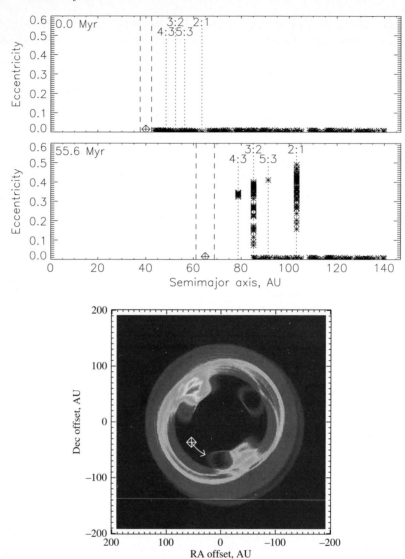

Fig. 2.9. Structure imposed on an initially axisymmetric planetesimal disk by the outward migration of a planet [86]. This model was proposed to explain the clumpy structure of the Vega disk and involves a Neptune mass planet which migrated from 40 to 65 AU over 56 Myr. *Top* Dynamical structure of the planetesimal disk, eccentricity versus semimajor axis, at the beginning and end of the migration. The planet is shown with a *diamond*, and the location of its resonances with *dotted lines*. The chaotic region of resonance overlap is shown with *dashed lines*. *Bottom* Spatial distribution (surface density) of planetesimals at the end of the migration. The planet is shown with a *diamond* and the *arrow* shows its direction of orbital motion

2.5 Interaction Between Planets and Dust

The preceding Sect. (2.4) dealt specifically with the structures imposed by planets on the planetesimal distribution. This is an important first step, since the dust we see is derived from those planetesimals. However, as discussed in Sect. 2.3, dust dynamics can be significantly different to that of the planetesimals, and so it is not obvious to what extent the dust distribution will follow that of the planetesimals. A model which takes into account the production in collisions of dust with a range of sizes and the subsequent dynamical evolution of that material is usually beyond the scope of current computing (and analytical) capabilities. Such models will likely become more common-place as more detailed observations demonstrate that more sophisticated models are necessary to explain the observations. For now, the types of structure which dust dynamics would produce can be understood by considering the dynamical evolution of dust grains released from a given planetesimal distribution. Those grains might then be considered to evolve in the absence of collisions, or in an idealised situation where the only collisions which matter are those with grains of similar size (and so which all have the same spatial distribution).

Here, I consider the effect of dust grain dynamics on the structures seen in a planetesimal belt in which some of the planetesimals are in resonance with a planet (Sect. 2.5.1), and the structures caused by trapping of dust into planetary resonances (Sect. 2.5.2).

2.5.1 Dust Produced from Resonant Planetesimals

Only the largest dust grains released from a resonant planetesimal remain in that resonance, and the reason is the effect of radiation pressure. First of all, consider the orbits of grains which remain bound to the star ($\beta < 0.5$). Radiation pressure has two effects: it changes the semimajor axis of the dust grain (so that $a' = a[1 - \beta]/[1 - 2\beta]$ for initially circular orbits), and also the location of the resonance (the resonance for dust is at a lower semimajor axis by a factor $(1 - \beta)^{1/3}$ than given in (2.14)). This means that the larger a particle's β (which typically means the smaller its size, although see Sect. 2.3.3 and chapter by Li) the further it starts from the resonance, and while resonant forces can accommodate this offset by increasing the libration width for large grains, there comes a size at which particles are no longer in resonance. Numerical simulations showed that for the 3:2 resonance, the critical size is that for which $\beta > \beta_{\mathrm{crit}}$, where

$$\beta_{\mathrm{crit}} = 2 \times 10^{-3}(M_{\mathrm{pl}}/M_\star)^{0.5}, \tag{2.17}$$

with a similar threshold for the 2:1 resonance [89]. Since the geometry of Fig. 2.8 is no longer valid for non-resonant grains, such grains have an axisymmetric distribution.

Next, consider the orbits of dust grains which are released onto hyperbolic orbits (i.e., $\beta > 0.5$). For $\beta = 1$ dust no force acts on the grains, and such grains leave the system with a constant velocity (that of the orbital motion of the parent planetesimal) which rapidly approaches radial motion. Assuming that these grains are created at a constant rate, this corresponds to a surface density distribution which falls off $\propto r^{-1}$. Since no force acts on the grains, one might naively expect no asymmetry in their distribution. However, their distribution can be non-axisymmetric if they are not produced from an axisymmetric distribution of parent bodies. Collision rates are highest between resonant planetesimals when they are in the clumps, and this means that a greater fraction of the $\beta > 0.5$ grains created from planetesimal collisions have trajectories which originate in the clumps. The distribution of such grains should thus exhibit spiral structure which emanates from the clumps (since while the motion of the dust is nearly radial, the source region, the clumps, are in orbital motion around the star). However, not all $\beta > 0.5$ grains are created in collisions between planetesimals with a clumpy resonant distribution; some originate in collisions between non-resonant grains with $\beta_{crit} < \beta < 0.5$, and so would have an axisymmetric distribution.

This motivates a division of the dust produced in a resonant planetesimal disk into four populations with distinct spatial distributions: (I) large grains $\beta < \beta_{crit}$ with a clumpy distribution, (II) intermediate grains $\beta_{crit} < \beta < 0.5$, with an axisymmetric distribution, (IIIa) small grains $\beta > 0.5$ from population (I) particles with extended spiral structure, (IIIb) small grains $\beta > 0.5$ from population (II) particles with extended axisymmetric structure. These distributions have been worked out numerically for the model presented in Fig. 2.9, and the structures expected for the four populations are shown in Fig. 2.10.

Aside from ascertaining the distribution of different grain sizes, it is important to determine which grain sizes actually contribute to the observation in question. This chapter will not deal specifically with such issues, for which a knowledge of the optical properties of the particles is needed, and for which the reader is referred to the chapter by Li in this book. However, the type of result that is obtained with such an analysis is illustrated in Fig. 2.10. This shows how observations in different wavebands are sensitive to different grain sizes and to different grain populations, with the shortest wavelengths probing the smallest grains; i.e., the disk would be expected to look different when observed at different wavelengths. For the Vega disk, this is indeed seen to be the case [51], although this does not mean the dynamics of this disk is completely understood, since the prediction for the spiral structure at the shortest wavelengths [89] has yet to be confirmed, and the large observed mass loss rate remains to be explained [73].

The fact that disk structure is expected to be (and is seen to be) a strong function of both grain size and wavelength of observation is good because it means that multiple wavelength observations of the same disk provide a means to test different models for the origin of structure formation. However, it also means that the models are becoming more complicated, and this means that the interpretation of observed structure is no longer straight forward,

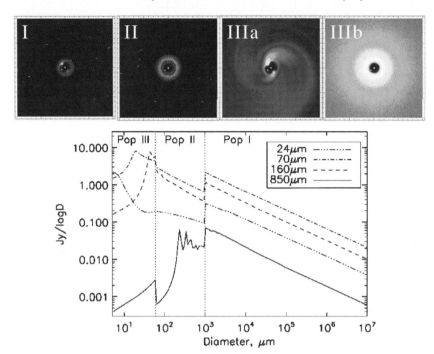

Fig. 2.10. Prediction for the structure of Vega's debris disk [89]. (*Top*) Spatial distribution of dust in different populations: (I) large grains, (II) intermediate grains, (IIIa) small grains (created from large grains), (IIIb) small grains (created from intermediate grains). All panels cover the same region (±100 arcsec from the star which is shown by an *asterisk* at the centre); the location of the planet is shown with a plus. (*Bottom*) Contribution of different grain sizes (and so different populations) to observations in different wavebands. The y axis is flux per log particle diameter, so that the area under the curve is the total flux, and the relative contribution of different grain sizes to that flux is evident from the appropriate region. For the size distribution shown here, the mid- to far-IR wavelength observations are dominated by population III grains, while sub-millimeter observations are dominated by population I grains

since there are multiple physical processes that have to be accounted for. For example, it should also be noted that the model described above only took account of the effect of radiation pressure on the dust orbits, and the relative velocity imparted to collisional fragments may also be important [40].

2.5.2 Resonant Trapping of Dust by P–R Drag

Planetary resonances can also sculpt a dust disk even if the parent planetesimals are not trapped in resonance, since the drag forces which act on dust to make it migrate inwards (see Sect. 2.3.3) mean that the dust may have the opportunity to encounter a planet's resonances. Resonant forces can then halt the migration causing a concentration of dust along the planet's orbit known as a *resonant ring*. For the same geometrical reasons as outlined in Sect. 2.4.2, these resonant rings are clumpy.

There are some important subtle differences in the structure of this type of resonant ring compared with the resonant planetesimal rings. One of these is the fact that the libration of ϕ is offset from 180° so that resonant forces can impart angular momentum to the particles to counteract that lost by P–R drag. This means that the loopy patterns in Fig. 2.8 are not symmetrical about the planet in such a way that the loop which is immediately behind the planet is closer to the planet than that in front of it. The magnitude of this effect is dependent on particle size (β). The concentration of all the loops from the different resonances and particle sizes behind the planet causes a clump to follow the planet around its orbit. This is sometimes referred to as a trailing wake. This effect was responsible for the discovery of the first resonant ring, since the zodiacal cloud was found to always be brighter in the direction behind the Earth's motion than in front of it [14]. This was interpreted as dust trapped in $q = 1$ resonances close to the Earth (i.e., with $p > 3$) (see top left panel of Fig. 2.11). The structure of the Earth's trailing wake will soon be known in great detail, as the infrared satellite Spitzer is currently flying directly through the middle of it. There is no evidence for a resonant

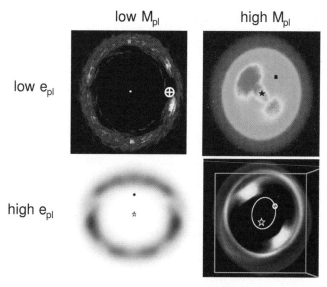

Fig. 2.11. Spatial distribution of dust which has migrated into the resonance of a planet forming a *resonant ring*. The structure of the ring depends on the mass and eccentricity of the planet [41], and examples of the four types of structure are taken from published models: low $M_{\rm pl}$, low $e_{\rm pl}$ (model for the Earth's resonant ring [14]); high $M_{\rm pl}$, low $e_{\rm pl}$ (model for the Vega dust ring [60]); low $M_{\rm pl}$, high $e_{\rm pl}$ (model for the ε Eridani dust ring [63]); high $M_{\rm pl}$, high $e_{\rm pl}$ (model for the Vega dust ring [83])

ring associated with Mars [42], but recent evidence shows that Venus has a resonant ring [44].

The structure of a resonant ring depends on the mass of the planet, because the resonant forces from a more massive planet are stronger meaning that dust can be trapped into resonances that are further from the planet, e.g., the 3:2 and $p = 1$ resonance such as the 2:1 and 3:1 resonances (see top right panel of Fig. 2.11). The ring structure is also dependent on the planet's eccentricity [41] (bottom panels of Fig. 2.11). However, one of the most important factors which determines that structure is the spatial distribution of source planetesimals and the size distribution of particles encountering the different resonances, since it is that determines which resonances are populated. It is not easy to ascertain the expected structure of a resonant ring, since a complete resonant ring model would have to consider the competition between production and destruction in collisions and removal by P–R drag, on top of which some fraction of the particles are trapped in different resonances for varying durations. Needless to say, current models make some approximations, and typically ignore collisions and consider only a relatively narrow range of particle sizes that are assumed to evolve independently [13, 55, 57].

One important point to consider is that for a resonant ring to form in this way the dust must migrate inwards on a timescale that is shorter than the timescale for it to be destroyed in collisions. As discussed in Sect. 2.3.4, the role of P–R drag in affecting the orbits of dust in the disks which are known about at present is negligible, since the collision timescale is short. Thus, while stellar wind drag forces may increase the drag rate for late type stars, the expectation is that resonant rings of this type are not present in the known disks [87, 40]. This serves as a caution that it is dangerous to apply our knowledge of the dynamical structures in the solar system's dust cloud [14, 47] directly to extrasolar systems without first having considered the dust dynamics. However, the example of the solar system also demonstrates that, once we are able to detect more tenuous debris disks, perturbations from Neptune-mass planets will be readily detectable, and it will even be possible to detect structures associated with planets as small as the Earth. In much the same way as it is not yet possible to detect the putative planets around stars like Vega, the dust structures associated with terrestrial planets may also be easier to detect than the planets themselves.

2.6 Conclusions

This chapter has considered the types of structures seen in the dusty debris disks of nearby stars (Sect. 2.2) and how those structures can be used to determine the layout of their planetary systems, in terms of the distributions of both planetesimals and planets. The text has dwelled on the successes of the models at explaining the observed structures, because this illustrates the elements that are essential to any debris disk model if the observations are to be successfully explained (Sect. 2.3), and because we are confident that we understand how a planet would perturb a planetesimal belt in an idealised system comprised of just one planet (Sect. 2.4) and to some extent how to extrapolate that to consider how the planet would affect the observed dust disk (Sect. 2.5). To summarize what we have learned: (i) the axisymmetric structure of debris disks can mostly be explained by a model in which dust is created in collisions in a narrow planetesimal belt and is subsequently acted on by radiation forces; (ii) the asymmetric structure of debris disks can mostly be explained by secular and resonant gravitational perturbations from unseen planets acting on the planetesimal belt and dust derived from it.

Knowing the radial location of the planetesimal belts is important because this demonstrates where in a protoplanetary disk grain growth must have continued to kilometer-sized planetesimals [90], and by analogy with the solar system there is a reason to believe that the location of the planetesimal belts tells us indirectly the whereabouts of unseen planets, although it is worth bearing in mind that there may be alternative explanations for gaps in the planetesimal distribution related to the physics of the protoplanetary disk. Nevertheless, it appears that where we have the capability to look for detailed disk structure, there is good correspondence between the asymmetric structures observed with those expected if there are planets in these systems. The modeling is also sufficiently advanced that the disk structure can be used to infer information on the properties of the perturbing planets (such as the planet's mass, orbit, and even evolutionary history). The planet properties which have been inferred in this way are particularly exciting when compared with those of exoplanets discovered using the radial velocity and transit techniques. Figure 2.12 shows how the debris disk planets are similar to Uranus and Neptune in the solar system, occupying a unique region of parameter space. This is possible because the large size of debris disks means that the planets perturbing them are most often at large orbital radii, and it is easy for planets as small as Neptune to impose structure on a debris disk. There is also the tantalizing possibility that in the future debris disk structures can be used to identify planets analogous to the Earth and Venus in extrasolar systems.

However, while it is incontrovertible that if there are planets present they would impose structure on a disk, the question of whether we have already seen these structures in extrasolar systems is still a matter for debate. In many cases, the presence of an unseen planet is the only explanation for the observed structures, but that does not mean that it has to be the right explanation. The problem is that it is hard to confirm that the planets are there, since they lie beyond the reach of radial velocity studies (see Fig. 2.12). Direct imaging could detect planets at this distance if they were a few times Jupiter mass [54], but not if they are Neptune mass. Thus the onus is on the models to make other testable predictions, and some of these have already been made (such as the orbital motion of the clumpy structures, and the disk structures expected to be seen at different wavelengths) and will be tested in the coming years. If these planets are confirmed, their addition onto plots like that shown in Fig. 2.12 will be invaluable for constraining planet formation models [31].

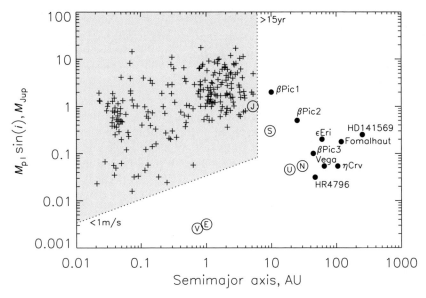

Fig. 2.12. Distribution of planet masses and semimajor axes. Solar system planets are plotted as *open circles*, and those known from radial velocity and transit studies with a plus (taken from the list on http://exoplanets.eu dated May 24, 2007). The *shaded region* shows the current limits of radial velocity surveys for sun-like stars. Debris disk planets inferred from disk structure (all awaiting confirmation) are shown with *filled circles*. References for the plotted planet parameters are: HR4796 [91], ε Eridani [60], Vega [86], HD141569 [88], η Corvi [93], Fomalhaut [64], β Pictoris [19], although it should be noted that these parameters, particularly planet mass, are often poorly constrained

It is also important to remember that this theory cannot yet predict the quantities of small grains we would expect to see in any given disk. There are too many uncertainties regarding the dust production mechanisms, and it is possible that these processes may differ among stars with, e.g., different dust compositions. Applying dynamical models of the kind presented in Sect. 2.3 to a greater number of resolved disk observations will help to understand these differences. However, there is still the possibility that the problem is more fundamental in a way which is best illustrated by the archetypal debris disk Vega. The observed mass loss rate from β meteoroids in this system is $2M_{\oplus}/\text{Myr}$, which indicates that this must be a transient, rather than a steady state, component [73]. It is thus possible that the small grain population in debris disks is inherently stochastic, perhaps influenced by input from recent massive collisions [77]. Fortunately, it appears that the large grain component of the majority of debris disks is evolving in steady state [95] and so can be understood within the framework described in this chapter, and the same is likely also true for the small grain component (it is just the relative quantities of the different components that is less certain).

However, the possibility must be considered that in some systems the observed dust is transient in such a way that its origin will require a significant overhaul to the models presented here. For example, there are a few cases of sun-like stars surrounded by hot dust (e.g., Sect. 2.2.1) which cannot be maintained by steady-state production in a massive asteroid given the age of the stars [94]. It is not clear what the origin of the transient event producing the dust is. However, it is known that the quantity of planetesimals in the inner solar system has had a stochastic component, notably involving a large influx ~ 700 Myr after the solar system formed in an event known as the late heavy bombardment, the origin of which is thought to have been a dynamical instability in the architecture of the giant planets [22]. So perhaps these systems are telling us about the more complex dynamics of their planetary systems. Given the complexity of planetary systems it seems inevitable that the models presented in this chapter are just the start of a very exciting exploration of the dynamics of extrasolar planetary systems.

References

1. O. Absil, et al.: Astron. Astrophys. **452**, 237 (2006)
2. D. R. Ardila, et al.: Astrophys. J. **617**, L147 (2004)
3. J. -C. Augereau and H. Beust: Astron. Astrophys. **455**, 987 (2005)
4. J. -C. Augereau, R. P. Nelson, A. M. Lagrange, J. C. B. Papaloizou and D. Mouillet: Astron. Astrophys. **370**, 447 (2001)
5. H. H. Aumann, et al.: Astrophys. J. **278**, L23 (1984)
6. D. Barrado y Navascues, J. R. Stauffer, L. Hartmann and S. C. Balachandran: Astrophys. J. **475**, 313 (1997)
7. D. E. Backman and F. Paresce: Debris disks. In: *Protostars and Planets III*, E. H. Levy and J. I. Lunine (Eds.) (Univ. Arizona Press, Tucson, 1993) pp 1253–1302
8. H. Beust and A. Morbidelli: Icarus **143**, 170 (2000)
9. J. A. Burns and P. L. Lamy, S. Soter: Icarus **40**, 1 (1979)
10. G. Bryden, et al.: Astrophys. J. **636**, 1098 (2006)
11. S. Charnoz, C. C. Porco, E. Déau, A. Brahic, J. N. Spitale, G. Bacques and K. Baillie: Science **310**, 1300 (2005)
12. M. Clampin, et al.: Astron. J. **126**, 385 (2003)
13. A. T. Deller and S. T. Maddison: Astrophys. J. **625**, 398 (2005)
14. S. F. Dermott, S. Jayaraman, Y. L. Xu, B. A. S. Gustafson and J. C. Liou: Nature, **369**, 719 (1994)
15. C. P. Dullemond and C. Dominik: Astron. Astrophys. **434**, 971 (2005)
16. P. Faber and A. C. Quillen: Mon. Not. Royal Astron. Soc. **382**, 1823 (2006)
17. R. S. Fisher, C. M. Telesco, R. K. Piña, R. F. Knacke, and M. C. Wyatt: Astrophys. J. **532**, L141 (2000)
18. D. A. Fischer and J. A. Valenti: Astrophys. J. **622**, 1102 (2005)
19. F. Freistetter, A. V. Krivov and T. Löhne: Astron. Astrophys. **466**, 389 (2007)
20. D. A. Golimowski, et al.: Astron. J. **131**, 3109 (2006)
21. D. A. Golimowski, et al.: In: *Spirit of Lyot 2007*, http://www.lyot2007.org (2007)

22. R. Gomes, H. F. Levison, K. Tsiganis and A. Morbidelli: Nature **435**, 466 (2005)
23. J. R. Graham, P. G. Kalas and B. C. Matthews: Astrophys. J. **654**, 595 (2007)
24. J. S. Greaves, et al.: Astrophys. J. **506**, L133 (1998)
25. J. S. Greaves, M. C. Wyatt, W. S. Holland and W. R. F. Dent: Mon. Not. Roy. Astron. Soc. **351**, L54 (2004)
26. J. S. Greaves, et al.: Astrophys. J. **619**, L187 (2005)
27. J. M. Hahn and R. Malhotra: Astron. J. **117**, 3041 (1999)
28. S. R. Heap, D. J. Lindler, T. M. Lanz, R. H. Cornett, I. Hubeny, S. P. Maran and B. Woodgate: Astrophys. J. **539**, 435 (2000)
29. W. S. Holland, et al.: Nature **392**, 788 (1998)
30. W. S. Holland, et al.: Astrophys. J. **582**, 1141 (2003)
31. S. Ida and D. N. C. Lin: Astrophys. J. **604**, 388 (2004)
32. M. Jura: Astrophys. J. **603**, 729 (2004)
33. P. Kalas: Astrophys. J. **635**, L169 (2005)
34. P. Kalas and D. Jewitt: Astron. J. **110**, 794 (1995)
35. P. Kalas, J. R. Graham and M. Clampin: Nature **435**, 1067 (2005)
36. P. Kalas, J. R. Graham, M. C. Clampin and M. P. Fitzgerald: Astrophys. J. **637**, L57 (2006)
37. P. Kalas, M. P. Fitzgerald and J. R. Graham: Astrophys. J. **661**, L85 (2007)
38. D. W. Koerner, A. I. Sargent and N. A. Ostroff: Astrophys. J. **560**, L181 (2001)
39. J. E. Krist, D. R. Ardila, D. A. Golimowski, M. Clampin and H. C. Ford: Astron. J. **129**, 1008 (2005)
40. A. V. Krivov, M. Queck, T. Löhne and M. Sremcevic: Astron. Astrophys. **462**, 199 (2007)
41. M. J. Kuchner and M. J. Holman: Astrophys. J. **588**, 1100 (2003)
42. M. J. Kuchner and W. T. Reach and M. E. Brown: Icarus **145**, 44 (2000)
43. P. O. Lagage and E. Pantin: Nature **369**, 629 (1994)
44. C. Leinert and B. Moster: Astron. Astrophys. **472**, 335 (2007)
45. A. Li and J. I. Lunine: Astrophys. J. **590**, 368 (2003)
46. S.-Y. Lin, N. Ohashi, J. Lim, P. T. P. Ho, M. Fukugawa and M. Tamura: Astrophys. J. **645**, 1297 (2006)
47. J. C. Liou and H. A. Zook: Astron. J. **118**, 580 (1999)
48. M. C. Liu: Science **305**, 1442 (2004)
49. R. Malhotra: Astron. J. **110**, 420 (1995)
50. V. Mannings and M. J. Barlow: Astrophys. J. **497**, 330 (1998)
51. K. A. Marsh, C. D. Dowell, T. Velusamy, K. Grogan and C. A. Beichman: Astrophys. J. **646**, L77 (2006)
52. R. G. Martin, S. H. Lubow, J. E. Pringle and M. C. Wyatt: Mon. Not. Royal Astron. Soc. **378**, 1589 (2007)
53. B. C. Matthews, P. G. Kalas and M. C. Wyatt: Astrophys. J. **663**, 1103 (2007)
54. S. A. Metchev, L. A. Hillenbrand and R. J. White: Astrophys. J. **582**, 1102 (2003)
55. A. Moro-Martín and R. Malhotra: Astron. J. **124**, 2305 (2002)
56. A. Moro-Martín and R. Malhotra: Astrophys. J. **633**, 1150 (2005)
57. A. Moro-Martín, S. Wolf and R. Malhotra: Astrophys. J. **621**, 1079 (2005)
58. R. A. Murray-Clay and E. I. Chiang: Astrophys. J. **619**, 623 (2005)
59. C. D. Murray and S. F. Dermott: *Solar System Dynamics* (Cambridge University Press, Cambridge, 1999)
60. L. M. Ozernoy, N. N. Gorkavyi, J. C. Mather and T. A. Taidakova: Astrophys. J. **537**, L147 (2000)

61. P. Plavchan, M. Jura and S. J. Lipscy: Astrophys. J. **631**, 1161 (2005)
62. C. J. Poulton, J. S. Greaves and A. C. Cameron: Mon. Not. Royal Astron. Soc. **372**, 53 (2006)
63. A. C. Quillen and S. Thorndike: Astrophys. J. **578**, L149 (2002)
64. A. C. Quillen: Mon. Not. Royal Astron. Soc. **372**, L14 (2006)
65. A. Roberge, A. J. Weinberger and E. M. Malumuth: Astrophys. J. **622**, 1151 (2005)
66. F. Roques, H. Scholl, B. Sicardy and B. A. Smith: Icarus **108**, 37 (1994)
67. G. Schneider, et al.: Astrophys. J. **513**, L127 (1999)
68. G. Schneider, M. D. Silverstone and D. C. Hines: Astrophys. J. **629**, L117 (2005)
69. G. Schneider, et al.: Astrophys. J. **650**, 414 (2006)
70. B. A. Smith and R. J. Terrile: Science **226**, 1421 (1984)
71. K. R. Stapelfeldt, et al.: Astrophys. J. Suppl. **154**, 458 (2004)
72. L. E. Strubbe and E. I. Chiang: Astrophys. J. **648**, 652 (2006)
73. K. Y. L. Su, et al.: Astrophys. J. **628**, 487 (2005)
74. K. Y. L. Su, et al.: Astrophys. J. **653**, 675 (2006)
75. H. Tanaka, S. Inaba and K. Nakazawa: Icarus **123**, 450 (1996)
76. C. M. Telesco, et al.: Astrophys. J. **530**, 329 (2000)
77. C. M. Telesco, et al.: Nature **433**, 133 (2005)
78. P. Thébault and J. C. Augereau: Astron. Astrophys **472**, 169, astro-ph/07060344 (2007)
79. P. Thébault, J. C. Augereau and H. Beust: Astron. Astrophys. **408**, 775 (2003)
80. C. A. Trujillo and M. E. Brown: Astrophys. J. **554**, L95 (2001)
81. A. J. Weinberger, R. M. Rich, E. E. Becklin, B. Zuckerman and K. Matthews: Astrophys. J. **544**, 937 (2000)
82. J. P. Williams, J. Najita, M. C. Liu, S. Bottinelli, J. M. Carpenter, L. A. Hillenbrand, M. R. Meyer and D. R. Soderblom: Astrophys. J. **604**, 414 (2004)
83. D. J. Wilner, M. J. Holman, M. J. Kuchner and P. T. P. Ho: Astrophys. J. **569**, L115 (2002)
84. J. Wisdom: Astron. J. **85**, 1122 (1980)
85. M. C. Wyatt: Signatures of Planets in Circumstellar Disks. Ph.D. Thesis, University of Florida, Gainesville (1999)
86. M. C. Wyatt: Astrophys. J. **598**, 1321 (2003)
87. M. C. Wyatt: Astron. Astrophys. **433**, 1007 (2005)
88. M. C. Wyatt: Astron. Astrophys. **440**, 937 (2005)
89. M. C. Wyatt: Astrophys. J. **639**, 1153 (2006)
90. M. C. Wyatt and W. R. F. Dent: Mon. Not. Roy. Astron. Soc. **334**, 589 (2002)
91. M. C. Wyatt, S. F. Dermott, C. M. Telesco, R. S. Fisher, K. Grogan, E. K. Holmes and R. K. Piña: Astrophys. J. **527**, 918 (1999)
92. M. C. Wyatt, W. S. Holland, J. S. Greaves and W. R. F. Dent: Earth Moon Planets **92**, 423 (2003)
93. M. C. Wyatt, J. S. Greaves, W. R. F. Dent and I. M. Coulson: Astrophys. J. **620**, 492 (2005)
94. M. C. Wyatt, R. Smith, J. S. Greaves, C. A. Beichman, G. Bryden and C. M. Lisse: Astrophys. J. **658**, 569 (2007)
95. M. C. Wyatt, R. Smith, K. Y. L. Su, G. H. Rieke, J. S. Greaves, C. A. Beichman, G. Bryden: Astrophys. J. **663**, 365 (2007)

3

Asteroids and Their Collisional Disruption

A.M. Nakamura[1], and P. Michel[2]

[1] Graduate School of Science, Kobe University, 1-1 Rokkodai-cho, Nada-ku, Kobe, Japan,
amnakamu@kobe-u.ac.jp
[2] Laboratoire Cassiopée, Observatoire de la Côte d'Azur, CNRS, Université de Nice Sophia-Antipolis, boulevard de l'Observatoire, 06300 Nice, France,
michel@oca.eu

Abstract The collisional process between small bodies is one of the key processes in the formation and evolution of a planetary system. Asteroids are the remnants which kept the memory of collisional processes that took place in different regions of our Solar System during its past and present history. Telescopic observations have collected data of such records, such as the ones provided by asteroid dynamical families and their related dust bands. Meteorites, micrometeorites, and interplanetary dust particles (IDPs), which are pieces of asteroids or comets collected on Earth, provide also information on the material properties of these small bodies, although they may only tell us about the strongest components capable of surviving the entry in Earth's atmosphere. In order to understand the collisional process, impact experiments have been performed in laboratory, using as targets terrestrial rocks whose mechanical properties are similar to those of some meteorites. The results of experiments together with numerical simulations and theoretical considerations have led to the conclusion that most asteroids smaller than several tens of kilometers in size have experienced major impact events, during which they have been at least severely shattered so that cracks and voids could be formed in their interior. For those who underwent a catastrophic disruption as a result of a collision at high impact energy, the outcome has been the formation of an asteroid family, some of which are still identifiable in the main asteroid belt. During such an event, the largest fragments that originate from the parent body can be large enough to undergo gravitational re-accumulations, so that at the end of the process, the cluster of fragments larger than a few hundreds of meters resulting from such a disruption is mostly composed of gravitational aggregates or rubble piles. Spacecraft explorations of multi-kilometer asteroids, namely 951 Gaspra, 243 Ida, and 433 Eros, who belong to the S taxonomic class — the dominant class in the inner Solar System — revealed that the surface of these bodies are shaped by impact processes, and that the bulk density (2.6 and 2.67 g/cm^3 for Ida and Eros, respectively) is generally lower than the supposed grain density of their material. However, direct evidence of a rubble pile structure has not been obtained, as the only information on their internal structure are inferred mostly from their surface properties. Conversely, in spite of its small self-gravity, the sub-kilometer asteroid 25143 Itokawa explored by the JAXA Hayabusa spacecraft in 2005 is the first S-class asteroid whose porosity is estimated

Nakamura, A.M., Michel, P.: *Asteroids and Their Collisional Disruption.* Lect. Notes
Phys. **758**, 71–97 (2009)
DOI 10.1007/978-3-540-76935-4_3   © Springer-Verlag Berlin Heidelberg 2009

to be as high as 40% (with a bulk density of 1.9 g/cm^3) and thus is considered to be a gravitational aggregate formed by reaccumulation of smaller pieces. The boulders on Itokawa have shapes and structures similar to those of laboratory rock fragments, suggesting some universal character of the disruption process. Since more and more asteroids are believed to have substantial porosity, current studies on the collisional disruption of solid bodies are to be extended to porous bodies, taking into account microporosity effects which have been neglected so far. Such porous bodies are not only present in the asteroid populations (Near-Earth Objects, main belt, and Trojan asteroids) but they are also supposed to constitute the populations evolving in the outer Solar System (Kuiper belt objects) and beyond (long-period comets). Thus, understanding the collisional process for different kinds of material appears crucial to determine its influence in the history of different populations of small bodies.

3.1 Introduction

Since the end of the accretion phase of the Solar System which led to the formation of our planets approximately 4.5 Gyr ago, the outcomes of mutual direct collisions between small bodies have mostly been disruptive and played a major role in the formation, evolution, and shape of small body populations. The major collisional disruption zones in the Solar System are the main asteroid belt [15], located between the orbits of Mars and Jupiter, and the Edgeworth–Kuiper belt [17] beyond Neptune's orbit. Mutual collisional velocities are typically of the order of several kilometers per second in the main asteroid belt [5], whereas they are about 1 km/s or less in the Kuiper belt (e.g., [17]). The collisional process is very complex and is still poorly understood, despite recent progresses. Thus, it is a major area of research which is studied by three interrelated and complimentary approaches, namely, laboratory experiments, numerical modeling, and scaling theories [27]. Observations of small bodies, both ground- and space-based, provide constraints to these studies and evidence that the collisional process is still active. Asteroids have been the targets of ground-based telescopic observations for more than two centuries [20] and have become more recently the targets of space missions (fly-by, in situ, and sample return). The observational data of asteroids are, therefore, more and more abundant and have statistical significance in many aspects. Additionally, we have the very samples from asteroids, i.e., meteorites, for in depth microscopic analyses. Therefore, the asteroids are so far the most studied objects in terms of physical properties connected to the collisional process.

In this chapter, a brief introduction to the current state of our knowledge on asteroids is first exposed, allowing further reading of more detailed reviews in recent literatures (e.g., [7]). Emphases are put on the evidence and processes related to collisional disruption events. Basic equations and quantities that are used to describe high-velocity collisions between solid bodies are then given and the impact response of solid bodies is illustrated by addressing the laboratory analog experiments. Finally, evidence of impact processes found on the small asteroid 25143 Itokawa explored by the JAXA (Japan Aerospace Exploration Agency) Hayabusa spacecraft are introduced; open problems and perspectives are then indicated.

3.2 Asteroids

3.2.1 Orbital and Size Distribution

Asteroids are small rocky bodies whose large majority do not have any signature of volatile activities. Only very recently, a few of them have been observed to have some sporadic activities [29]. Most of these bodies are orbiting between Mars and Jupiter in a region called the Main Belt (MB hereafter). The structure of the MB is dynamically constrained by the gravitational perturbations of planets, and more particularly the biggest planet Jupiter. Mean motion resonances with Jupiter, which occur at locations where the orbital period of a small body is proportional to that of Jupiter, correspond to empty zones in the MB called Kirkwood gaps, after the name of their discoverer. Such gaps are a signature of the efficiency of such resonances to destabilize the trajectory of small bodies located into them. Figure 3.1 shows the distribution of semi-major axis and inclination of the bodies in the MB. The gaps are visible and are associated to those dynamical mechanisms such as mean motion resonances with Jupiter, and secular resonances which occur when the frequency of the longitude of perihelion of a small body is equal to one of the proper frequencies of the Solar System generally associated to the average frequency of the longitude of perihelion of a planet (such as the ν_6 secular resonance associated to Saturn). Asteroids located in those resonances are rapidly transported from the MB to the Near-Earth space due to the fast increase of their eccentricity caused by the resonant dynamics. Thus, the ν_6 secular resonance at the inner edge of the MB and the 3 : 1 mean motion resonance with Jupiter at 2.5 AU are among the most effective resonances that supply asteroids to the near Earth object (NEO) population. The timescale for increasing the eccentricity from MB value to Earth-crossing orbits is only a few million years in these resonances [23, 42]. The median lifetime of the NEO population is about 9 Myr [24], and the end-state is either a collision with the Sun for 60% of them, an ejection outside Jupiter's orbit for 30% of them, or a collision with a planet for the rest of the population. However, this population has been kept more or less in a steady-state number over more than 3 Byr [35], as indicated by the dating and counting of Moon's craters, which suggest that the flux of impactors has been kept constant on average (apart from some short fluctuations) over this period. This is due to the fact that while some NEOs reach their end-state, collisions occur in the MB. These collisions generate new fragments whose evolutions lead them to a resonance and eventually to the Near-Earth space.

There are also populations of small bodies that lie farther away from the Sun than the MB. The Trojan asteroids evolve in the 1 : 1 mean motion resonance with Jupiter on the same orbit as the planet, 60° before and after the planet on the Lagrangian points L_4 and L_5 of the three-body problem. Small bodies with perihelia greater than Jupiter's semimajor axis (5.2 AU) and semimajor axes smaller than that of Neptune (30.1 AU) are called Centaurs. These objects originally come from the population of trans-Neptunian objects (TNOs), also called Kuiper belt objects (KBOs), located beyond Neptune's orbit. In all these populations, small bodies with well-determined orbits have been given a permanent number and a name, such as the small NEO 25143 Itokawa, the largest asteroid in the MB called 1 Ceres, the Centaur 5145 Pholus, and the first discovered TNO (15760) 1992 QB$_1$.

The clusters of asteroids that are easily identified in the *proper orbital element* space in the MB are dynamical asteroid families. *Proper elements* are quasi-integrals of motion which are more stable than the osculating elements, although some mechanisms such as the Yarkovsky thermal effect can cause slow variations over time. In such a proper element space, the points related to real objects have a stronger link to their original place than in the osculating element space, which allows the identification of connected groups related to asteroid families. In Fig. 3.1, three prominent asteroid families, Themis, Eos, and Koronis families that were first discovered in 1918 [32] are indicated. The origin of an asteroid family is the catastrophic disruption of a large asteroid (called the parent body of the family) as a result of a collision with a smaller body. The largest *children* formed during such events, by reaccumulation of small fragments of the parent body are the known family members (see Michel's chapter in this issue), which have been identified thanks to their large enough size, spectra similarities, and small dispersion by ground-based observations. The finest portion of the ejecta from the parent body spreads further and sometimes can be identified as dust bands associated to the family (see Ishiguro and Ueno's chapters in this issue). The age of a young asteroid family is determined by directly tracking the orbital evolutions of the family members backwards in time all the way to their starting orbits, provided that the time since the birth event is not long enough to reach the time of unpredictability of the orbital motions due to chaos and highly non-linear phenomena. The Karin cluster is such a young group of asteroids produced by the disruption of a ≈ 30 km-sized body, only 5.75 ± 0.2 Myr ago [50]. Because the proper orbital elements of family members undergo some diffusion due to high-order resonances and the Yarkovsky effect, which leads to a semimajor axis drift depending on the object's spin, orbit, and material properties [7], it is, as we just said, impossible to determine the age of old families by direct orbital integration backward in time based on purely dynamical considerations.

The number of asteroids increases with the absolute magnitude H. The amount of the reflected sunlight from an asteroid is proportional to the square of the diameter; the absolute magnitude H and the diameter D of an object are related by the following expression:

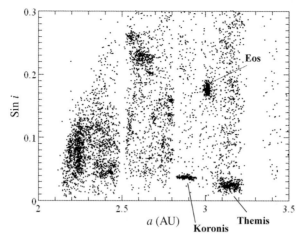

Fig. 3.1. The distribution of 5335 numbered asteroids with respect to proper semi-major axis a and inclination i (see [40]; data from PDS Small Body Node). The inner edge is determined by the ν_6 secular resonance. The gap at 2.5 AU corresponds to the location of the 3 : 1 mean motion resonance with Jupiter. Themis, Eos, and Koronis asteroid families are indicated. See also Fig. 15 in Jewitt's chapter in this issue

$$D = \frac{1329}{\sqrt{p_v}} 10^{-\frac{H}{5}} \tag{3.1}$$

where p_v is the geometric albedo, whose value ranges from 0.04 to 0.4 in the different taxonomic classes of the asteroid population. Thus, when the albedo is not known, the diameter of an asteroid can roughly be estimated with an ambiguity within a factor of a few. The cumulative size distribution of asteroids having a diameter larger than a given diameter D, called $N(> D)$, increases with decreasing D. A power-law is generally used to fit this distribution with a power of about -2 in the range of 1–10^3 km (see also Sect. 3.2.4), although the slope is still uncertain in the kilometer-sized range. As for asteroid families, power-law exponents mostly in the range -2 to -4 are found for the size distribution of large members [70].

3.2.2 Asteriod Composition

The composition of asteroids is continuously investigated by remote-sensing techniques, such as spectroscopy of the reflected sunlight from the surface, and in laboratory by the analyses of meteorites. Establishing the link between asteroids and meteorites is crucial as it can provide clues on the composition and internal structure of asteroids and on the orbital properties of the meteorites' parent bodies. Moreover, an identified link would give answers to the question of where in the Solar System those meteorites experienced the different chemical and physical processes.

The dynamical link between asteroids and meteorites can be studied both by model calculations of the orbital evolutions of main belt asteroids and by observationally determined orbits of meteorite falls [59]. Material connection between asteroids and meteorites can be estimated from the visible and the near-infrared (NIR) spectroscopic ground-based observations of asteroids and X-ray and gamma-ray spectroscopic observations made in situ by spacecrafts visiting an asteroid. The diagnostic features of silicates in the visible and NIR reflectance spectra are two relatively narrow and symmetric absorption features of pyroxene (at wavelengths of 0.9 and 1.9 μm) and a broad asymmetric 1 μm absorption feature of olivine. Different types of meteorites have been related to different taxonomic classes of asteroids by a resemblance in color and albedo. The taxonomic classification of asteroids has been extended from the original one made in the 20th century. Asteroids were initially classified into C and S (i.e., carbonaceous and stony)-classes [12], and then into several classes, including M (metal)-class, but because some subtle differences have been identified within each class, those initial classes have been split into sub-classes [11]. Then, new classes have been defined, such as, for instance the W-class, which is composed of M-class asteroids for which a 3 μm water absorption band has been found in their spectra [57].

The most common taxonomic class in the inner MB is the S-class. S-class asteroids typically have a geometric albedo of ~ 0.15 [61] and the meteorites believed to be associated to this class are the ordinary chondrites, the most abundant meteorites collected on Earth's surface. However, until recently, it was not clear whether S-class asteroids are really the parent bodies of ordinary chondrites, because a detailed comparison of the reflectance spectra shows a mismatch in the spectral slope, albedo, and absorption band depth: the spectra of S-class asteroids are redder and darker than those of ordinary chondrites and the NIR silicate absorption bands are shallower [14]. This discrepancy has then been understood as an effect of the *space weathering* which affects the top thin surface of small bodies. Surface modification processes due to the space environment, such as solar wind ion implantation, sputtering, and micrometeorite bombardment and their resulting optical effects all correspond to what is called *space weathering*. Indeed, X-ray fluorescence spectroscopy performed on the asteroids 433 Eros and 25143 Itokawa shows that these bodies have elemental compositions consistent with that of ordinary chondrites [52, 62].

C-class asteroids have low geometric albedo, typically ~ 0.05, and monotonic spectra in the visible wavelength range with ultraviolet absorption feature. They are believed to be the parent bodies of carbonaceous chondrites and are abundant in the middle zone of the MB. Darker and redder colored asteroids, which are considered to have more organics, are dominant in the outer region of the MB and among Trojan asteroids (called D-class asteroids). The orbital distribution of the different asteroid classes, which shows some groupings, is an indication that the degree of thermal metamorphism and consequently the aqueous alteration depends on the heliocentric distance from the Sun.

3.2.3 Internal Structure

Mass estimates of asteroids were very limited until the beginning of space-craft explorations towards asteroids and the precise determination of asteroid orbits. The masses of the largest ones have been determined from mutual close approaches and by their gravitational perturbations onto smaller objects. Moreover, the masses of visited asteroids could also be determined by their perturbation on the observing spacecraft. Then, the discovery by the Galileo spacecraft [13] of a satellite, called Dactyl, around the asteroid 243 Ida stimulated search efforts for binary systems. Now photometric light-curve observations, radar-imaging observations, and direct imaging using adaptive optics attached on the largest telescopes are used to detect binary systems. Once the asteroid mass is determined, the bulk density can be calculated using a shape model.

The current internal structure of asteroids is a consequence of their collisional history. It can range from monolithic, fractured, and shattered due to moderate impacts to gravitationally reagglomerated rubble piles. The bulk density of an asteroid gives a first-order indication of its internal structure. For instance, the C-class asteroid 253 Mathilde visited by the NEAR spacecraft in 1997 has a bulk density of 1.3 g/cm^3 [68]. This is much lower than the bulk density of carbonaceous chondrites. For example, CI and CM chondrites have average bulk densities of 2.1 and 2.2 g/cm^3, whereas the average bulk densities of CR, CO, and CV chondrites are greater than 3 g/cm^3 [9]. The low bulk density of Mathilde is interpreted as an indication of high porosity inside the body. Note that the type of porosity measured in meteorites is microporosity, which is related to fractures, voids, and pores on the scale of tens of micrometers. Based on their grain density (~ 2.7 g/cm^3), CM chondrites have a degree of microporosity of about 12%. If Mathilde has material components and microstructures similar to CM chondrites, its total porosity is $\sim 52\%$ and its macroporosity accounts for $\sim 40\%$. The typical macroporosity of S-class asteroids is about 20%, based on the bulk density measured for Ida and Eros, 2.6 and 2.67 g/cm^3, respectively, although the 500 m S-class asteroid 25413 Itokawa, with a bulk density of 1.9 g/cm^3, has a macroporosity of about 40%, similar to the one of the C-class Mathilde. This shows the great diversity of asteroids' properties, even in a same spectral class.

These information about the internal structure are extremely important. The outcome of a collisional disruption of an asteroid depends highly on its mechanical properties [27, 38, 55] which constrain the transmission efficiency of the impact energy throughout the body. For instance, the presence of five big craters larger than 5 km in diameter on the surface of the asteroid Mathilde is considered as an indication that porous asteroids are more robust against impacts (absorb more greatly the impact energy). Indeed, based on our current understanding, a non-porous object of similar size would not have survived the impact events needed to create such craters. Actually, the transfer of the shock wave throughout the body, more efficient in a non-porous material, would have destroyed it all.

As one can see, the information that we have on the internal structure of small bodies are all indirect. Thus, only in situ investigations by space missions will allow us to have a more direct characterization. However, even in this case, challenging techniques will have to be used, such as radar tomography, to characterize the properties deep inside an object.

3.2.4 Timescales of Processes During Asteroid Evolution Histories

The evolution history of asteroids can be illustrated by key timescales. The collision lifetime τ_{col} of an asteroid of diameter D has been estimated assuming constant collisional velocities and impact rates over time, i.e., assuming the orbital structure of asteroids does not change much during the population history. Its expression is given by:

$$\frac{1}{\tau_{col}(D)} = \int_{D_p^*} \frac{P_i}{\pi} \sigma_{col}\, n(D^*)\mathrm{d}D^* \tag{3.2}$$

where D_p^*, P_i, σ_{col}, and $n(D)$ are the threshold diameter of a body that can *catastrophically* break up the body of diameter D, the intrinsic collisional probability in units of km^{-2} yr^{-1} (P_i is defined as the probability that a single member of the impacting population will hit the target over a unit of time), the collision cross-section, and the number of asteroids per unit diameter, respectively ([16]).

To compute P_i and the average collisional velocity, Bottke et al. [5] took a representative sample of MB asteroids (all 682 asteroids with $D > 50$ km) and calculated the collision probabilities and impact velocities between all possible pairs of asteroids, assuming fixed values of semimajor axis, eccentricity, and inclination (a, e, i). Note that so-called Öpik-like codes like that in Bottke et al. assume the orbits can be integrated over uniform distributions of longitudes of apsides and nodes; this approximation is considered reasonable because secular precession randomizes the orientations of asteroid orbits over $\approx 10^4$ yr timescales. However, it fails while an object is in a resonance. After all possible orbital intersection positions for each projectile–target pair were evaluated and weighted, it is found that MB objects striking one another have $P_i = 2.86 \times 10^{-18}$ km^{-2}/yr^{-1} and a collisional velocity equal to 5.3 km/s. These values have then been corroborated by different authors and methods. Gravitational focusing is generally neglected because escape velocities from asteroids are $\approx$m/s, whereas asteroid impact velocities are of the order of several kilometers per second. Thus, the collisional cross-section is expressed as:

$$\sigma_{col} = \frac{\pi(D + D^*)^2}{4}. \tag{3.3}$$

The size distribution of asteroids $n(D)$ can undergo some slight changes over time due to the dynamical and collisional evolution of the population, but the current size distribution can still be approximated by analytical forms (e.g., [25]):

$$n(D)dD = 2.7 \times 10^{12} D^{-2.95} \, dD \text{ for } D > 100 \text{ m} \tag{3.4}$$

Then (3.2) can be approximated to ([18]),

$$\frac{1}{\tau_{\text{col}}(D)} \sim \frac{P_i D^2}{4} N(> D) \tag{3.5}$$

$$N(> D) = \int_{D_p^*} n(D^*)dD^* \tag{3.6}$$

The threshold diameter D^* is determined by the impact physics and will be discussed in the next section. A typical value for D^* used in the current studies for a 10 km-diameter asteroid is $600 - 700$ m [27]. Because the number of asteroids larger than $600 - 700$ m in diameter calculated by the (3.6) is $\sim 5 \times 10^6$, $\tau_{\text{col}}(D = 10 \text{ km}) \sim 3$ Gyr in agreement with current estimates of the collision lifetime of 10 km-sized asteroid in the MB by the latest collisional evolution models [8, 51].

The orbit of small asteroids is changed by the Yarkovsky effect on timescales smaller than their collisional lifetime. The semimajor axis displacement δa that a stony body having a diameter of 0.01 km can undergo is estimated to be about 0.001 AU/Myr, while it is about $2-5 \times 10^{-5}$ AU/ Myr for 2–4 km diameter asteroids [6]. Thus, the mobility δa decreases with the size of the body. These values of δa have important dynamical consequences. For instance, 0.01–0.1 AU are typical distances that a MB meteoroid might have to travel to reach a powerful resonance, and diffusing on such distances would require 10–100 Myr only. Therefore, meteorites do not have to be directly injected into a resonance as a result of their parent body disruption. They can rather be created further away from the resonance and then be injected into it as a result of the Yarkovsky effect after a few 10 Myr. Then, although the transport to the Earth once inside the resonance is short (only a few million years), this scenario explains why the cosmic ray exposure age (CRE, see further) measured for most meteorites is greater than the short transport timescale by resonances (see, e.g., [41]). Similarly, the orbital distribution of members of old families, such as Koronis, can be explained by a combination of the Yarkovsky effect and diffusion in high order resonances, starting from the initial spreading produced by the disruption of the parent body, which is likely narrower than the observed one.

On the course of their dynamical and collisional evolutions, the surface of asteroids is processed in the interplanetary environment. The already mentioned space weathering is one of the material processing that can be remotely detected. The timescale of this process is not well established at present [53]. The degree of the damage by cosmic ray is analyzed on meteorites. CRE ages of the most common meteorites, i.e., the ordinary chondrites, range from a few million years to ~ 100 Myr [36]. CRE ages give the length of time a body has been exposed in space as a meter-scale object or near the surface of a larger body and therefore can be used to estimate the time between the liberation from the parent body and the arrival on Earth's ground.

3.3 Disruption by Hypervelocity Impact

3.3.1 Initial Shock Pressure and Propagation

When two small bodies collide with each other with a velocity of the order of kilometers per second, shock waves are generated and propagate in these bodies. Figure 3.2 shows a schematic view of a collision. The shock wave velocity U_s is well approximated by a linear relationship with the particle velocity u_p for the considered velocity range (see Fig. 3.3). Its expression is given by:

$$U_{sk} = C_{Bk} + s_k u_{pk} \qquad (k = p, t) \qquad (3.7)$$

where the subscripts p and t denote each of the two colliding bodies, i.e., the projectile (conventionally considered as the smaller body involved) and the target. Here C_B and s are the bulk sound velocity and a dimensionless constant of the order of unity, respectively. In a planar impact approximation, the pressure at contact P is:

$$P_k - P_{0k} = \rho_{0k} U_{sk} u_{pk} \qquad (k = p, t) \qquad (3.8)$$

where P_0 is the pressure before the compression, which can be neglected in most cases, and ρ_0 is the bulk density. This is one of the Hugoniot equations (see e.g. [37]). The pressure is determined under the following two boundary conditions at the surface of contact between the two bodies: (1) the pressure balance and (2) the equality of particle velocity in both bodies. The first condition expresses as:

$$P_t = P_p \qquad (3.9)$$

and the second condition is:

$$u_{pt} = V_i - u_{pp} \qquad (3.10)$$

where V_i is the impact velocity.

If the two bodies are made of the same material, then

$$u_{pt} = u_{pp} = \frac{V_i}{2} \qquad (3.11)$$

and

$$P_p = P_t = \rho_0 \left(C_B + s \frac{V_i}{2} \right) \frac{V_i}{2}. \qquad (3.12)$$

The initial shock pressure during the collision is of the order of tens of GPa in the MB, where typical collision velocities are of several kilometers per

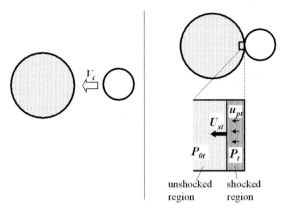

Fig. 3.2. Schematic view of a collision

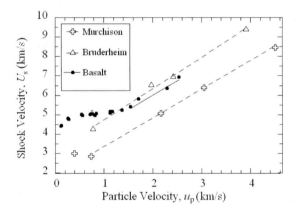

Fig. 3.3. Relationship between shock and particle velocities for two meteorites, namely Murchison (carbonaceous chondrite, CM2) and Bruderheim (ordinary chondrite, L6) [2], and Kinosaki basalt [48]. The values of the bulk density ρ_0 are 2.2, 3.3, 2.7 g/cm^3, respectively; those of the bulk sound velocity C_B are 1.87, 3.11, 3.0 km/s, respectively, and those of the constant s are 1.48, 1.62, and 1.5, respectively

second [5]. This is about (or more than) 100 times the compressive strength of terrestrial rocks. Since brittle materials are weaker by an order of magnitude in tension than in compression, such mutual collisions naturally lead to the breakup of the small rocky bodies involved in the collision. The level of initial pressure is only marginal with respect to the level required for partial rock melting, which is typically in the range between 50 and 100 GPa [37]. However, some shock effects are identified on minerals and meteorites at a pressure level less than 50 GPa.

The pressure of the shock wave decays in the body with the propagation distance from the impact point, with a rate depending on the material and on the pressure level [26]. The pressure rapidly decays in a *strong* shock

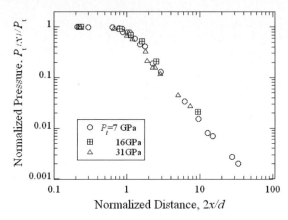

Fig. 3.4. Experimentally determined shock pressure decay in Kinosaki basalt [49]. The pressure $P_t(x)$ at the propagation distance x is normalized by the initial peak pressure P_t. The radius $d/2$ of the projectile was 4.5, 4.5, and 14 mm for experiments with initial $P_t = 31$, 16, and 7 GPa, respectively. The projectile's shape was a disk (with thickness about one-ninth of the diameter) for the 31 and 7 GPa experiments and a cylinder (with height equal to the diameter) in the 16 GPa experiment

regime where the particle velocity is much larger than the sound velocity, i.e., $u_p \gg C_B$. Figure 3.4 shows a typical decay curve of non-porous rocks in the *intermediate regime*, where $u_p \sim C_B$, and illustrates that the initial shock pressure decreases by three orders of magnitude when the wave travels a distance of a few tenths of the projectile's radius.

3.3.2 Hyper-Velocity Impact Experiment Conditions

In a typical set-up, a two-stage light-gas gun is used to accelerate a millimeter– or centimeter–sized projectile to the velocities involved in asteroidal collisions (several kilometers per second). Powders are fired in the first stage and the combustion gas gives pressure to a plastic piston which in turn compresses a light gas in the second stage. The projectile is accelerated by the light gas, i.e., H or He, and the velocity is limited by the sound velocity of the light gas. Projectiles are usually composed of plastics or metals. To avoid that the projectile breaks in the gun muzzle, a sabot is used to accelerate the projectile. As for the targets, a wide variety of materials have been used, including rocks, ices, and even sands in a thin paper bag [67].

Material properties of the targets, which have important influence on the collisional and cratering processes, are the bulk density, the yield strength, the compressive, shear, and tensile strengths ([3], see also Michel's chapter in this issue), the porosity, the longitudinal and transversal wave velocities, and the Hugoniot parameters (e.g., C_B and s in (3.7)). The measurements of these material properties have been performed for terrestrial rocks, although some of them require sophisticated instrumentations. Other material parameters used in numerical simulations of impact disruption are the Weibull parameters, which represent the distribution of incipient flaws in the material. These parameters have recently been measured for one of the materials used in previous impact disruption experiments (see Sect. 3.3.3). Among all these material parameters, only the bulk density and inferred porosity have been determined for some asteroids. Other material properties have not been measured directly for asteroids and comets, so all our understanding of the collisional process relies on terrestrial rocks and meteorites.

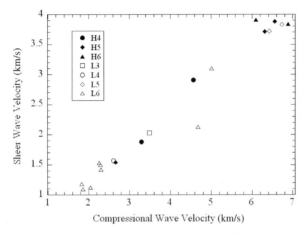

Fig. 3.5. Elastic wave velocities in L-type and H-type ordinary chondrites [69] and terrestrial rocks used in previous analog impact experiments [60]

Meteorites are the only materials available in laboratory which are more or less directly connected to some asteroids. However, meteorites have only been used in a few impact experiments, due to the small volumes of sample available. Therefore, terrestrial rocks and synthesized materials are more commonly used. Figure 3.5 shows elastic wave velocities in ordinary chondrites [69] and in materials used in previous analog impact experiments [60]. Figures 3.3 and 3.5 indicate that the terrestrial rocks used in previous analog experiments have some of their mechanical properties which are similar to those of ordinary chondrites. However, there are three distinct problems that prevent to make a direct link between the material constituing meteorites and asteroid material. First, we do not have samples for all the asteroid taxonomic classes. In particular, our collection is likely to be biased toward material coming from the inner part of the asteroid belt. Second, our collection has probably suffered greatly from a selection effect against weaker materials, as only the strongest material can survive the dynamical pressure undergone during atmospheric entry and transit. For example, analyses of the trajectory of a meteorite's fall have suggested that meter-class, stony, near-Earth asteroids (NEAs) have tensile strengths more than an order of magnitude lower than those measured for ordinary chondrites [10]. Third, there is a great difference between asteroids and meteorites in size scale. It is known that the strength of brittle materials decreases with size, because larger rocks have statistically a higher probability to contain a weaker part (larger incipient flaws) in their volume than smaller rocks. This will be discussed in the next section. Moreover, asteroids are found to have a smaller bulk density than taxonomically corresponding meteorites, indicating higher internal porosity in those bodies than in meteorites.

Therefore, although meteorites are rich in information (in particular, concerning the chronology of our Solar System), their physical properties may not be representative of those of the material that evolves in space. Thus, our understanding of the material properties of asteroids is still very limited. Similarly, our understanding of material behavior under impacts is limited to terrestrial rocks, and we can just hope that our findings can still provide us information that are relevant to asteroid disruption events.

3.3.3 Distribution of Incipient Flaws in a Rock: The Weibull Parameters

Natural materials intrinsically have non-uniform physical properties. This non-uniformity often becomes the major source of scattering in the outcome of collisional disruptions of rocks in laboratory experiments under same initial conditions. For example, everything else being equal in experimental conditions, targets cut out from a same large block of rock break differently one another, due to the one by one different strength of these targets. The statistical behavior of the material strength is usually expressed [66] by a Weibull distribution:

$$n(\sigma) = K \left(\frac{\sigma}{\sigma_N} \right)^m \tag{3.13}$$

where $n(\sigma)$ is the density number of flaws in a rock that activate at a stress not greater than σ [66], K is constant, and m and σ_N are the constant Weibull parameters (note that in the literature, the Weibull parameters are indicated to be m and k, where k is related to K and σ_N). The probability of failure $P_{\text{probability}}$ of a specimen of a given volume at the stress σ is represented by:

$$P_{\text{probability}}(\sigma) = 1 - \exp \left[- \left(\frac{\sigma}{\sigma_N} \right)^m \right] \tag{3.14}$$

where m and σ_N are called shape and scale, respectively. The value of m gives an indication of the degree of homogeneity within the material. Higher values of m indicate that flaws are more evenly distributed throughout the material, and consequently, the strength is nearly independent of the volume of the specimen. Lower values of m indicate that flaws are fewer and less evenly distributed, causing greater scatter in strength. The characteristic stress σ_N is the stress at which 63.2% of similar units subjected to stress will have failed.

It is only recently that the Weibull parameters of the target material used in impact experiments have been experimentally determined [46]. Figure 3.6

shows the X-ray CT-scan image of a basalt target used in previous impact disruption experiments and the measured result of the probability of failure under static loading. The value of the shape parameter m is in the range of 15–17 for this material and is higher than the value used in numerical simulations of the impact disruption of this material (using about 9–9.5).

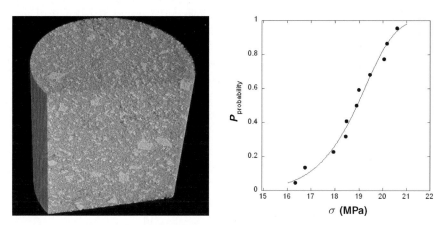

Fig. 3.6. *Left*: X-ray CT-scan image (diameter is 4.2 mm, taken by XMS-BS9, Microscopic Scan Co.) of a basalt target used in some impact experiments with in-depth data [43, 44]. *Right*: Probability of failure $P_{\mathrm{probability}}$ in a diametral compression test of a basalt specimen of diameter 10 mm and thickness 5 mm [46]

The dependency of material strength on target's size is directly related to the Weibull distribution of incipient flaws. Indeed, the minimum stress σ_{min} at which a flaw in a target of volume V activates is derived from $n(\sigma_{\mathrm{min}}) = 1/V = K(\sigma_{\mathrm{min}}/\sigma_{\mathrm{N}})^{\mathrm{m}}$. Thus, we have:

$$\sigma_{\mathrm{min}} = \sigma_{\mathrm{N}}(KV)^{-1/\mathrm{m}}. \tag{3.15}$$

The threshold for failure σ_{min} thus goes with the $-3/m$ power of the target's size, and therefore, larger targets start to break at lower stresses than smaller ones. Hence, for $m = 6$, the strength is proportional to $r^{-1/2}$, where r is the body's radius, and such a decrease of strength with body's size is often used in collisional evolution models.

3.3.4 Outcome of Collisional Disruption

Largest Fragment

One of the most fundamental result of an impact disruption is the degree of fragmentation as a function of the impact initial conditions. The mass fraction of the largest fragment to the original target is commonly used as the indicator of the degree of fragmentation. When the stress level of the wave in the target at the antipodal point from the impact point is larger than the tensile strength of the target, the antipodal surface is removed (broken off) due to tensile failure and the largest fragment corresponds to an internal part of the target. It is usually called in this case a *core* fragment. Fine fragments are rather generated in the vicinity of the impact point. Such tensile failure at the target surface is called *spallation* and is also seen in laboratory cratering impacts on brittle materials. Figure 3.7 shows an example of a *core* fragment. On the other hand, when the stress level of the wave in the target is low due to either poor transmissivity of the shock wave by the target material or low initial pressure, spallation does not occur and cone-shaped largest fragments with their tops heading at the impact point are created.

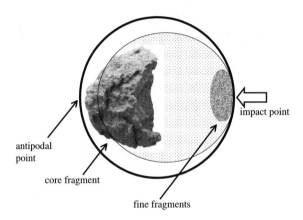

Fig. 3.7. A core fragment from a 6 cm diameter basalt disruption experiment [44]

The most direct indication of the impact condition is the energy density, also called specific impact energy Q, defined as the kinetic energy of the projectile (the smaller of the two colliding bodies) divided by the mass of the target in the frame where the target is at rest. Because the initial pressure and the following pressure decay are different depending on the impact velocity or the impacting materials (e.g., a iron–rock collision involves much higher initial pressure than a rock–rock collision), they cannot be used as an intrinsic indicator for impact events with different impact velocities or involving very different materials from solid rocks. Conversely, the energy density is commonly used as a reference not only in laboratory experiments but also in numerical simulations of the collisional disruption and evolution of asteroids. The threshold value for catastrophic disruption, Q^*, is defined as the impact energy required to produce a largest remaining piece whose mass corresponds to 50% of the mass of the original body [21, 27]. Then assuming that both bodies have the same bulk density, the projectile diameter D^* needed for catastrophically disrupting an object of diameter D with an impact velocity V_i is

$$D^* = \left(\frac{2Q^*}{V_{\mathrm{i}}^2}\right)^{(1/3)} D \qquad (3.16)$$

Generally, it is determined for a fixed impact velocity, and its trend is analyzed as a function of the target's size by adjusting the projectile's mass. Usually, impact velocities consistent with the ones in the MB (about 5 km/s) are considered, but then, one should note that the curves of Q^* as a function of diameter published for this velocity, in principle, cannot be used for collisions made at other velocities.

The specific impact energy to shatter an object, Q_{S}^*, is defined as the threshold value at which the largest remaining intact piece immediately following the fragmentation contains 50% of the mass of the original body. We refer to it as the shattering energy. At large body sizes (above a few hundred meters), the shattered pieces may reaccumulate due to their mutual gravitational attractions, depending on their velocity relative to their mutual escape velocity. Therefore, at those large sizes, in the so-called gravity regime, a higher impact energy threshold Q_{D}^* is defined and corresponds to the specific energy such that the largest fragment (which may be produced by reaccumulation of smaller ones) contains 50% of the mass of the original body. This is called the threshold energy for disruption. Note that in the strength regime where gravity is negligible, $Q_{\mathrm{S}}^* = Q_{\mathrm{D}}^*$. A typical value of Q_{S}^* for a centimeter-scale rocky target is $\approx 10^7$ erg/g at impact velocities consistent with MB values. The value of Q_{S}^* is extrapolated at larger target's sizes from the experimental values at centimeter-sized by numerical simulations or using some scaling laws. However, due to our poor understanding of the process of fragmentation, depending on the assumptions and models used, the extrapolated value of Q_{S}^* can vary by several orders of magnitude at a given size [27]. Although the gravity regime should be better understood, as Newton's law of gravity is well-known compared to the concept of solid strength, the body's size at which the transition occurs between the strength and gravity regime (where Q_{S}^* starts to differ with Q_{D}^*) is still a subject of debates and is somewhere in the range between a few hundred of meters to a few kilometers according to various studies. Small body populations are composed of bodies with a wide range of material properties, and several evidence point toward the presence of bodies with a high degree of porosity (e.g., [9, 22]). As a consequence, determining the outcome of the disruption of a porous body is now considered to be extremely important for studying the origin and collisional evolution of small body populations, and for determining efficient mitigation strategies against a threatening NEO. Generally, the impact energy threshold for disruption depends on material strength and gravity when the body is large enough [27]; however, porosity plays a complicated role. The static compressive strength (S) of porous material is usually lower than that of dense material, as indicated by an empirical formula known for ceramics [56], given by $S = c_1 \mathrm{e}^{-c_2(1-\phi)}$, where c_1 and c_2 are constant, and ϕ denotes the filling factor, $(1-\phi)$ being the porosity. On one hand, with increasing porosity, the target body has a weaker tensile strength. On the other hand, the increasing volume of void space decreases the transmission efficiency of stress waves in the target body, so that porous bodies appear stronger against impact.

The impact response of a small body, and consequently, the mass of the largest remnant from its disruption are thus highly sensitive to material properties and especially to the degree of porosity. Therefore, it is crucial to have a better knowledge of the collisional process as a function of the internal structure of the bodies, and to determine those properties for the real objects belonging to the different populations of small bodies. This last part will require the development of in situ space missions toward different bodies, as from the ground, only a limited knowledge can be inferred from the light emitted from their surface.

Fragment Size, Shape, Velocity, and Spin Distributions in Small-Scale Laboratory Experiments

In laboratory experiments, the masses of the largest fragments are measured. The masses (sizes) of largest fragments are highly dependent on the impact conditions such as the energy density and the geometry of the collision characterized by the target's shape and the impact angle of the projectile. The size distribution of smaller fragments is derived from either the masses of individual fragments or from the total mass of fragments in each size range binned by sieves with different opening sizes [60]. The size distribution of smaller fragments whose sizes are typically equal to or smaller than ~ 1 mm in usual laboratory centimeter-scale disruption experiments, is not highly sensitive to the impact conditions. The differential size distribution of fragments from a rocky body can often be well fitted with a power index between -3 and -3.5. Interestingly, a power-law exponent of -3.5 is also used for the size distribution of interstellar dust grains (MRN-distribution [34]). The size distribution of even smaller fragments (fine fragments) with sizes less than $10\,\mu$m is investigated at specific directions from the impact point using witness plates [45]. The size distribution of fine fragments from basalt targets shows similar power-law index to that of $\sim 10\,\mu$m -1 mm fragments. The ones for chondrites also show a similar tendency [19].

The shape of fragments is investigated using high-speed images taken during impact experiments or is directly measured using a slide caliper. The shape is charaterized by the axial ratios, B/A and C/A, in an ellipsoid approximation (with $A > B > C$). One method starts by measuring the largest dimension of the fragment, A, and the other method starts by measuring the smallest dimension of the fragment, C. These two main methods of measurement of the axes can lead to different results. Fragments from catastrophic disruptions rarely have B/A or C/A ratios below 0.3 and 0.2, respectively. The mean values cluster around 0.7 and 0.5 over widely different experimental conditions [21]. In cratering events, larger fragments are presumably spall fragments and have plate-like shapes. It has been reported that most of the spall fragments from craters excavated in Gabbro had B/A values greater than 0.6 and C/A values less than 0.25 [54].

The surface roughness of fragments is investigated by fractal analysis. The fractal dimension is not dependent on the degree of fragmentation and the value determined by a divider method is $\sim 2.1 - 2.4$ (see [31]).

In general, fragment velocities and rotational frequencies are the highest near the impact point and decrease with increasing distance from this site. Surface fragments tend to have higher velocities than fragments from the interior of the target. The upper bounds of the fragment velocity and rotational frequency distributions usually decrease with increasing fragment's size ([27]).

3.4 Impact Process on Asteorid 25413 Itokawa

In 1991, a MB S-class asteroid 951 Gaspra was visited during a fly-by of the NASA Galileo spacecraft [63]. Gaspra has an irregular shape of $19 \times 12 \times 11$ km and has a great number of small craters and grooves on its surface as well as color variations suggesting some space weathering effect. In 1993, the Galileo spacecraft made a fly-by to another MB S-class asteroid called 243 Ida [4]. Ida is a body of 58 km long. It belongs to the Koronis family and was found to have a 1.4 km diameter moon, called 243(1) Dactyl. Owing to the discovery of this tiny satellite, the bulk density of Ida was determined to be 2.6 g/cm^3. Next, a MB C-class asteroid 253 Mathilde was visited during a fly-by of the NASA NEAR spacecraft [64]. Mathilde is a $66 \times 48 \times 46$ km body with a very slow rotational period of 17.4 days. The bulk density of the body was found to be 1.3 g/cm^3. This small value led to the conclusion that the asteroid contains 40% or more macroporosity. There are at least five craters larger than 5 km in size on Mathilde and this was also interpreted as a piece of evidence that Mathilde has a high fraction of vacuum space inside, which allowed the whole body to survive the impacts that created those craters. The high porosity effectively attenuates the shock waves generated by such impacts and helps to maintain the whole body intact. Otherwise, it was believed that the impact energy needed to form such large craters should have disrupted the body if there was no dissipation mechanism of this energy. Although Mathilde has such a high porosity, it is not clear if Mathilde is a reaccumulated body from a major impact event. Mathilde may be as likely a primitive porous body. In February 2000, the NEAR spacecraft was inserted into orbit around 433 Eros, the second largest NEO, and was renamed the NEAR-Shoemaker spacescraft [65]. Eros is a S-class NEO whose size is $33 \times 13 \times 13$ km. Its bulk density is 2.67 g/cm^3, which is close to the value estimated for Ida. During the $\sim$1-year mission in orbit around Eros, more than a hundred of thousands images were taken. The presence of global ridges and grooves and higher order gravitational data all indicate that Eros has at least partially a cohesive, homogeneous interior in spite of $\sim$ 20% of macroporosity. This percentage of porosity is inferred

from spectral data which suggest that Eros has a similar composition to ordinary chondrites. Thus, the total porosity can be estimated from the ratio
of the bulk density of Eros (2.67 g/cm^3) and the grain density of the component minerals (3.75 g/cm^3, when assuming L-chondrites composition) to
be $(1 - 2.67/3.75) \times 100 = 28.8\%$. Since the microporosity of L-chondrites is
on average 10.8%, assuming a same level of microporosity in Eros leads to a
macroporosity of 18% within this asteroid [9]. Therefore, S-class bodies are
usually not interpreted as being extremely porous and their porosity is rather
considered as macroporosity. Conversely, the porosity indicated for dark-type
(e.g., C-class) bodies is considered as being composed of microporosity at
higher levels in addition to large voids and fractures. The material constituing these bodies is thus believed to be porous at micro-scale, which makes
them behave differently than non-porous (or macro-porous) bodies during
impact phenomena.

In 2005, the JAXA Hayabusa spacecraft performed a sample return mission to the S-class NEO 25143 Itokawa, whose length is about 500 m (see
Fig. 3.8). Although the return of the sample is not guaranteed yet, the mission is already considered as a great success in the scientific community, given
the wealth of new information and surprises that it provided by the images and
measurements made during the visit of the spacescraft. Itokawa is the smallest
asteroid that has ever been visited by an artificial satellite. The largest boulder
(Yoshinodai) lying on its surface is one-tenth the size of the whole asteroid.
Yoshinodai and other ten meter-size boulders are too large to be generated
from any impact crater candidates on Itokawa. The bulk density of Itokawa
is 1.9 g/cm^3 [1] and is far below the values found for other S-class asteroids.
These facts collectively strongly suggest that Itokawa is a gravitationally reaccumulated asteroid from a major impact disruption of a parent asteroid [22].
Therefore, images of the boulders on Itokawa's surface with pixel resolution

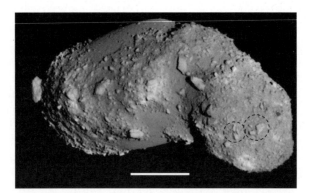

Fig. 3.8. An overview of the asteroid 25143 Itokawa ((c) ISAS/JAXA). Scale horizontal bar= 100 m. The boulder at the left end is the largest boulder (Yoshinodai).
Boulders in *circles* appear as if they were originated from a single larger boulder or
two large boulders which underwent cratering impact on Itokawa's surface [47]

of ~ 6 mm to ~ 70 cm taken by the Asteroid Multi-band Imaging CAmera (AMICA) provide us with the actual outcome of the collisional disruption event of the parent body of a subkilometer-sized asteroid [58].

The boulders on Itokawa exhibit a wide spectrum of angularities and irregularities. There are thin, flat-looking boulders, angular and conical boulders, and irregularly shaped (wavy-shaped) boulders. A similar variety is also found for fragments in laboratory. Figure 3.9 shows a comparison of boulders on Itokawa and laboratory fragments from centimeter-sized targets collected after impact experiments. Cracks and fractures are observed both on boulders and experimental fragments.

Although there is a difference of many orders of magnitude in the scale and complexity of the physical processes, as well as in the environment (gravity), the similarities in shapes and structures of the boulders and the laboratory fragments establish a bridge between disruption in laboratory of solid bodies (governed by the growth and coalescence of microscopic flaws) and the natural collisional disruption process at larger scales in space. These similarities suggest a universal character of at least some parts of the process throughout these scales.

In principle, the impact process related to the boulders on Itokawa's surface is expected to be erosive because of the tiny gravitational attraction of the body. The fraction of ejecta from a cratering process that have velocities less than the escape velocity is dependent upon the strength of the surface [28] if the crater is excavated on a cohesive surface. In the case of Itokawa, the internal structure cannot be inferred with great details, but it is likely that the thickness of the boulder layer does not exceed a few tens of meters. When the impactor is large enough to penetrate into the boulder layer and the bedrock, the fraction of the falling-back ejecta is governed by the strength of the bedrock. Similarly, when the impactor is smaller than the size of individual boulders, the fraction of the ejecta from a crater excavated on a boulder itself that can fall back onto the surface is also controlled by the strength of the boulder material. Since the escape velocity from the surface of Itokawa is only 10–20 cm/s, the fraction of the ejecta that can fall back on the surface is expected to be very small [39], unless there is a great dissipation of energy during cratering events. Therefore, most of the boulders on the surface should rather come from the collisional disruption event that formed Itokawa and may also correspond to original boulders of the parent body of Itokawa. The slope of the cumulative size distribution of boulders on Itokawa's surface is, however, close to -3 [39] which is steeper than the size distributions of fine fragments in laboratory disruptions. One possible explanation of this discrepancy is that some boulders may not be genuine in the sense that some of them may have originated from a single larger boulder which was broken up during its fall back on the surface, or later by a cratering event. In fact, boulder pairs or groups (or families) have been indentified on Itokawa's surface as shown in Fig. 3.8. These families of boulders might well be the remnants of the impact disruptions of larger boulders on

Fig. 3.9. Irregularly shaped fragments from laboratory impact experiments and boulders on the surface of Itokawa ((c) ISAS/JAXA, [47])

the surface [47]. The other explanation is that at least some of the boulders on the surface of Itokawa are original boulders belonging to the parent body of Itokawa.

3.5 Summary and Perspectives

More than 200 years have passed since the first asteroid 1 Ceres was found in 1801 in Palermo by G. Piazzi [20], and our knowledge on asteroids has been changed qualitatively and continuously during the last decades. Asteroids appear more active and complicated than ever thought. Many asteroid-satellite and binary systems are being discovered and provide information on asteroid's masses and bulk densities. The bulk density, which is lower than typical bulk densities of small rocks, tells us that asteroid interiors contain generally substantial void spaces. Moreover, the dynamical evolution of asteroids appears more complex as it is controlled not only by Kepler's laws, but also by interaction with photons, e.g., the Yarkovsky thermal effect. The interaction with photons also affects the spin state of these bodies when their shape is not spherical, which is the case of almost all of them, and when they are small enough, and now there are a few observational evidence showing that this so-called YORP effect has changed some real asteroid spin periods [30, 33]. Another observation shows the evidence of volatile activity on the surface of bodies which reside in the main belt [29].

Although our understanding of the collisional evolution of asteroids suggests that most of the ones whose size is at most a few tens of kilometers have experienced major catastrophic disruption events and have fractured or rubble pile internal structures (in agreement with measured bulk densities), only one sub-kilometer asteroid, the smallest ever explored 25413 Itokawa, has shown direct evidence that it is a gravitationally reaccumulated body. However, the detailed internal structure and birth scenario of Itokawa have not been revealed yet. Space missions aiming at making a sample return and/or investigating deep into the interior using techniques such as radar tomography are thus strongly required.

Studies of the collisional disruption of asteroids are performed by three different approaches: laboratory experiments, numerical simulations, and theoretical scaling considerations. Laboratory disruption experiments of rocky materials have provided a first-order understanding of the collisional disruption process of solid bodies. Now our study of collisinal disruption of such bodies is to be extended to porous ones. Such extension, and its extrapolation at larger sizes by numerical models will lead to a greater understanding of the collisional response and evolution of the small bodies, in particular those which evolve in the outer part of the main belt and further, e.g., Trojans asteroids, Kuiper belt objects, and comets. It will then also be possible to address the fundamental problem of collisional accretion which, during the early history of our Solar System, led to the formation of our planets.

References

1. S. Abe, T. Mukai, N. Hirata, O. S. Barnouin-Jha, A. F. Cheng, H. Demura, R. W. Gaskell, T. Hashimoto, K. Hiraoka, T. Honda, T. Kubota, M. Matsuoka, T. Mizuno, R. Nakamura, D. J. Scheeres, and M. Yoshikawa: *Mass and local topography measurements of itokawa by hayabusa*, Science **312**, 1344 (2006)
2. W. W. Anderson and T. J. Ahrens: *Shock Wave Equations of State of Chondritic Meteorite*, In: Shock Compression of Condensed Matter – 1997, S. C. Schmidt, et al. (Eds.) (AIP Press, Woodbury, NY, 1998) pp. 115–118.
3. W. Benz and E. Asphaug: *Impact simulations with fracture. I-Method and tests*, Icarus **107**, 98 (1994)
4. M. J.S. Belton, C. R. Chapman, K. P. Klaasen, A. P. Harch, P. C. Thomas, J. Veverka, A. S. McEwen, and R. T. Pappalardo: *Galileo's Encounter with 243 Ida: an Overview of the Imaging Experiment*, Icarus **120**, 1 (1996)
5. W. F. Bottke Jr., M. C. Nolan, R. Greenberg, and R. A. Kolvoord: *Velocity distributions among colliding asteroids*, Icarus **107**, 255 (1994)
6. W. F. Bottke Jr., D. Vokrouhlicky, D. R. Rubincam, and M. Broz: *The Effect of Yarkovsky Thermal Forces on the Dynamical Evolution of Asteroids and Meteoroids*, In: Asteroids III, W. F. Bottke Jr, A. Cellino, P. Paolicchi and R. P. Binzel (Eds.) (Univ. Arizona Press, Tucson, 2002) pp. 395–408.
7. W. F. Bottke Jr., A. Cellino, P. Paolicchi, and R. P. Binzel, Eds. *Asteroids III*, (Univ. Arizona Press, Tucson, 2002)
8. W. F. Bottke, Jr., D. D. Durda, D. Nesvornỳ, R. Jedicke, A. Morbidelli, D. Vokrouhlicky, and H. F. Levison.: *Linking the collisional history of the main asteroid belt to its dynamical excitation and depletion*, Icarus **179**, 63 (2005)
9. D. T. Britt, D. Yeomans, K. Housen, and G. Consolmagno: *Asteroid Density, Porosity, and Structure*, In: Asteroids III, W. F. Bottke Jr, A. Cellino, P. Paolicchi and R. P. Binzel (Eds.) (Univ. Arizona Press, Tucson, 2002) pp. 485–500.
10. P. Brown, D. Pack, W. N. Edwards, D. O. Revelle, B. B. Yoo, R. E. Spalding, and E. Tagliaferri: *The orbit, atmospheric dynamics, and initial mass of the Park Forest meteorite*, Meteorit. Planet. Sci. **39**, 1781 (2004)

11. S. J. Bus, F. Vilas, and M. A. Barucci: *Visible-Wavelength Spectroscopy of Asteroids*, In: Asteroids III, W. F. Bottke Jr, A. Cellino, P. Paolicchi and R. P. Binzel (Eds.) (Univ. Arizona Press, Tucson, 2002) pp. 169–182.
12. C. R. Chapman, D. Morrison, and B. Zellner: *Surface properties of asteroids: A synthesis of polarimetry, radiometry, and spectrophotometry*, Icarus **25**, 104 (1975)
13. C. R. Chapman, J. Veverka, P. C. Thomas, K. Klaasen, M. J. S. Belton, A. Harch, A. McEwen, T. V. Johnson, P. Helfenstein, M. E. Davies, W. J., Merline, and T. Denk: *Discovery and Physical Properties of Dactyl a Satellite of Asteroid 243 Ida*, Nature **374**, 783 (1995)
14. B. E. Clark, B. Hapke, C. Pieters, and D. Britt: *Asteroid Space Weathering and Regolith Evolution*, In: Asteroids III, W. F. Bottke Jr, A. Cellino, P. Paolicchi, and R. P. Binzel (Eds.) (Univ. of Arizona Press, Tucson, 2002) pp. 585–599.
15. D. R. Davis, S. J. Weidenschilling, P. Farinella, P. Paolicchi, and R. P. Binzel: *Asteroid collisional history: Effects on sizes and spins*, In: Asteroids II, R. P. Binzel, T. Gehrels, and M. Matthews (Eds.) (Univ. of Arizona Press, Tucson, 1989) pp. 805–826.
16. P. Farinella, D. R. Davis, A. Cellino, and V. Zapplà: *The collision lifetime of asteroid 951 Gaspra*, Astron. Astrophys. **257**, 329 (1992)
17. P. Farinella and D. R. Davis: *Short-period comets: primordial bodies or collisional fragments?*, Science **273**, 938 (1996)
18. P. Farinella, D. Vokrouhlický, W. K. Hartmann: *Meteorite delivery via Yarkovsky orbital drift*, Icarus **132**, 378 (1998)
19. G. J. Flynn, D. D. Durda, J. W. Kreft, I. Sitnitsky, M. Strait: *Catastrophic disruption experiments on the murchison hydrous meteorite*, Lunar and Planetary Science XXXVIII, March 2007, Texas, LPI Contribution **1338**, p.1744.
20. G. Fodera Serio, A. Manara, and P. Sicoli: *Giuseppe Piazzi and the Discovery of Ceres*, In: Asteroids III, W. F. Bottke Jr, A. Cellino, P. Paolicchi, and R. P. Binzel (Eds.) (Univ. of Arizona Press, Tucson, 2002) pp. 17–24.
21. A. Fujiwara, P. Cerroni, D. R. Davis, E. V. Ryan, M. Di Martino, K. Holsapple, and K. Housen: *Experiments and scaling laws for catastrophic collisions*, In: Asteroids II, R. P. Binzel, T. Gehrels, and M. Matthews (Eds.) (Univ. of Arizona Press, Tucson, 1989) pp. 240–268.
22. A. Fujiwara, J. Kawaguchi, D. K. Yeomans, M. Abe, T. Mukai, T. Okada, J. Saito, H. Yano, M. Yoshikawa, D. J. Scheers, O. Barnouin-Jha, A. F. Cheng, H. Demura, R. W. Gaskell, N. Hirata, H. Ikeda, T. Kominato, H. Miyamoto, A. M. Nakamura, R. Nakamura, S. Sasaki, and K. Uesugi: *The rubble-pile asteroid Itokawa as observed by Hayabusa*, Science **312**, 1330 (2006)
23. B. J. Gladman, F. Migliorini, A. Morbidelli, V. Zappalà, P. Michel, A. Cellino, Ch. Froeschlé, H. F. Levison, M. Bailey, and M. Duncan: *Dynamical lifetimes of objetcs injected into asteroid belt resonances*, Science **277**, 197 (1997)
24. B. J. Gladman, P. Michel, and Ch. Froeschlé: *The near-earth object population*, Icarus **146**, 176 (2000)
25. R. Greenberg, W. F. Bottke, and M. Nolan: *Collisional and dynamical history of Ida*, Icarus **120**, 106 (1996)
26. K. A. Holsapple: *The scaling of impact processes in planetary sciences*, Annu. Rev. Earth Planet. Sci. **21**, 333 (1993)
27. K. Holsapple, I. Giblin, K. Housen, A. Nakamura, and E. Ryan: *Asteroid Impacts: Laboratory Experiments and Scaling Laws*, In: Asteroids III, W. F. Bottke

Jr, A. Cellino, P. Paolicchi, and R. P. Binzel (Eds.) (Univ. of Arizona Press, Tucson, 2002) pp. 443-462.

28. K. R. Housen, R. M. Shumidt, and K. A. Holsapple: *Crater ejecta scaling laws - fundamental forms based on dimensional analysis*, J. Geophys. Res. **88**, 2485 (1983)

29. H. H. Hshieh and D. Jewitt: *A population of comets in the main asteroid belt*, Science **312**, 561 (2006)

30. M. Kaasalainen, J. Ďurech, B. D. Warner, Y. N. Krugly, N. M. Gaftonyuk: *Acceleration of the rotation of asteroid 1862 Apollo by radiation torques*, Nature **446**, 420 (2007)

31. T. Kadono, J. Kameda, K. Saruwatari, H. Tanaka, S. Yamamoto, and A. Fujiwara: *Surface roughness of alumina fragments caused by hypervelocity impact*, Planet. Space Sci. **54**, 212 (2006)

32. Y. Kozai: *Kiyotsugu Hirayama and His Families of Asteroids* In: Proc. Seventy-five (75) years of Hirayama asteroid families: The role of collisions in the solar system history, Y. Kozai, R. P. Binzel, and T. Hirayama (Eds.), Astronomical Society of the Pacific Conference Series **63**, 1 (1994)

33. S. C. Lowry, A. Fitzsimmons, P. Pravec, D. Vokrouhlický, H. Boehnhardt, P. A. Taylor, J.-L. Margot, A. Galád, M. Irwin, J. Irwin, and P. Kusnirák: *Direct detection of the asteroidal YORP effect*, Science **316**, 272 (2007)

34. J. S. Mathis, W. Rumple, and K. H. Nordsieck: *The size distribution of interstellar grains*, Astrophys. J. **217**, 425 (1977)

35. A. S. McEwen, J. M. Moore, and E. M. Shoemaker: *The phanerozoic impact cratering rate: evidence from the farside of the moon*, J. Geophys. Res. **102**, 9231 (1997)

36. K. Marti and T. Graf: *Cosmic-ray exposure history of ordinary chondrites*, Annu. Rev. Earth Planet. Sci. **20**, 221 (1992)

37. H. J. Melosh: *Impact Cratering A Geologic Process*, (Oxford University Press, New York, 1989)

38. P. Michel, W. Benz, and D. C. Richardson: *Modelling collisions between asteroids: from laboratory experiments to numerical simulations: disruption of fragmented parent bodies as the origin of asteroid families*, Nature **421**, 608 (2003)

39. T. Michikami, A. M. Nakamura, N. Hirata, R. W. Gaskell, R. Nakamura, T. Honda, C. Honda, K. Hiraoka, J. Saito, H. Demura, M. Ishiguro, and H. Miyamoto: *Size-frequency statistics of boulders on global surface of asteroid 25143 Itokawa*, Earth Planet. Space **60**, 13–20 (2008)

40. A. Milani and Z. Knezevic: *Asteroid proper elements and the dynamical structure of the asteroid belt*, Icarus **107**, 219 (1994)

41. A. Morbidelli and B. J. Gladman: *Orbital and temporal distributions of meteorites originating in the asteroid belt*, Meteoritics Planetary Sci **33**, 999 (1998)

42. A. Morbidelli, W. F. Bottke Jr., Ch. Froeschle, and P. Michel: *Origin and Evolution of Near-Earth Objects*, In: Asteroids III, W. F. Bottke Jr, A. Cellino, P. Paolicchi, and R. P. Binzel (Eds.) (Univ. of Arizona Press, Tucson, 2002) pp. 409–422.

43. A. Nakamura and A. Fujiwara: *Velocity distribution of fragments formed in a simulated collisional disruption*, Icarus **92**, 132 (1991)

44. A. M. Nakamura: *Laboratory studies on the velocity of fragments from impact disruptions*, The Institute of Space and Aeronautical Science Report, **651** (1993)
45. A. M. Nakamura, A. Fujiwara, T. Kadono: *Velocity of finer fragments from impact*, Planet Spacs Sci. **42**, 1043 (1994)
46. A. M. Nakamura, P. Michel, and M. Setoh: *Weibull parameters of yakuno basalt targets used in documented high-velocity impact experiments*, J. Geophys. Res. (Planets) **112**, 10.1029/2006JE002757 (2007)
47. A. M. Nakamura, T. Michikami, N. Hirata, A. Fujiwara, R. Nakamura, M. Ishiguro, H. Miyamoto, H. Demura, K. Hiraoka, T. Honda, C. Honda, J. Saito, T. Hashimoto, and T. Kubota, *Impact Process of Boulders on the Surface of Asteroid 25143 Itokawa– Fragments from Collisional Disruption*, Earth Planet. Space **60**, 7–12 (2008)
48. S. Nakazawa, S. Watanabe, M. Kato, Y. Iijima, T. Kobayashi, and T. Sekine: *Hugoniot equation of state of basalt*, Planet. Space Sci. **45**, 1489 (1997)
49. S. Nakazawa, S. Watanabe, Y. Iijima, and M. Kato: *Experimental investigation of shock wave attenuation in basalt*, Icarus **156**, 539 (2002)
50. D. Nesvorný, W. F. Bottke Jr., L. Dones, and H. F. Levison: *The recent breakup of an asteroid in the main-belt region*, Nature **417**, 720 (2002)
51. D. P. O'Brien and R. Greenberg: *The collisional and dynamical evolution of the main-belt and NEA size distributions*, Icarus **178**, 179 (2005)
52. T. Okada, K. Shirai, Y. Yamamoto, T. Arai, K. Ogawa, K. Hosono, and M. Kato:*X-ray Fluorescence Spectrometry of Asteroid Itokawa by Hayabusa*, Science **312**, 1338 (2006)
53. P. Paolicchi, S. Marchi, D. Nesvorny, et al.: *Towards a general model of space weathering of S-complex asteroids and ordinary chondrites*, Astron. Astrophys. **464**, 1139 (2007)
54. C. A. Polanskey and T. J. Ahrens: *Impact spallation experiments – Fracture patterns and spall velocities*, Icarus **87**, 140 (1990)
55. D. C. Richardson, Z. M. Leinhardt, H. J. Melosh, W. F. Bottke Jr., and E. Asphaug.: *Gravitational Aggregates: Evidence and Evolution*, In: Asteroids III, W. F. Bottke Jr, A. Cellino, P. Paolicchi, and R. P. Binzel (Eds.) (Univ. of Arizona Press, Tucson, 2002) pp. 501–515.
56. E. Ryshkewitch: *Compression strength of porous sintered alumina and zirconia*, J. Am. Ceram. Soc., **36**, [2], 65 (1953)
57. A. S. Rivkin, E. S. Howell, F. Vilas, and L. A. Lebofsky: *Hydrated minerals on asteroids: The astronomical report*, In: Asteroids III, W. F. Bottke Jr, A. Cellino, P. Paolicchi, and R. P. Binzel (Eds.), (Univ. of Arizona Press, Tucson, 2002), pp. 235–253.
58. J. Saito, H. Miyamoto, R. Nakamura, M. Ishiguro, T. Michikami, A. M. Nakamura, H. Demura, S. Sasaki, N. Hirata, C. Honda, A. Yamamoto, Y. Yokota, T. Fuse, F. Yoshida, D. J. Tholen, R. W. Gaskell, T. Hashimoto, T. Kubota, Y. Higuchi, T. Nakamura, P. Smith, K. Hiraoka, T. Honda, S. Kobayashi, M. Furuya, N. Matsumoto, E. Nemoto, A. Yukishita, K. Kitazato, B. Dermawan, A. Sogame, J. Terazono, C. Shinohara, H. Akiyama: *Detailed images of asteroid 25143 itokawa from hayabusa*, Science **312**, 1341 (2006)
59. P. Spurný, J. Oberst, and D. Heinlein: *Photographic observations of Neuschwanstein, a second meteorite from the orbit of the Příbram chondrite*, Nature **423**, 151 (2003)

60. Y. Takagi, H. Mizutani, and S.-I. Kawakami: *Impact fragmentation experiments of basalts and pyrophyllites*, Icarus **59**, 462 (1984)

61. D. J. Tholen and M. A. Barucci: *Asteroid Taxonomy* In: Asteroids II, R. P. Binzel, T. Gehrels, and M. S. Matthews (Eds.) (Univ. of Arizona Press, Tucson, 1989) pp. 298–315.

62. J. I. Trombka, S. W. Squyres, J. Bruckner, W. V. Boynton, R. C. Reedy, T. J. McCoy, P. Gorenstein, L. G. Evans, J. R. Arnold, R. D. Starr, L. R. Nittler, M. E. Murphy, I. Mikheeva, R. L. McNutt, T. P. McClanahan, E. Mc-Cartney, J. O. Goldsten, R. E. Gold, S. R. Floyd, P. E. Clark, T. H. Burbine, J. S. Bhangoo, S. H. Bailey, M. Petaev: *The elemental composition of asteroids 433 Eros: results of the NEAR-Shoemaker X-ray spectrometer*, Science **289**, 2101 (2000)

63. J. Veverka, M. J. S. Belton, K. P. Klaasen, and C. R. Chapman: *Galileo's encounter with 951 Gaspra: Overview*, Icarus **107**, 2 (1994)

64. J. Veverka, P. Thomas, A. Harch, B. Clark, J. F. Bell, B. Carcich, J. Joseph, S. Murchie, N. Izenberg, C. Chapman, W. Merline, M. Malin, L. McFadden, M. Robinson: *NEAR encounter with asteroid 253 Mathilde: overview*, Icarus **140**, 3 (1999)

65. J. Veverka, M. Robinson, P. Thomas, S. Murchie, J. F. Bell, N. Izenberg, C. Chapman, A. Harch, M. Bell, B. Carcich, A. Cheng, B. Clark, D. Domingue, D. Dunham, R. Farquhar, M. J. Gaffey, E. Hawkins, J. Joseph, R. Kirk, H. Li, P. Lucey, M. Malin, P. Martin, L. McFadden, W. J. Merline, J. K. Miller, W. M. Owen, C. Peterson, L. Prockter, J. Warren, D. Wellnitz, B. G. Williams, D. K. Yeomans: *NEAR at Eros: imaging and spectral results*, Science **289**, 2088 (2000)

66. W. Weibull: *A statistical theory of the strength of materials* ingvetensk. Akad Handl. **151**, 1 (1939)

67. M. Yanagisawa and T. Itoi: *Impact Fragmentation Experiments of Porous and Weak Targets*. In: Proc. of Seventy-five (75) years of Hirayama asteroid families: The role of collisions in the solar system history, Y. Kozai, R. P. Binzel, and T. Hirayama (Eds.), Astronomical Society of the Pacific Conference Series **63**, 243 (1994).

68. D. K. Yeomans, J.-P. Barriot, D. W. Dunham, R. W. Farquhar, J. D. Giorgini, C. E. Helfrich, A. S. Konopliv, J. V. McAdams, J. K. Miller, W. M. Owen, Jr., D. J. Scheeres, S. P. Synnott, and B. G. Williams: *Estimating the mass of asteroid 253 Mathilde from tracking data during the NEAR flyby*, Science **278**, 2106 (1997)

69. K. Yomogida and T. Matsui: *Physical properties of ordinary chondrites*, J. Geophys. Res. **88**, B11 9513 (1983)

70. V. Zappalà, A. Cellino, A. Dell'Oro: *Physical and Dynamical Properties of Asteroid Families*, In: Asteroids III, W. F. Bottke Jr, A. Cellino, P. Paolicchi, and R. P. Binzel (Eds.) (Univ. of Arizona Press, Tucson, 2002) pp. 619–631.

On the Strength and Disruption Mechanisms of Small Bodies in the Solar System

P. Michel

Laboratoire Cassiopée, Observatoire de la Côte d'Azur, CNRS, Université de Nice Sophia-Antipolis, boulevard de l'Observatoire, 06300 Nice, France,
michel@oca.eu

Abstract During their evolutions, the small bodies of our Solar System are affected by several mechanisms which can modify their properties. While dynamical mechanisms are at the origin of their orbital variations, there are other mechanisms which can change their shape, spin, and even their size when their strength threshold is reached, resulting in their disruption. Such mechanisms have been identified and studied, by both analytical and numerical tools. The main mechanisms that can result in the disruption of a small body are collisional events, tidal perturbations, and spin-ups. However, the efficiency of these mechanisms depends on the strength of the material constituing the small body, which also plays a role in its possible equilibrium shape. As it is often believed that most small bodies larger than a few hundreds meters in radius are gravitational aggregates or rubble piles, i.e., cohesionless bodies, a fluid model is often used to determine their bulk densities, based on their shape and assuming hydrostatic equilibrium. A representation by a fluid has also been often used to estimate their tidal disruption (Roche) distance to a planet. However, cohesionless bodies do not behave like fluids. In particular, they are subjected to different failure criteria depending on the supposed strength model. This chapter presents several important aspects of material strengths that are believed to be adapted to Solar System small bodies and reviews the most recent studies of the different mechanisms that can be at the origin of the disruption of these bodies. Our understanding of the complex process of rock failure is still poor and remains an open area of research. While our knowledge has improved on the disruption mechanisms of small bodies of our Solar System, there is still a large debate on the appropriate strength models for these bodies. Moreover, material properties of terrestrial rocks or meteorites are generally used to model small bodies in space, and only space missions to some of these bodies devoted to precise in situ analysis and sample return will allow us to determine whether those models are appropriate or need to be revised.

4.1 Introduction

In our Solar System, there are several populations of small bodies, which differ both by their locations and by their physical properties. While most

Michel, P.: *On the Strength and Disruption Mechanisms of Small Bodies in the Solar System.*
Lect. Notes Phys. **758**, 99–128 (2009)
DOI 10.1007/978-3-540-76935-4_4   © Springer-Verlag Berlin Heidelberg 2009

asteroids evolve in the main belt, a region located between the orbits of Mars and Jupiter, some of them originating from this region cross the orbits of the terrestrial planets (the so-called Near-Earth Objects or NEOs), while another population evolves on the same orbit as Jupiter on the L4 and L5 lagrangian points (the so-called Trojan asteroids). Then, another population of small bodies called Kuiper Belt Objects (or KBOs) evolves beyond the orbit of Neptune and is at the origin of the Jupiter-Family Comets (JFCs). Finally, the Long Period Comets (LPCs) come temporarily in the Solar System from an external location called the Oort Cloud.

Small bodies of our Solar System are all affected by planetary gravitational perturbations. Thus, their orbits are more or less stable, depending on their locations. For instance, most NEOs are transported to Earth-crossing orbits from the main belt as a result of their injection into mean motion resonances with Jupiter, or secular ones with Saturn. These resonances increase their orbital eccentricity such that their perihelion distance becomes eventually shorter than 1 AU on only a few million years timescale (e.g., [5, 9]). The population of JFCs is also transported from KBO orbits through resonant channels, and finally the LPCs are believed to come from the Oort cloud due to some stellar perturbations or galactic tides. Thus, these populations are all dynamically active.

In addition to these changes in their trajectories, small bodies of our Solar System can also undergo dramatic changes in their physical properties due to different mechanisms. Lightcurves obtained by ground-based observations, and images obtained from space missions, all show that these bodies can have very irregular shapes and heavily cratered surfaces, indicating a quite intense collisional activity. Moreover, spin rates give important clues about the composition and strength of these bodies. Then, the presence of binary objects, which represent about 15% of the main belt and NEO populations, indicates that some processes are efficient to form such systems.

So, what are the mechanisms that can modify the physical properties of a small body? We know at least three mechanisms which can be effective enough to change the shape or disrupt a small body, depending on its strength. The first most intuitive one is the collisional process. It is well known that populations of small bodies evolve collisionally. Witnesses of these collisional events are, for instance, the asteroid families in the main belt. About 20 asteroid families have been identified, and each corresponds to a group of small bodies, who share the same orbital and spectral properties. From these characteristics, reproduced recently by numerical simulations (see, e.g., [30] and references therein), it is now established that an asteroid family is the outcome of the disruption of a large asteroid due to an impact with another small asteroid. As a consequence, a large asteroid is transformed into a group of smaller bodies, and the shapes, sizes, spins, and orbits of these objects depend on several parameters of the collision, one of them being the strength of the parent body. The second mechanism which can lead to a change of the physical properties of an object is the increase of its spin due to a thermal effect called the YORP effect [43]. When a rotating body has an irregular shape, it can reemit the light received from the Sun in a different direction than the one from which it received, and such difference in direction can lead to a change of its spin rate. Although acting on long timescales, such effect has recently been observed [22]. When an acceleration occurs, depending on the strength and internal properties of the object, the spin can reach the threshold above which the shape of the body is not in equilibrium anymore, so that either the shape readjusts into another equilibrium or the body breaks up. The third mechanism which can produce similar effects is due to tidal encounters with a massive object (a planet). It is well known that below a certain limit distance, tidal forces can cause the deformation or the disruption of an object. This distance is known as the Roche limit for fluid bodies [38], but as we will see, it can take different values and the bodies can take a wide range of shapes at this distance when solid materials (with and without cohesion) are considered.

The efficiency of all the mechanisms described above relies at least partially on the assumed strength of the small body in which these mechanisms act. This is why it is important that the definition of strength is clearly understood, and this chapter addresses this problem. In Sect. 4.2, the definition of strength is given for different kinds of materials. Section 4.3 summarizes the most recent study on the spin limits of small bodies and what the observed spins tell us on the strength and internal structure of these objects. The latest results on the limit distances of small bodies to a planet as a function of their strength are then presented in Sect. 4.4. Several reviews have already been devoted to our current understanding of the collisional disruption of small bodies based on numerical simulations (see, e.g., [29, 30]), therefore this problem is briefly discussed in Sect. 4.5, concentrating only on the some important issues and open areas. Discussions, conclusions, and perspectives are then given in Sect. 4.6.

4.2 The Strength of Materials

4.2.1 What Do We Mean by Strength?

The behavior of a small solid body subjected to different forces is a wide area of research, and the results depend at least partially on the definition used for the strength of the material. In this section, we expose some important concepts which can help better understand the meaning of strength of a small body.

There is no doubt that the term "strength" is often used in imprecise ways. Given the implications of this concept in different areas of study, we believe that it is important to present it in different places to ensure that a same language is spoken among researchers dealing with it. The description presented here is largely inspired from different works by Holsapple, and Holsapple and Michel [15, 16, 31]. Materials such as rocks, soils, and ices, which are the main constituents of small bodies of our Solar System, are complex and characterized by several kinds of strength.

Generally, the concept of "strength" is a measure of an ability to withstand stress. But stress, as a tensor, can take on many different forms. One of the simplest is a uniaxial tension, for which one principal stress is positive and the two others are zero. The tensile strength, i.e., the value of this stress at which the specimen breaks, is often (mis)used to characterize material strength as a whole. Thus, while it is common to equate "zero tensile strength" to a fluid body, that is not correct. In fact, a body can both be solid and have no tensile strength. For instance, dry sand has no tensile strength. However, contrary to a fluid, dry sand and granular materials, in general, can withstand considerable shear stress when they are under pressure: that is why we can walk on dry sand and not on water. Here comes into play a second kind of strength: the shear strength which measures the ability to withstand pure shear. The shear strength in a granular material under confining pressure comes from the fact that the interlocking particles must move apart to slide over one another, and the confining pressure resists that. A third kind of strength, the compressive strength, governs the ability to withstand compressive uniaxial stress. Thus, in general, a material has tensile strength, shear strength at zero pressure (technically the "cohesion"), and compressive strength. In geological materials, such as soils and rocks, the failure stresses depend strongly on the confining pressure; as a result, these three strength values can be markedly different. Then, contrary to a common assumption, a cohesionless body is simply a solid body whose cohesion (shear strength at zero pressure) is null, but that does not mean that it does not have any shear strength under confining pressure. For instance, there are strong evidence that probably most asteroids greater than a few kilometers in diameter are rubble piles or gravitational aggregates (see [36] for a definition of those terms). For such bodies, cohesion can be ignored but they should not be represented by a fluid. In their case, the confining pressure at the origin of the shear strength is played by their self-gravity. Hence, a body can be cohesionless but nevertheless solid.

4.2.2 Failure Criteria of Solid Bodies

Once the strength of a material has been defined consistently, a failure law is required to determine imminent failure states of stress. Failure criteria for geological materials parallel the yield criteria for metals. Recall that the maximum stress at which a load can be applied without causing any permanent deformation defines the *elastic limit*. It is also called the *yield point*, for which it marks the initiation of plastic or irreversible deformation.

There are two common yield criteria for metals: the Tresca criterion and the von Mises criterion. The Tresca criterion states that yield occurs when the maximum shear stress on any plane reaches a critical value. The von

Mises criterion replaces the shear stress with the square root of the second invariant J_2 of the deviatoric stress tensor (non-diagonal components of the stress tensor), which depends on all shear stresses. A common assumption of these criteria is that the average stress (pressure), given in terms of principal stresses σ_i as $P = (\sigma_1 + \sigma_2 + \sigma_3)/3$, has no effect. Then, in a plane in principal stress space perpendicular to the pure pressure axis, the von Mises criterion is a circle, while the Tresca criterion is a hexagon (see, for example, [7] for a good discussion of various yield and failure criteria).

For geological materials, failure can also be described by two such criteria, but with an important addition: because the allowable shear depends on the confining pressure, the size of either of the circle or hexagon depends on the pressure or normal stress. The Mohr–Coulomb criterion (MC) assumes that the maximum shear stress on any plane (τ_{max}) depends linearly on the normal stress (σ_n) on the plane:

$$\tau_{max} = Y - \sigma_n \tan(\phi) \tag{4.1}$$

where the constant of proportionality is the tangent of the *angle of friction* ϕ and the constant Y is called the *cohesion* (shear strength at zero pressure); both are material constants determined by experiments. This defines an envelope (limit curve) of maximum shear stress. Thus, compressive stress (negative) increases the allowable shear. In a three-dimensional principal stress space, this criterion defines a hexagonal cylinder that increases linearly in size for increasing pressure [7]. The MC criterion can be considered a Tresca criterion generalized to account for the normal stress effect.

Another criterion called Drucker–Prager (DP) is also common model for geological materials. The DP criterion can be considered a modification of the von Mises criterion, which now assumes that the allowable shear stress depends linearly on the confining pressure. The shear stress magnitude is measured by the square root of the second invariant J_2 of the deviator stress (see (4.3)). Thus, the DP criterion is similar to models for linear friction and is defined by two constants: one characterizes the "cohesion" (shear strength at zero pressure), and the second characterizes the dependence on the confining pressure and is related to the angle of friction. Those two constants determine the tensile and compressive strengths. When the cohesion is zero, so is the tensile strength, but not the compressive strength. Physically, the pressure dependence is, as already explained, the consequence of the interlocking of the granular particles and not the friction of the surfaces of the particles. In fact, a closely packed mass of uniform rigid *frictionless* spherical particles has an angle of friction about $23°$. So the term angle of friction is somewhat a misnomer and angle of *interlocking* would be more correct. However, we will keep using the usual name angle of friction. Figure 4.1 gives a representation of the DP model. Using the three principal stresses σ_1, σ_2, σ_3 (positive in tension) of a general three-dimensional stress state, the pressure (positive in tension) is given as:

$$P = \frac{1}{3}(\sigma_1 + \sigma_2 + \sigma_3) \qquad (4.2)$$

and the square root of the second invariant of the deviator stress is:

$$\sqrt{J_2} = \frac{1}{\sqrt{6}} \sqrt{\left[(\sigma_1 - \sigma_2)^2 + (\sigma_2 - \sigma_3)^2 + (\sigma_3 - \sigma_1)^2 \right]} \qquad (4.3)$$

Then, the DP failure criterion is generally given as:

$$\sqrt{J_2} \le k - 3sP \qquad (4.4)$$

which is illustrated as a straight line with slope $3s$ and intercept k in Fig. 4.1. Clearly, negative pressure (compression) increases the allowable $\sqrt{J_2}$ when s is positive.

For the special case of a pure shear stress only, $\sqrt{J_2}$ is just that shear stress and the pressure P is zero. In Fig. 4.1, the uniaxial tension strength σ_T has $\sqrt{J_2} = 3^{-1/2}\sigma_T$ and $P = \sigma_T/3$. The uniaxial compression strength σ_c has $\sqrt{J_2} = -3^{-1/2}\sigma_C$ and $P = \sigma_C/3$.

The DP criterion can be made to match the MC one in all combinations of pressure plus uniaxial compression if the parameters s and k are related to the cohesion and the angle of friction ϕ used in the MC model. In particular, the slope s is related to the angle of friction ϕ of the MC model by:

$$s = \frac{2 \sin \phi}{\sqrt{3}(3 - \sin \phi)}. \qquad (4.5)$$

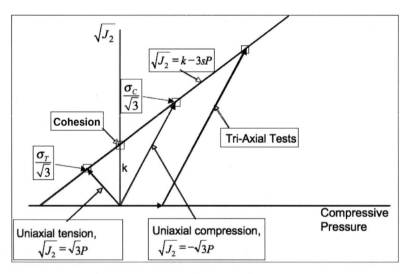

Fig. 4.1. The Drucker–Prager failure criterion. The abscissa is positive in compression. The four small squares indicate the failure condition in, respectively from the left: tension, shear, compression, and a confined compression or tri-axial test. The intercept at zero pressure at the value k is called the cohesion and the slope of the line passing through k is $3s$. From [16] and [31]

The intercept k of the DP model is also the shear stress τ for failure in pure shear. Technically, the term "cohesion" means the intercept value of shear stress at zero pressure. When the cohesion is zero, so is the tensile strength, and vice versa; both cases would have the envelope starting at the origin in Fig. 4.1. For instance, an appropriate failure criterion for rubble piles is a criterion for which those two measures are zero, while the plot shows the more general case where they are non-zero.

The tensile stress σ_T for failure is located at the intersection of the tensile line shown in Fig. 4.1 slopping to the left with the straight line representing the criterion. Its value is given by:

$$\sigma_T = \frac{\sqrt{3}}{\sqrt{3}s + 1}k. \tag{4.6}$$

Similarly, the compressive failure occurs when the compression line intercepts the failure line, and the resulting compressive stress is given by:

$$\sigma_P = \frac{\sqrt{3}}{\sqrt{3}s + 1}k. \tag{4.7}$$

Hence, the ratio of compressive strength to tensile strength is given by:

$$\frac{\sigma_C}{\sigma_T} = \frac{\sqrt{3}s + 1}{\sqrt{3}s - 1} \tag{4.8}$$

which defines the slope of the DP criterion, i.e., the friction coefficient s.

To give an order of idea of the difference between those strengths, a common friction angle for rocks is $45°$, so that $s = 0.356$, and the ratio of compressive strength to tensile strength is $-4.22{:}1$. In this case, from (4.6), the shear stress for failure k would be 0.93 of σ_T, i.e., the shear and tensile strengths are roughly equal.

This envelope is usually determined experimentally by a test known as a "confined compression" or a "tri-axial" test. In such a test, a uniform confining pressure is applied to a specimen in all directions, and the axial stress is increased in compression until failure occurs when the envelope is reached. This path is illustrated in Fig. 4.1.

Finally, from a practical point of view, the two criteria (MC and DP) can offer different advantages. The MC model defines a maximum shear stress directly, which is determined by the difference of the maximum and minimum principal stresses. As a consequence, to use this criterion in algebraic manipulations involving general stress states, one must first determine the principal stresses, then which is the largest and which is the smallest. The result is a difference in the algebra of the results in six different regimes, where the three principal stress components take on different orderings. An example of the six possible cases of the ordering of the stress magnitudes is given in Fig. 4 of [12]. Moreover, there are "corners" in the curves shown where the ordering of the principal stresses changes. In contrast, the DP criterion has a single algebraic relation for all stress states. Thus, although the algebraic form of that relation is more complicated than the MC criterion, there is no need to consider the six different possibilities of the ordering of the stress magnitudes. The algebraic complexity of the DP model is of little consequence when an algebraic manipulation program such as *Mathematica* is used. For instance, Holsapple and Michel used the DP failure criterion to characterize the tidal disruption limit distance of a cohesionless ellipsoid to a planet, noting that the differences between the two models are small (see [15] and Sect. 4.4).

4.2.3 Strength Dependence on Object's Size and Loading Rate

It is generally believed that the effective static cohesive and tensile strengths decrease with increasing body size. The origin of this assumption comes from indications that a distribution of incipient flaws is present within the volume of a solid body. Because larger bodies are more likely to contain larger natural flaws than smaller bodies, the strength is expected to decrease with the body's size. Thus, the use of a strength measure that decreases with size is now a common feature of the studies of disruption of small bodies by impacts (see, e.g., [13, 30]) and has even been demonstrated experimentally [19].

A common model for a distribution of incipient flaws in a solid body is a power-law Weibull distribution [45]. Such a distribution is used in numerical simulations of catastrophic disruption of solid bodies (see Sect. 4.5) to generate the initial flaws in the bodies involved. A two-parameter Weibull distribution is usually assumed to describe the network of incipient flaws in any material, expressed as:

$$N(\varepsilon) = k\varepsilon^{\mathrm{m}} \tag{4.9}$$

where ε is the strain and N is the number density of flaws that activate (i.e., start their propagation) at or below the value of strain. The Weibull parameters m and k are material constants which have been measured for a number of geological and industrial materials, although data are quite scarce for some important rocks (see [24]). In particular, the parameter k varies widely between various rock types and the exponent m ranges typically between 6 and 12, but can have a wider range of values. Recently, it was measured for the first time for the same basalt material as the one used in some impact experiments, and its value was found to be around 17 in static loadings [33].

From the Weibull distribution, it is easily shown that the most probable static strength S of a specimen of volume V (diameter D) decreases with increasing size as:

$$S \propto V^{-1/\mathrm{m}} \propto D^{-3/\mathrm{m}}. \tag{4.10}$$

As explained above, such a decrease in strength is simply because larger specimens are more likely to have larger cracks.

The values of the Weibull parameters represent important material properties. Large values of m describe homogeneous rocks with uniform fracture threshold, while small values apply to rocks with widely varying flaw activation thresholds. The existence of incipient flaws within any rock is understood to originate from its cooling history and from crystal lattice imperfections. Due to the initial presence of these flaws, when a finite strain rate $\dot{\varepsilon}$ is applied, a stress increase occurs in time, which is compensated by the propagation of active flaws causing a stress release. Thus, a competition takes place between the stress increase due to loading and the stress release due to flaw activation and propagation, until a temporary equilibrium is reached at the time of peak stress. Then, the stress decreases to zero as active flaws propagate rapidly through the rock.

From these explanations, it is obvious that the crack growth velocity c_g is an important parameter since it governs the stress release due to an active flaw. Experiments indicate that it relates to the speed c_l of longitudinal waves in a rock by $c_g \approx 0.4c_l$, and this is usually the value used in numerical simulations of fragmentation. Since cracks propagate at this fixed velocity, under moderate conditions, the weakest flaws (those which activate at lower values of ε) suffice to accomodate the growing stresses. Therefore, the peak stress at failure is low and fragments are relatively large (see, e.g., Fig. 1 in [30]). Conversely, more resistant flaws have time to activate at high strain rates. In this case, the peak failure stress is high and fragments are small. This process depends strongly on the assumed value of the crack growth velocity c_g. In particular, fragment sizes scale with c_g. For instance, a higher velocity would enhance the efficiency at which a crack relieves stress, since stress release is proportional to crack length cubed. As a consequence, fewer flaws would be required to relieve a given increase in stress.

Thus, the concept of material strength reaches another level of complexity as it can also depend on the dynamical context. From the explanations above, one may conclude that defining the material strength as *the stress at which sudden failure occurs* is not rigorously adequate. Material strength could rather be defined as the stress at which the first flaw begins to fail, thereby initiating an inelastic behavior characterized by irreversible deformation. But in practice, the adopted definition is the peak stress which the rock undergoes prior to failure. It is then not a material constant, since as explained above, the peak stress is a function of the loading history of the rock. This is the reason why a distinction is made between static strength and dynamic strength on the basis of the loading rate. For extremely small loading rates, elastic stresses increase in equilibrium until the onset of catastrophic failure. This occurs at loading rates that are typically smaller than $\approx 10^{-6}$ strains per second. Static tensile strength decreases with increasing size of the rock due to the greater probability of finding a weaker (larger) flaw. At high enough loading rates, stresses can continue to build while catastrophic rupture has begun. In this case, it is more appropriate to speak of dynamic failure. For most rocks, dynamic strain rates are of the order of 1 s^{-1} and decrease with increasing rock size. Therefore, the peak stress that the rock suffers prior to failure is rigorously called the material's dynamic strength at that strain rate. This dynamic strength increases with strain rate and is always greater than the static strength.

All hypervelocity impacts into small targets are in the dynamic regime, but some impacts on large bodies can still be in a regime close to the static one. In this case, part of the event, close to the impact point, can be dynamic, but some important aspects can also be understood in terms of quasi-static failure. Therefore, in all studies that are described in the following sections, apart from the problem of catastrophic disruption, the tensile strength will generally be the static one.

4.3 Rotation Rates and Implications on the Strength of Small Bodies

The spin rates of small bodies of the Solar System give an important clue about the composition and strength of those bodies. Indeed, the greatest spin that a body can take without being deformed or disrupted depends directly on those properties. For instance, a simple analysis based on the property of zero tensile stresses at the body's poles [11] led to the conclusion that an object whose assumed typical mass density is 2.5 g/cm^3 has a period limit of 2.1 h. This value is smaller than the measured rotation period of all large asteroids. Thus, it was suggested that most asteroids must be gravitational aggregates or rubble piles with no tensile strength. This value was later revised [12, 14] by a complete stress analysis of spinning, self-graviting, ellipsoidal bodies using the MC failure criterion for cohesionless solid bodies (see Sect. 4.2.2). From this analysis, it was concluded that the spin limits are not determined by tensile failure, but by shear failure. Consequently, it was found that the spin limits depend on the angle of friction (see Sect. 4.2.2) of the material of the body. A typical minimum period was found to be about 2.6 h, which is higher than the previous estimate [11], but still smaller than the rotation period of large asteroids. Numerical experiments of spinning rubble piles (modeled as hard spheres maintained together by gravity) found that such rubble piles behave in a manner consistent with those last theoretical expectations [37].

It was thus tempting to conclude on this basis that most asteroids are rubble piles, because none of them was found to rotate faster than the limit above which a rubble pile would break, in principle. However, recent data for small asteroids indicated that some of them rotate at a rate which is much greater than those previous limits, which suggests that they have some cohesive and tensile strength. This raised the question whether the spin limits observed for large asteroids really rule out that their material is strengthless.

These questions have been addressed in a recent study [16], and this section summarizes its principle and main results. It is an extension of previous studies [12, 14] and considers spinning bodies with cohesive (and therefore

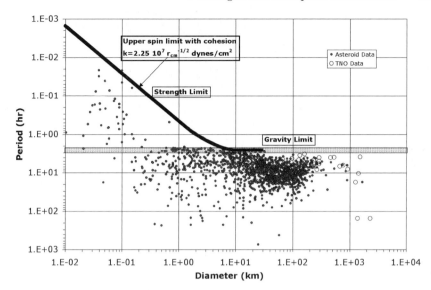

Fig. 4.2. Spin limits and data for small Solar System bodies. The *dark sloped line* assumes a size-dependent strength; it becomes asymptotic to the horizontal red band for materials without cohesion. On the left, the spin limit for cohesive bodies is determined by the cohesive/tensile strength and defines a strength regime. The horizontal asymptote on the right characterizes a gravity regime, where tensile/cohesive strength is of no consequence. Gravity regime values do depend on shape and angle of friction, so average values have been assumed to represent them on the plot. The data in the upper left triangular region are the fast spinning near-Earth asteroids. The triangular points for the large diameter bodies on the right are trans-Neptunian objects (from [16] and [18])

tensile) strength. The fundamental approach consisted of calculating the internal stress state in an ellipsoidal spinning body as a function of size, shape, and spin of the body, and comparing the stress state with the limit failure state (provided by the DP model; see Sect. 4.2.2) to determine the spin limits at which the failure occurs. It must be noted that rather than solving for the stress state by assuming linear elasticity from some actual prior history of the material, the limit states are solved in the spirit of limit analyses of plasticity theories. Those limit states correspond to the situation when the body has reached a final state at which collapse is imminent, and give the final and greatest loadings for failure. The great advantage of this approach is that the prior history of the material is not a factor, and the analyses give a limiting loading state with a greater spin than any found from first failure using a linear elastic approach. This is important because small bodies of our Solar System are formed by complex processes and undergo disruptions, reaccumulations, heating, cratering events, tidal forces, etc. Those processes inevitably introduce residual stresses that cannot be known, so the assumption of linear elastic behavior from a virgin stress-free state is not reasonable given the history of these bodies. There exist several explicit examples (see [15]) that illustrate the important differences between stress analyses using elasticity theory, and those using the limit state approach. This limit state approach will also be used to determine the tidal disruption limit distances of solid bodies ([15, 18, 31]; see Sect. 4.4).

A significant complexity is added once the cohesive terms are included in the strength measure. Indeed, for a cohesionless failure criterion, it was found [14] that the limit states had simultaneous failure at all points in the body and the algebra to find that state was rather simple. When the material has cohesion, failure is rather attained only at certain points and planes within the body, and the algebra that determines the stress state with certain failure locations seems insurmountable (even for a symbolic program such as *Mathematica*). Therefore, the approach chosen in his study was to construct the volume-averaged stresses and to use those stresses to compare to the failure criterion [16]. In other words, spin states are determined which, on average, cause stresses to equal the failure threshold (see Sect. 4.2.2). It is clear that for a body having a certain degree of cohesion, more or less critical stress states than the average may exist at some locations within the body, so in reality failure may occur at a lower spin than that found with this approach. But in the particular case of a cohesionless material, failure occurs at all locations in the body at the limit spin. Therefore in this case, both the average and the exact methods give identical results. In other words, in the gravity regime (in which self-gravity dominates over strength, so that cohesion can be ignored), the results are exact; while in the strength regime (in which self-gravity is negligible), they are only approximate.

Once the average stresses have been expressed [16], they can be inserted into the DP criterion (4.4) to find the combinations of spin and shape that satisfy this criterion. Then for a given shape, represented by the aspect ratios of an ellipsoid, the spin at a given limit state can be found. Defining a scaled spin as:

$$\Omega^2 = \frac{\omega^2}{\pi \rho G}, \tag{4.11}$$

where ω is the actual spin, ρ is the bulk density, and G is the gravitational constant, the results can be put into the form of relations for the scaled spin at failure states as functions of the cohesion, the angle of friction ϕ (see Sect. 4.2.2), the average body radius $\bar{r} = (abc)^{1/3}$ (where a, b, and c are the semi-axes of the ellipsoid representing the object with $a > b > c$) and the two aspect ratios $\alpha = c/a$ and $\beta = b/a$:

$$\Omega = F(k, \phi, \bar{r}, \alpha, \beta). \tag{4.12}$$

The spin limits are indicated in Fig. 4.2. Both constant strength and decreasing strength with body size have been considered. For the size-dependent strength (see Sect. 4.2.3), the value assumed for the Weibull parameter m ((4.9)) is $m = 6$, which seems to fit many data for crack distributions in samples from micron to kilometer sizes [19]. From (4.10), for $m = 6$, the strength of a body decreases with the body's diameter to the power of $-1/2$. Therefore, the strength (cohesion) expresses as:

$$k = \kappa \bar{r}^{-1/2} \tag{4.13}$$

where the strength coefficient κ is the strength of an object of 1 cm in radius (here k is used for the cohesion and should not be confused with the Weibull parameter having the same name). The spin limits represented in Fig. 4.2 have been estimated assuming $\kappa = 2.25 \times 10^7$ dynes cm$^{3/2}$, which is one order of magnitude below the measurements of the tensile strength of Georgia Keystone granite specimens [19]. The sloped line corresponding to this size-variable strength (Fig. 4.2) gives an extremely good upper envelope for the current data over the entire range of small body sizes. Measured strengths are still scarce, so it is not necessarily surprising that the best fit is provided by the line corresponding to a strength value smaller than the measured one.

Note that these estimates of spin limits [16] assume that the bodies have ideal ellipsoidal shapes (with aspect ratios of 0.7 and prolate for the representation in the figure) and a fixed friction angle. The reality is obviously more complex, which may explain that the observed spin limits are smaller than the ones assuming the measured tensile strength. In particular, small bodies do not have ideal shapes, their actual friction angle is not known, they may have weaker materials and other non-ideal properties. Moreover, determinations of the observed body's sizes and shapes contain their own error bars, which may shift the points in Fig. 4.2 toward higher or lower values. Note finally that the magnitude of the dependence of the estimated spin limit on shape and friction angle is well within a factor of 2, except in extreme cases. Also, the average method used to make those estimates gives an upper bound to the limit spin in the strength regime.

Thus, a detailed investigation of the spin limits as a function of the mentioned parameters, including strength, has been performed recently [16] and the conclusion of this investigation is rich in implications. In particular, it is found that the presence of tensile and cohesive strength for a large body (> 10 km in diameter) makes no difference in its spin limit. Therefore, the observed spin limit (also called the spin barrier) for large bodies cannot be interpreted as evidence of a zero-strength (cohesive/tensile) rubble pile structure. It is the gravity that limits the spin in those cases, even if they have some cohesion. So, large asteroids may be rubble piles, but not on the pure basis of the so-called spin barrier (however, other evidence may point to a rubble pile structure). On the other hand, the strength that allows the higher spins of the smaller and fast spinning kilometer-sized bodies is only of the order of 10–100 kPa, which is very small compared to the strength of small terrestrial rocks. So, these small asteroids do not have to be very strong to be able to rotate so fast. They could be some kind of rubble piles that have accumulated slight bondings between constituents.

As a conclusion, the spin data of small bodies can give us some indications on the internal structure of these bodies. However, based on the current observed spins and our current understanding on the strength of large bodies, they are not sufficient to indicate whether these bodies are rubble piles or monolithics. If some small asteroids were found to spin above the theoretical limits provided by the described approach [16], then this would be a first indication of the potential existence of strong (monolithic) rocky bodies in the Solar System.

4.4 Tidal Disruption of Small Bodies

When a small body has a close encounter with a planet, depending on the approach distance, it can be subjected to tidal forces that may change its shape or even disrupt it. Such mechanism was at the origin of the observed disruption of Comet Shoemaker-Levy 9, which was fragmented into 21 pieces during a first passage close to Juipter, and which collided with the giant planet during its next passage in 1994. Tidal disruption has often been proposed as a formation mechanism for binary asteroids, which represent 15% of the NEO population, and as an explanation of crater chains and doublet craters on planetary surfaces. The strength of a small body is an important parameter in the determination of its limit distance to a planet for tidal disruption (or shape readjustment).

The investigation of the limit distance for tidal disruption started in the 19th century, using a fluid to represent the small body. This led to the concept of the Roche limit [38], which is still often used nowadays. A great number of studies followed until know, which accounted for important parameters in different manners from one study to the other. The last theoretical studies on this problem provided a continuum theory which allows the determination of this limit distance for cohesionless bodies, and a lower bound of this distance for small bodies with cohesion [15, 31, 18]. We summarize here the theory and results provided by these studies, and the reader interested in other previous studies on these problems can refer to these last publications in which a history of previous works and their differences are well exposed.

Although the Roche limit [38] for tidal disruption of orbiting satellites assumes a fluid body, a length to diameter of exactly 2.07:1, and a particular orientation of the body, it is often used in studies of Solar System satellites and small asteroids or comets encountering a planet. Clearly, these bodies are neither fluid, nor generally are that elongated, so more appropriate theories are needed and have been developed since this first work. Recently, exact analytical results for the distortion and disruption limits of solid spinning ellipsoid bodies subjected to tidal forces, using the DP model with zero cohesion (see Sect. 4.2.2) have been presented [15]. The study used the same approach as the one exposed in Sect. 4.3 to study the spin limits for solid ellipsoidal bodies. It was followed by a study along the same lines, in which the cohesion of the small bodies was now considered, which, due to the added complexity, could not provide exact but only approximate results [18], for the same reasons as the ones already exposed in the case of spin limits (see Sect. 4.3). Thus, a static theory was developed that predicts conditions for breakup, the nature of the deformation at the limit state, but it does not track the dynamics of the body as it comes apart. At the end of the section, we will briefly expose results from dynamical investigations.

In the case of cohesionless bodies, as already indicated in previous sections, the strength is essentially characterized by a single parameter associated with an angle of friction ranging from $0°$ to $90°$. The case with a null angle of friction has no shear strength whatsoever, so it corresponds to the case of a fluid or a gas (and the limit distance corresponds to the Roche limit). The case of $90°$ represents a material that cannot fail in shear, but still has zero tensile strength. Typical dry soils have angles of friction of $30°$–$40°$. As most satellites are spin-locked with the planet around which they evolve, both the spin-locked case and the zero spin case, a possible case for passing stray body, have been considered to characterize the limit distance.

The equilibrium problem of an ellipsoid body has been described [12]. Three stress equilibrium equations must be satisfied by the stresses σ_{ij} in any body in static equilbrium with body forces b_i, which are given as (using repeated index summation convention):

$$\frac{\partial}{\partial x_j}\sigma_{ij} + \rho b_i = 0 \qquad (4.14)$$

where ρ is the bulk density of the body. An $(x, \ y, \ z)$ coordinate system aligned with the ordered principal axes of the ellipsoid is used. In the problems here, the body forces arise from mutual gravitational forces, centrifugal forces, and/or tidal forces; they all have the simple linear forms $b_x = k_x x$, $b_y = k_y y$, $b_z = k_z z$. The full expressions of k_x, k_y, and k_z are explicitly presented [15]. Then, for the limit states sought, the stresses must satisfy the DP failure criterion (see Sect. 4.2.2) at all points x, y, and z. Also, the surface tensions are zero on the surface points of the ellipsoidal body surface defined by: $\frac{x^2}{a} + \frac{y^2}{b} + \frac{z^2}{c} - 1 = 0$. This problem has been solved [12], showing that the distribution of stresses in that limit state just at uniform global failure has the simple form:

$$\sigma_x = -\rho k_x a^2 \left[1 - \left(\frac{x}{a}\right)^2 - \left(\frac{y}{b}\right)^2 - \left(\frac{z}{c}\right)^2\right],$$

$$\sigma_y = -\rho k_y b^2 \left[1 - \left(\frac{x}{a}\right)^2 - \left(\frac{y}{b}\right)^2 - \left(\frac{z}{c}\right)^2\right],$$

$$\sigma_z = -\rho k_z c^2 \left[1 - \left(\frac{x}{a}\right)^2 - \left(\frac{y}{b}\right)^2 - \left(\frac{z}{c}\right)^2\right], \qquad (4.15)$$

and the shear stresses in this coordinate system are all zero. The body force constants k_x, k_y, k_z depend on the body forces, so those forces must be such that the DP failure criterion is not violated. That condition determines the limit states. Putting the expressions of these components into the

DP criterion ((4.4)), one can see that the common functional dependence $1 - \left(\frac{x}{a}\right)^2 - \left(\frac{y}{b}\right)^2 - \left(\frac{z}{c}\right)^2$ will cancel out of (4.15). That is because the limit stress state has simultaneous failure at all points. Thus, we can omit that functional dependence and focus on finding the combinations of the leading multipliers of the three terms of (4.15) that satisfy the failure criterion. We define the dimensionless distance by:

$$\delta = \left(\frac{\rho}{\rho_p}\right)^{1/3} \frac{d}{R} \qquad (4.16)$$

where ρ_p is the bulk density of the primary (the planet). Then, failure will occur when [15]:

$$\frac{1}{6}[(c_x - c_y)^2 + (c_y - c_z)^2 + (c_z - c_x)^2] = s^2[c_x + c_y + c_z]^2 \qquad (4.17)$$

where, for arbitrary spin and when the long axis points towards the Earth:

$$c_x = \left(-A_x + \frac{1}{2}\Omega^2 + \frac{4}{3}\delta^{-3}\right),$$
$$c_y = \beta^2\left(-A_y + \frac{1}{2}\Omega^2 - \frac{2}{3}\delta^{-3}\right),$$
$$c_z = \alpha^2\left(-A_z - \frac{2}{3}\delta^{-3}\right) \qquad (4.18)$$

and A_x, A_y, and A_z are the components of the self-gravitational potential of a homogeneous ellipsoidal body of uniform mass density ρ in the body coordinate system expressed as: $U = \pi\rho G(A_0 + A_x x^2 + A_y y^2 + A_z z^2)$ (e.g., [6]). Ω is the scaled spin already expressed in Sect. 4.3. A similar form when the long axis points along the trajectory at its closest approach can be obtained [15]. The criterion expressed in these forms can then be used to solve for the dimensionless distances δ at the failure condition as a function of the aspect ratios α and β (which determine the A_x, A_y, and A_z), the mass ratio p of the secondary to the primary, and for any value of the constant s related to the angle of friction. The solution always has the dimensionless form:

$$\delta = \frac{d}{R}\left(\frac{\rho}{\rho_p}\right)^{1/3} = F[\alpha, \beta, p, \phi, \Omega] \qquad (4.19)$$

so that the bulk density ratio only occurs with this cube root. Note that in the spin-locked case, the spin is given by:

$$\omega = \frac{G(M + m)}{d^3} \qquad (4.20)$$

where M and m are the masses of the primary and secondary, respectively. The number of independent variables is then reduced by one when the scaled spin is zero or the spin-locked value, and by another one when $p = 0$, i.e., when the mass of the secondary is negligible compared to that of the primary, which is the case for an asteroid flying by a planet or a small satellite of a giant planet.

Note that this limit distance to the primary corresponds to the distance below which a secondary cannot exist with its assumed shape, because the failure criterion would be violated. However, it does not mean that below this distance, the secondary would disrupt. A flow rule is required to indicate the nature of any readjustment (or disruption). Then, if those changes lead to a new configuration that is within failure at the given distance, a shape change is indicated. Otherwise, if the new shape still violates the failure criterion, a global disruption is indicated. Such analysis has been done [15], but goes beyond the scope of this review.

The results provided by this static theory show that a spin-locked spherical body can approach a planet as close as $d/R = 1.23168(\rho_p/\rho)^{1/3}$ if its angle of friction is 90°, and the orbit distance decreases smoothly as the angle of friction increases. For a generic rock value, say $\phi = 30°$, the closest orbit for a spherical satellite is about $d/R = 1.5(\rho_p/\rho)^{1/3}$. The fluid case with zero angle of friction has a distance of infinity, as there is no solution for a spherical body in this case. Other general ellipsoid shapes have then been fully investigated, and it was found that for each combinations of aspect ratios α and β, there is a range of permissible orbital distances for any angle of friction $\phi > 0°$. For instance, a prolate body of negligible mass with aspect ratios of 0.8 and $\phi = 20°$ can orbit as close as $d/R = 1.78261(\rho_p/\rho)^{1/3}$, center to center, and if $\phi = 40°$, it can orbit as close as $d/R = 1.15141(\rho_p/\rho)^{1/3}$. Then, an elongated prolate body with $\alpha = 0.4$ and $\phi = 40°$ can orbit as close as $d/R = 1.92929(\rho_p/\rho)^{1/3}$. Figure 4.3 illustrates the same application to a stray body with zero spin, $p = 0$ and $\alpha = 0.8$.

Thus, all of these are noticeably closer than the fluid Roche limit of $d/R = 2.455(\rho_p/\rho)^{1/3}$, and for a solid, even cohesionless, the shapes are not limited to fluid shapes. This is why it is important to make no confusion between a fluid and a cohesionless body. This is particularly important in the context of the study of satellites of giant planets. In fact, many planetary satellites are inside their Roche limit, and do not have the aspect ratios required for a fluid at this limit (see, e.g., Table 1 in [15]). The same holds true for stray (zero-spin) cohesionless bodies (the case of a fluid body in this configuration is called the Jean's problem, [20]).

The difference with a fluid body is even more striking when cohesion is added to the strength of a small body. This has been investigated recently by [18], who extended their previous analysis of limit distances of cohesionless ellipsoids to the limit distances of ellipsoids with cohesion. Recall that when the cohesion term k is zero in the DP criterion (4.4), the general form of the stresses satisfying the equilibrium condition, the boundary values and the failure criterion at all points in the body can be found exactly in closed form with a quadratic dependence on the coordinates. Conversely, when the

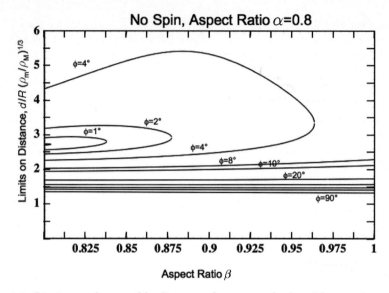

Fig. 4.3. Limits on the possible distances for a stray body with no spin and for various aspect ratios α and β, and different angles of friction Φ. Note that for small angles of friction, the region of permissible distances is a *closed curve*. This means that the body cannot approach too close, because tidal forcs would be too strong, nor range too far as some tidal forces are still necessary for its equilibrium. The limit case of such behavior is the Roche fluid case, for which there is a single point at the center of the closed curves, near the prolate bodies on the left. For typical angles of friction $(20°–40°)$, there is no large variation with the aspect ratios. As in the spin-locked situation, the miminum distances for cohesionless stray bodies are all well below the Roche limit, and a much wider range of aspect ratios can exist

cohesion is not zero, there is no solution to this general equilibrium problem with simultaneous failure at all locations. Therefore, an exact answer cannot be determined, but an approximate solution can be found by averaging the stresses across the body, as explained in Sect. 4.3 for the spin limits. The difference here with the spin limit study is that the tidal terms are added. It has been proven that the loads for which the average stresses are at failure are equal to or greater than the actual limit loads [17]. Thus, using average stresses to look for failure gives an upper bound on actual limit loads. In other words, the analysis provides a lower bound to the closest approach for collapse. But recall again that in the special cases with zero cohesion the results are exact. In fact, a study along the same lines devoted to tidal encounters of granular bodies has been done using the volume average approach and, although the calculations of limit distances were not investigated with the same level of details, it was consistent with the results above using the closed form for the stresses [39].

The limit distance for a small cohesive body can be expressed in terms of seven non-dimensional parameters (one dependent and six independent variables), leading for each orientation of the body to the reduced form:

$$\delta = \frac{d}{R}\left(\frac{\rho}{\rho_{\rm p}}\right)^{1/3} = G\left[\alpha, \beta, p, \Omega, \phi, \frac{k}{\rho^2 G r^2}\right] \qquad (4.21)$$

using the previously defined aspect ratios α, β, mass ratio p, and a scaled cohesion term $k^* = k/(\rho^2 G r^2)$, where r is the body's average radius. This scaled term corresponds to the ratio of the cohesive strength to a gravity pressure. For any given value of these six parameters, numerical results can then easily be obtained, using appropriate programs (see [18] for significant examples and illustrations of important dependencies). In this analysis, the cohesion is also assumed to decrease with the body's size, and the same size dependency as in Sect. 4.3 is used.

For bodies larger than a few kilometers in diameter, the limit distance is the one provided by the theory for cohesionless bodies, as the cohesion is so small that gravity is dominant (due to the decrease of cohesion with size). Thus, the new approach in which cohesion is included, is only relevant for those bodies whose size is below a few kilometers. For them, the limit distance becomes much closer to the primary, and depending on the values of the six parameters (or on the density ratio), the distance can even be smaller than the primary's radius, which means that these small bodies cannot be disrupted by tidal forces. An interesting application relates to planetary satellites, and a preliminary analysis showed that some of them must have non-zero friction angles or cohesion to evolve at the observed distance from their primary (see [18]). Thus, the approach developed to determine limit distances can provide some indirect indications on the possible internal structure of real objects, which is a very interesting aspect. Also, it can tell us whether some shape readjustment or disruption will occur during the close approach with the Earth of a real asteroid. For instance, if we assume that the asteroid Apophis (2004 MN4) is a cohesionless body, then the static theory developed by Holsapple and Michel indicates that its close approach with the Earth at 5.6 Earth's radii in 2029 will correspond to its tidal limit distance only if its bulk density is smaller than 0.25 g/cm^3 or unless it is highly elongated, or its angle of friction is less than about 5° [15, 31].

Thus, according to these recent studies, both cohesionless bodies and cohesive ones with expected properties of geological solids can exist in arbitrary ellipsoidal shapes and much closer to a primary than a fluid body. The main limitation of these studies is that they are not based on a dynamic approach but a static one, which is used to determine the onset of disruption or shape readjustment. Actually, in case of readjustment, the angular moment of inertia would change, so the spin would change. Then the new state would not be in the spin-locked or zero spin configuration as supposed here. The effects of tidal torques on the asteroid's spin have been investigated in details [40, 41], and a complete analysis should incorporate these results, thereby accounting for the change in the limit distance as the small body's spin rate changes during a fly-by. Such analysis will have to be performed in the future.

Also, when a small body goes through the limit distance, the resulting motions are affected by how the body changes its shape or breaks up, and by the resulting dynamics. That is a much more complicated problem, which is left for future studies. Some semi-analytical studies have been recently done to address this problem [39], but numerical simulations are probably the best tools. In a pioneering numerical study [34], the break-up of bodies encountering a planet was considered. Small bodies were modeled as granular aggregates comprised of 247 smooth spheres that interacted with each other only through inelastic collisions and were held together by gravity. Numerical simulations were then used to determine the motion of individual spheres. Various parameters, such as the initial angular velocity vector and encounter variables, were changed to explore the consequences of different close approach configurations. But this study was limited by the low resolution (number of smooth spheres) constrained by the computer performances and numerical codes at that time. More recently, numerical simulations of tidal disruption of rubble piles at higher resolutions using a sophisticated N-body code named *pkdgrav* have been performed [44] in order to determine whether tidal disruptions can explain the presence of 15% of binaries in the NEO population. The results show that tidal encounters with a planet (Earth) can form binaries. However, when a small body experiences a tidal encounter, it is likely to experience another such encounter in its close future, so that the binary which is first formed is often eventually disrupted by the next encounter. Thus, although tidal encounters are efficient to form temporary binaries, they cannot be at the origin of the high fraction of binaries observed in the NEO population. Some other potential mechanisms must be found and investigated.[1] Also, these simulations only addressed the problem of tidal approaches of cohesionless bodies, and they will be extended to the case of cohesive bodies in a close future. In principle, the limit distances should be similar to the ones provided by the static approach (for some values of the angle of friction and cohesion), but a dynamical approach will also allow the determination of the behavior of the small body and its seperate pieces once it breaks up. The gravitational evolution of fragments from a disrupted body has already been studied but only in the specific case of the collisional disruption of a small body. This is the last mechanism of disruption which is briefly summarized in the following section.

4.5 Collisional Disruption of Small Bodies

In this section, we just summarize the most important concepts and issues concerning the catastrophic disruption of an asteroid due to a collision. The

---

[1] a recent study has indicated that the spin-up of small NEOs and main belt asteroids due to the YORP thermal effect is at the origin of small binaries and explains their observed properties (K. Walsh, D. C. Richardson and P. Michel: Rotational breakup as the origin of small binaries, Nature 454, 188 (2008)).

reader interested in more details can refer to a few articles which expose the main recent results concerning this process (see, e.g., [29, 30]).

Collisional processes occur frequently between the small bodies of our Solar System. The best witnesses of those events are *asteroid families* in the main belt. Each family originates from the break-up of a large body, which is now represented by a group of asteroids sharing the same spectral and orbital properties (see, e.g., [21, 47]). As in the case of spin limits (Sect. 4.3), two regimes of collisional disruption have been defined: the *strength* regime, in which the fragmentation of the body is the only process determining the outcome (this is the case of impact experiments in laboratory), and the *gravity* regime, in which not only the fragmentation but also the gravitational interactions of fragments have an influence on their final size and velocity distributions. The transition between the two regimes has been found to occur for body sizes in the hundred meter range by numerical simulations [3], while the transition between the two regimes derived from the spin limits occurs at higher diameters (kilometer range; see Sect. 4.3 and [16]).

The first numerical simulations which reproduced succesfully large-scale events represented by asteroid families [25, 26, 27, 28], showed that when a large parent body (several tens of kilometers in diameter) is disrupted by a collision with a projectile, the generated fragments interact gravitationally during their ejections, and some of them reaccumulate to form aggregates. The final outcome of such a disruption is thus a distribution of fragments, most of the large ones being aggregates formed by gravitational reaccumulations of smaller ones. The implication of these results is that most large family members should be rubble piles and not monolithic bodies. Moreover, it was found that collisional disruptions form naturally binary systems and satellites [8, 25], although the timescale of their stability still needs to be determined.

The physics of the gravitational phase during which generated fragments evolve under their mutual attractions relies on the fundamental laws of classical mechanics, which are well understood. However, the development of numerical simulations which account for all the processes that may occur during this phase is a difficult task. When a large body is fragmented, the number of generated sizeable fragments can be as large as a few millions. Therefore, the numerical difficulties come from the fact that the forces must be computed for a large number of particles, up to millions, and this requires the use of efficient numerical methods to reduce the CPU time to reasonable values. Moreover, during their evolutions, these fragments do not only evolve under distant interactions, they can also undergo physical collisions between them, which must also be dealt with. A numerical N-body code called *pkdgrav* has been developed (see [35]) to compute the evolutions of large numbers of particles. It is a parallel tree-code which is able to compute the gravitational evolutions of millions of particles and handles collisions between them. During the gravitational phase, such collisions are assumed to not cause fragmentation, but only mergers or bounces. This simplification is justified by the fact that the relative velocities between the ejected fragments are small enough that collisions between them are quite smooth. So, when collisions occur during this phase, depending on some velocity and spin criteria, the particles either merge into a single one whose mass is the sum of the particle masses and whose position is at the center of mass of the particles, or bounce with some coefficient of restitution to account for dissipation (see [26] for details). In this approximation, while the aggregates that are formed have "correct" masses, their shapes are all spherical because of the merging procedure. Of course, the final shape and spin distributions are also an important outcome of a disruption and a study is currently under way to improve the simulations: instead of merging into a single spherical particle, colliding particles will be able to stick together using rigid body approximations. Such improvement will allow the determination of the shapes and spins of aggregates formed during a catastrophic disruption.

The most poorly understood part of the collisional process is the fragmentation phase, following immediately the impact of the projectile. It usually lasts twice the time for the shock wave to propagate through the whole target (a few seconds for a kilometer-sized body). The process of rock fragmentation is still a widely open area of research, relying on a large number of assumptions based on a limited number of data. Moreover, not only the physics is badly understood, the numerical techniques used to perform the computation are also confronted to some difficulties. Indeed, the fragmentation process in a rock involves two kinds of approaches, which are generally incompatible. A high-velocity impact on a rock generates a shock wave, followed by a rarefaction wave which will activate the crack propagation. Thus, the rock can be seen as a continuum for the shock treatment. On the other hand, a rock contains some discrete elements (the initial cracks). This mixture of continuum and discrete features makes the development of a numerical scheme difficult. A numerical code used to compute the fragmentation phase is generally called a *hydrocode*, which emphasizes the fact that this process involves the physics of hydrodynamics, although it occurs in a solid. Indeed, the difference between a fluid and a solid is that the deviatoric (non-diagonal) part of the stress stensor is not null in the case of a solid, while in a fluid only the spherical (diagonal) part of the stress tensor representing the pressure plays a role (see Sect. 4.2 for a detailed explanation of the difference between a fluid and a solid). Thus, three kinds of waves (elastic, plastic, and shock) propagate through a rock during an impact. Elastic waves are well known and determined by linear realtionships between the stress and strain tensors. Plastic waves begin to develop when the material strength changes with the wave amplitude. Then, at wave amplitudes that are high enough and associated to shock waves, the body is treated as a fluid. Being non-linear, the transitory behaviors between these kinds of waves are difficult to determine analytically from constitutive models, and this probably motivated the development of numerical algorithms. The process has thus been studied by implementing the bulk properties of a given rock in a numerical model of continuous medium (a *hydrocode*), including a yielding criterion and an equation of state for the appropriate material. The main power of this method is that no assumption on the form of the stress wave that drives the fragmentation is required since the initial conditions evolve numerically based on a rheological model and a failure criterion. The appropriate regime (elastic, plastic, or shock) is determined by the computation.

The 3D Lagrangian hydrocode developed by [2] represents the state-of-the-art in numerical computations of dynamical fracture of brittle solids. It uses the method called Smooth Particle Hydrodynamics (SPH) (see [1] for a review of this method). Basically, the value of the different hydrodynamics quantities are known at finite numbers of points which move with the flow. Starting from a spatial distribution of these points called particles, the SPH technique allows the computation of the spatial derivatives without the necessity of an underlying grid. The 3D SPH hydrocode is thus able to simulate consistently from statistical and hydrodynamical points of view, the fragments that are smaller or larger than the chosen resolution (number of SPH particles). The resulting system has proven to predict successfully the sizes, positions, and velocities of fragments measured in laboratory experiments, without requiring the adjustment of too many free parameters [2]; moreover, associated with the N-body code *pkdgrav*, it has succesfully reproduced the main properties of asteroid families [25, 26, 27, 28].

For the sake of completeness, we recall here the basic equations that must be solved to compute the fragmentation process. Other important concepts, such as the equations of state, the model of brittle failure used to propagate damage, and the method to distribute appropriately incipient flaws in the modeled rock with a Weibull distribution are not reproduced here, as they have been described several times (see [2, 30]).

The basic equations that must be solved to compute the process are the well-known conservation equations of hydrodynamics that can be found in standard textbooks. The first equation represents the mass conservation. Its expression is:

$$\frac{d\rho}{dt} + \rho \frac{\partial}{\partial x_\alpha} 1 v_\alpha = 0 \tag{4.22}$$

where d/dt is the Lagrangian time derivative. Other variables have their usual meaning (i.e., ρ is the bulk density, v is the velocity, and x is the position) and the usual summation rule over repeated indices is used. The second equation describes the momentum conservation (in absence of gravity):

$$\frac{dv_\alpha}{dt} = \frac{1}{\rho} \frac{\partial}{\partial x_\beta} \sigma_{\alpha\beta} \tag{4.23}$$

where $\sigma_{\alpha\beta}$ is the stress tensor given by:

$$\sigma_{\alpha\beta} = -P\delta_{\alpha\beta} + S_{\alpha\beta} \tag{4.24}$$

where P is the isotropic pressure and $S_{\alpha\beta}$ is the traceless deviatoric stress tensor. The energy conservation is then expressed by the equation:

$$\frac{du}{dt} = -\frac{P}{\rho}\frac{\partial}{\partial x_\alpha}v_\alpha + \frac{1}{\rho}S_{\alpha\beta}\dot{\varepsilon}_{\alpha\beta} \qquad (4.25)$$

where $\dot{\varepsilon}_{\alpha\beta}$ is the strain rate tensor given by:

$$\dot{\varepsilon}_{\alpha\beta} = \frac{1}{2}\left(\frac{\partial}{\partial x_\beta}v_\alpha + \frac{\partial}{\partial x_\alpha}v_\beta\right). \qquad (4.26)$$

This set of equations is still insufficient in the case of a solid since the evolution in time of $S_{\alpha\beta}$ must be specified. The basic Hooke's law model is assumed in which the stress deviator rate is proportional to the strain rate:

$$\frac{dS_{\alpha\beta}}{dt} = 2\mu\left(\dot{\varepsilon}_{\alpha\beta} - \frac{1}{3}\delta_{\alpha\beta}\right) + S_{\alpha\gamma}R_{\beta\gamma} + S_{\beta\gamma}R_{\alpha\gamma} \qquad (4.27)$$

where μ is the shear modulus and $R_{\alpha\beta}$ is the rotation rate tensor given by:

$$R_{\alpha\beta} = \frac{1}{2}\left(\frac{\partial}{\partial x_\beta}v_\alpha - \frac{\partial}{\partial x_\alpha}v_\beta\right). \qquad (4.28)$$

This term allows the transformation of the stresses from the reference frame associated with the material to the laboratory reference frame in which the other equations are specified.

This set of equations can now be solved, provided an equation of state is specified, $P = P(\rho, u)$, linking the pressure P to the density ρ and internal energy u. The Tillotson equation of state for solid material [42] is generally used. Its expression and method of computation, as well as parameters for a wide variety of rocks are described in [23], appendix II. Other equations of states have been developed and all have different pros and cons and remain necessarily limited to materials studied in laboratory. This is one of the limits of any collisional model that necessarily relies on the behavior of known materials that do not necessarily represent the materials constituing an asteroid.

Perfectly elastic materials are well described by these equations. Plastic behavior beyond the Hugoniot elastic limit is introduced in these relations by using the von Mises yielding criterion (see Sect. 4.2). This criterion limits the deviatoric stress tensor to:

$$S_{\alpha\beta} = fS_{\alpha\beta} \qquad (4.29)$$

where f is computed from:

$$f = \min\left[\frac{Y_0^2}{3J_2}, 1\right], \qquad (4.30)$$

where J_2 is the second invariant of the deviatoric stress tensor (see Sect. 4.2) and Y_0 is a material-dependent yielding stress which generally depends on temperature, density, etc., in such a way that it decreases with increasing temperature until it vanishes beyond the melting point.

The von Mises criterion is adapted to describe the failure of ductile media such as metals. Brittle materials like rocks do not undergo a plastic failure but rather "break" if the applied stresses exceed a given threshold. Conversely, the yielding beyond the Hugoniot elastic limit does not prescribe any permanent change in the constitution of the material, since once stresses are reduced the original material remains behind, possibly heaten by the motion against the remaining stress, but otherwise not weakened. Therefore, it is not adapted to impact into rocks, as any yielding beyond the elastic limit invariably involves irreversible damage, and one needs to know how the rock is permanently altered by the event. A realistic fracture model is then clearly required to study the disruption of a solid body. The Grady–Kipp model of brittle failure is generally the model implemented in numerical codes aimed at simulating fragmentation processes in solid bodies [10].

Despite these recent successes of impact simulations to reproduce some experiments and asteroid family properties, there are still many issues and uncertainties in the treatment of the fragmentation phase. Some of the important ones are:

- **Material parameters**: One of the main limitations of all researches devoted to the fragmentation process comes from the uncertainties on the material properties of the objects involved in the event. For instance, 10

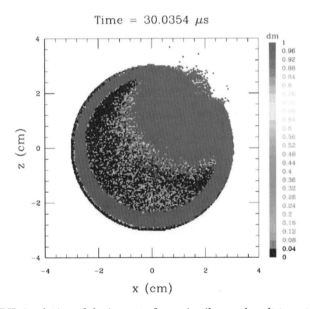

Time = 30.0354 μs

Fig. 4.4. SPH simulation of the impact of a projectile on a basalt target in the same conditions as a high-velocity experiment [32]. The plot shows how damage (labeled dm) propagated 30 μs after the impact (red zones are fully damaged). In particular, a core fragment can be identified which has the same mass and velocity as the one measured in the experiment

material parameters describe the usually adopted Tillotson equation of state (see e.g. [23], appendix II). Other sensitive material-dependent parameters are, for instance, the shear and bulk modulus, but the most problematic parameters are probably the two Weibull parameters m and k used to characterize the distribution of initial cracks in the target. In fact, as already mentioned (Sect. 4.2.3), data are still scarce about these parameters, due to the experimental difficulty to determine their values. This is a crucial problem because so far, the validation of numerical simulations by confrontation to experiments has been done by choosing freely those missing values so as to match the experiments [2]. This is not a totally satifactory approach for an ab initio method such as the one provided by SPH simulations. Unfortunately, this is often the only alternative which one has. A database including both the material parameters of targets and outcomes of impact experiments using these targets is thus required to perform a full validation of numerical codes. Such a project has started using the experimental expertise of japanese researchers from Kobe University, and the numerical expertise of french and swiss researchers from Côte d'Azur Observatory and Berne University. For instance, measurements of Weibull parameters of a Yakuno basalt used in impact experiments were made in this purpose [33].

- **Crack propagation speed in a solid**: The value of the crack growth speed is usually assumed to be 40% of the longitudinal sound speed in numerical simulations. This speed highly influences the number of cracks that can be activated for a given strain rate (see Sect. 4.2.3). It thus plays a major role in the number and sizes of fragments which are eventually created. The lack of measurements of this speed and its possible dependency on material type leave no choice but to use an intermediate value such as the one currently assumed. However, it is important to keep in mind that this may need to be revised once some cracks are found to propagate at higher/lower speed in a sufficient number of experiments.

- **Model of fragmentation:** Up to now, all published simulations of impact disruption have been done using in general the Grady–Kipp model of brittle failure [10]. In this model, damage increases as a result of crack activation, and microporosity (pore crushing, compaction) is not treated. However, several materials contain a certain degree of porosity (e.g., pumice, gypsum), and asteroids belonging to dark taxonomic types (e.g., C, D) are believed to contain a high level of microporosity. The behavior of a porous material subjected to an impact is likely to be different from the behavior of a non-porous one, as already indicated by some experiments (e.g., [19]). Therefore, a model for porous materials is required, in order to be able to address the problem of dark-type asteroid family formations, and to characterize the impact response of porous bodies in general (including porous planetesimals during the phase of planetary growth). Such models have been developed recently and inserted in numerical codes [4, 46]. However, so far their application has been limited to the cratering regime and their validity is still not guaranteed. Moreover, it will be important to check their validity in the disruptive regime by comparison to impact experiments, such as recent ones made on pumice by the group of Kobe led by A. Nakamura.

- **Rotating targets:** All simulations of catastrophic disruption have been performed starting with a non-rotating target. However, in the real world, small bodies are spinning (see Sect. 4.3), and the effect of the rotation on the fragmentation is totally unknown. Some preliminary experiments have been performed suggesting that, everything else being equal, a rotating target is easier to disrupt than a non-rotating one (K. Housen, private communication). If this is confirmed, this will be an important result as all models of collisional evolutions of small body populations use prescriptions that are provided by numerical simulations on non-rotating targets. In particular, the lifetime of small bodies may be shorter than expected if their rotation has an effect on their ability to survive collisions. It will thus be important to characterize the impact response of rotating bodies, both experimentally and numerically, although on both sides, starting with a rotating body is confronted to several difficulties.

4.6 Conclusion

Our understanding of the disruption mechanisms of small bodies of our Solar System has greatly improved in the last decades, thanks to the development of analytical theories and sophisticated simulations. However, there are still many uncertainties, and problems that need to be investigated. Concerning the spin limits and tidal disruptions, the theories that have been developed so far are all static. Nevertheless, they allowed us to understand that the spin barrier observed for large asteroids does not imply necessarily that these bodies are pure rubble piles, in contrast with the usual interpretation. On the other hand, the small fast rotators do not need to have much cohesion to spin at such high rates (see Sect. 4.3). Then, these theories allowed us to revisit the concept of Roche limit for solid ellipsoidal bodies with and without cohesion. They showed that, contrary to a fluid, a solid body can come much closer to a planet or a primary and with a wide variety of shapes (see Sect. 4.4). However, these theories rely on a model of material strength (Drucker–Prager or Mohr–Coulomb) which is not necessarily unique and they are limited to bodies whose shapes are idealized ellipsoids. Some other strength models or non-idealized shapes should certainly be considered. Then, a complete dynamical investigation of these problems is required to determine the outcome of rotational and tidal break-ups of small bodies.

Sophisticated numerical codes have been developed to study the process of impact disruption of a small body. The outcome of some impact experiments and the main properties of some asteroid families have been reproduced successfully with one of those codes, based on the smooth particle hydrodynamics technique and the Grady–Kipp model of brittle failure. However, as discussed in Sect. 4.5, there are still many issues, and the road is still long before being able to characterize with high accuracy, the impact response of a small body as a function of its material properties. This is a challenging topic which has many applications. Indeed, the collisional process plays a fundamental role in the different phases of the history of our Solar System, from the phase of planetary growth by collisional accretion to the current phase during which small bodies are catastrophically disrupted. Moreover, the determination of the impact response of a small body as a function of its physical properties is crucial in the definition of efficient mitigation strategies aimed at deflecting a potential threatening near-Earth asteroid whose trajectory leads it to the Earth.

Thus, researches devoted to these disruption mechanisms and to the concept of strength will certainly keep busy future generations of researchers, and they will also take advantage of future space missions devoted to in situ investigations and sample returns from small bodies.

References

1. W. Benz: *Smooth Particle Hydrodynamics – A Review*. In: Proceedings of the NATO Advanced Research Workshop on The Numerical Modelling of Nonlinear Stellar Pulsations Problems and Prospects, J. Robert Buchler (Ed.) (Kluwer Academic Publishers, Dordrecht, 1990)
2. W. Benz and E. Asphaug: *Impact simulations with fracture. I- Method and tests*, Icarus **107**, 98 (1994)
3. W. Benz and E. Asphaug: *Catastrophic disruptions revisited*, Icarus **142**, 5 (1999)
4. M. Jutzi, W. Benz and P. Michel: *Numerical simulations of impacts involving porous bodies: I. Implementing sub-resolution porosity in a 3D SPH Hydrocode*, Icarus (2008) in press
5. W. F. Bottke, A. Morbidelli, R. Jedicke, J. M. Petit, H. Levison, P. Michel and T. S. Metcalfe: *Debiased orbital and absolute magnitude distribution of the Near-Earth Object population*, Icarus **156**, 399 (2002)
6. S. Chandrasekhar: *Ellipsoidal Figures of Equilibrium* (Dover, New York, 1969)
7. W. F. Chen and D. J. Han: *Plasticity for Structural Engineers* (Springer, Berlin and New-York, 1988)
8. D. D. Durda, W. F. Bottke, B. L. Enke, W. J. Merline, E. Asphaug, D. C. Richardson, Z. M. Leihnardt: *The formation of asteroid satellites in large impacts: results from numerical simulations*, Icarus **170**, 243 (2004)
9. B. J. Gladman, F. Migliorini, A. Morbidelli, V. Zappalà, P. Michel, A. Cellino, Ch. Froeschlé, H. F. Levison, M. Bailey and M. Duncan: *Dynamical lifetimes of objetcs injected into asteroid belt resonances*, Science **277**, 197 (1997)
10. D. E. Grady and M. E. Kipp: *Continuum modeling of explosive fracture in oil shale*, Int. J. Rock Mech. Min. Sci. Geomech. Abstr. **17**, 147 (1980)
11. A. W. Harris: *The rotation rates of very small asteroids: Evidence for "rubble pile" structure*, Lunar Planet. Sci. **27**, 493 (1996)
12. K. A. Holsapple: *Equilbrium configurations of solid ellipsoidal cohesionless bodies*, Icarus **154**, 432 (2001)

13. K. A. Holsapple, I. Giblin, K. R. Housen, A. Nakamura and E. Ryan: *Asteroid impacts: Laboratory experiments and scaling laws.* In: Asteroids III, W. F. Bottke, A. Cellino, P. Paolicchi and R. P. Binzel (Eds.) (University of Arizona Press, Tuscon, 2002) pp. 443–462.
14. K. A. Holsapple: *Equilibrium figures of spinning bodies with self-gravity*, Icarus **172**, 272 (2004)
15. K. A. Holsapple and P. Michel: *Tidal disruptions: a continuum theory for solid bodies*, Icarus **183**, 331 (2006)
16. K. A. Holsapple: *Spin limits of Solar System bodies: From the small fast-rotators to 2003 EL61*, Icarus **187**, 500 (2007)
17. K. A. Holsapple: *Spinning rods, elliptical disks and solid ellipsoidal bodies: Elastic and plastic stresses and limit spins*, International Journal of Non-linear Mechanics, in press (2008)
18. K. A. Holsapple and P. Michel: *Tidal disruption II: a continuum theory for solid bodies with strength, with applications to the satellites of our Solar System*, Icarus, **193**, 283 (2008)
19. K. R. Housen and K. A. Holsapple: *Scale effects in strength-dominated collisions of rocky asteroids*, Icarus **142**, 21 (1999)
20. J. H. Jeans: *The motion of tidally-distorted masses, with special reference to the theories of cosmogony*, Mem. Roy. Astron. Soc. London **62**, 1 (1917)
21. Z. Kneževič, A. Lemaître and A. Milani: *The determination of asteroid proper elements.* In: Asteroids III, W. F. Bottke, A. Cellino, P. Paolicchi and R. P. Binzel (Eds.) (University of Arizona Press, Tucson, 2002) pp. 603–612.
22. S. C. Lowry, et al.: *Direct detection of the asteroidal YORP effect*, Science **316**, 272 (2007)
23. H. J. Melosh: *Impact cratering: a geologic process* (Oxford University Press, New York, 1989).
24. H. J. Melosh, E. V. Ryan and E. Asphaug: J. Geophys. Res. **97**, 14735 (1992)
25. P. Michel, W. Benz, P. Tanga and D. C. Richardson: *Collisions and gravitational reaccumulations: forming asteroid families and satellites*, Science **294**, 1696 (2001)
26. P. Michel, W. Benz, P. Tanga and D. C. Richardson: *Formation of asteroid families by catastrophic disruption: simulations with fragmentation and gravitational reaccumulation*, Icarus **160**, 10 (2002)
27. P. Michel, W. Benz and D. C. Richardson: *Fragmented parent bodies as the origin of asteroid families*, Nature **421**, 608 (2003)
28. P. Michel, W. Benz and D. C. Richardson: *Disruption of pre-shattered parent bodies*, Icarus **168**, 420 (2004)
29. P. Michel, W. Benz and D. C. Richardson: *Catastrophic disruption and family formation: a review of numerical simulations including both fragmentation and gravitational reaccumulation*, Planet. Space Sci. **52**, 1109 (2004)
30. Michel P.: Modelling collisions between asteroids: from laboratory experiments to numerical simulations. In: Souchay J. (ed.) *Dynamics of Extended Celestial Bodies*, Lect. Notes Phys. **682**, pp. 117–143 Springer, Berlin (2006)
31. P. Michel and K. A. Holsapple: *Tidal disturbances of small cohesionless bodies: limit on planetary close approache distances.* In: Nearth-Earth Objects, our celestial neighbors: Opportunity and Risks, A. Milani, G. B. Valsecchi and D. Vokrouhlicky (Eds.), IAU Symposium 236 (IAU, 2007) pp. 201–210
32. A. Nakamura and A. Fujiwara: *Velocity distribution of fragments formed in a simulated collisional disruption*, Icarus **92**, 132 (1991)

33. A. M. Nakamura, P. Michel and M. Seto: *Weibull parameters of Yakuno basalt targets used in documented high-velocity impact experiments*, J. Geophys. Res. **112**, E02001, doi:10.1029/2006JE002757 (2007)

34. D. C. Richardson, W. F. Bottke and S. G. Love: *Tidal distortion and disruption of Earth-crossing asteroids*, Icarus **134**, 47 (1998)

35. D. C. Richardson, T. Quinn, J. Stadel and G. Lake: *Direct large-scale N-body simulations of planetesimal dynamics*, Icarus **143**, 45 (2000)

36. D. C. Richardson, Z. M. Leinhardt, W. F. Bottke, H. J. Melosh and E. Asphaug: *Gravitational aggregates: Evidence and evolution.* In: Asteroids III, W. F. Bottke, A. Cellino, P. Paolicchi and R. P. Binzel (Eds.) (University of Arizona Press, Tucson, 2002), pp. 501–515.

37. D. C. Richardson, P. Elankumaran and R. Sanderson: *Numerical experiments with rubble piles: equilibrium shapes and spins*, Icarus **173**, 349 (2005)

38. E. A. Roche: Acad. Sci. Lett. Montpelier. Mem. Section Sci. **1**, 243 (1847)

39. I. Sharma, J. T. Jenkins and J. A. Burns: *Tidal encounters of ellipsoidal granular asteroids with planets*, Icarus **183**, 312 (2006)

40. D. J. Scheeres, S. J. Ostro, R. A. Werner, E. Asphaug and R. S. Hudson: *Effect of gravitational interactions on asteroid spin states*, Icarus **147**, 106 (2000)

41. D. J. Scheeres: *Changes in rotational angular momentum due to gravitational interactions between two finite bodies*, Celest. Mech. Dynam. Astron. **81**, 39 (2001).

42. J. H. Tillotson: *Metallic equations of state for hypervelocity impact*, General Atomic Report **GA-3216**, July (1962).

43. D. Vokrouhlický and D. Capek: *YORP-induced long-term evolution of the spin state of small asteroids and meteoroids: Rubbincam's approximation*, Icarus **159**, 449 (2002)

44. K. J. Walsh and D. C. Richardson: *Binary near-Earth asteroid formation: Rubble pile model of tidal disruptions*, Icarus **180**, 201 (2006)

45. W. A. Weibull: *Ingvetensk. Akad. Handl.* **151**, 5 (1939)

46. K. Wuennemann, G. S. Collins and H. J. Melosh: *A strain-based porosity model for use in hydrocode simulations of impacts and implications for transient crater growth in porous targets*, Icarus **180**, 514 (2006)

47. V. Zappalà, A. Cellino, A. Dell'Oro and P. Paolicchi: *Physical and dynamical properties of asteroid families.* In: Asteroids III, W. F. Bottke, A. Cellino, P. Paolicchi and R. P. Binzel (Eds.) (University of Arizona Press, Tucson, 2002), pp. 619–631.

5

Meteoroids and Meteors – Observations and Connection to Parent Bodies

S. Abe

Institute of Astronomy, National Central University, 300 Jhongda Road, Jhongli, Taoyuan 32001, Taiwan,
shinsuke.avell@gmail.com

Abstract Meteoroid are a small rocky bodies traveling through interplanetary space. Meteors are phenomena caused by the interaction of meteoroids with the Earth's upper atmosphere. In this chapter, the author will briefly discuss observational methods and then concentrate on optical observations of meteors. First, the basic properties of meteor phenomenon in the atmosphere and classification of meteoroids are introduced and then coincidental phenomena, e.g., wake, jets, and train, are mentioned. Scientific observations (imaging and spectroscopy) carried out using various observational techniques allow measuring characteristics of meteoroids, e.g., orbits, density, strength, compositions. All information are potentially useful for investigating parent bodies of meteoroids, such as comets and asteroids. Searching for organics-related CHON and water in meteoroids is of particular interest for astrobiology.

5.1 Introduction of Meteoroids and Meteors

Comets (solar system small bodies) are planetesimals that somehow did not grow into bodies as large as the major planets. They are thought to be remnants of planetesimals were formed in the protoplanetary disk and thus they reveal important information about the formation of our solar system. Dust grains about sub-micrometer to centimeter in diameter are ejected from these bodies and are moving around the Sun as meteoroids.

Meteoroids can be observed during atmospheric entry as a meteor phenomenon. Most meteoroids are weakly bound highly porous chunks of rocky material. These meteoroids enter the Earth's atmosphere at hypervelocities (approximately several tens of kilometer per second), so they reach very high temperatures (1,800–2,000 K) during entry that they ablate nearly fully. In other words, the terrestrial upper atmosphere is a natural detector of meteoroids. The groups of meteors that appear annually are called 'meteor showers' (see Fig. 5.1). On the night of a meteor shower, any naked-eye observer can recognize that the majority of the visible meteors seem to emanate from a

Abe, S.: *Meteoroids and Meteors – Observations and Connection to Parent Bodies*. Lect. Notes Phys. **758**, 129–166 (2009)
DOI 10.1007/978-3-540-76935-4_5   © Springer-Verlag Berlin Heidelberg 2009

specific point in the sky. The direction where the meteors comes from is traditionally called 'radiant.' All dust streams, whether of cometary or asteroidal origin, show this 'radiant' phenomenon. Many such streams have been identified as meteor showers throughout the year. The general nomenclature for a meteor shower and a meteoroid stream are using the Latin name of the constellation from which the radiant emanates, e.g., Geminids, Perseids, and Quadrantids.

Typical meteors are associated with meteoroid sizes between 0.05 mm and 20 cm in diameter, which are ejected from a small solar system bodies. A meteor brighter than approximately –3 magnitude is called a 'fireball.' An exceptionally bright fireball (–8 magnitude or brighter) is eventually called a 'bolide.' Some meteors are related with the fall of a meteorite. The term 'meteorite' should be exclusively used only for meteoroids recovered on the Earth's surface. The geocentric entry speed of the meteoroid varies from 11 to 72 km s^{-1}, which depends on gravity of the Earth (11 km s^{-1}; escape velocity from the Earth) and the parabolic velocity (42 km s^{-1}; escape velocity from the solar system) relative to the Earth's motion (30.3 km s^{-1}), respectively. For example, the head-on meteor shower 'Leonids' shows nearly the maximum speed of about 72 km s^{-1} [35].

A meteor shower occurs when the Earth's path crosses a dense dust trail generated by its parent comet, which has approximately the orbital period of the parent. The Leonid meteor shower is one of the most interesting meteor showers and has shown strong activity roughly every 33 years at least in the last 100 years. This corresponds to the orbital period of the comet, 55P/Tempel–Tuttle. Meteor streams that form as a result of cometary activity around perihelion are often referred to as trails. A trail is created at each perihelion passage of the parent comet. On the other hand, non-shower meteors are called 'sporadic meteors.'

Meteoroids are ejected from a cometary nucleus when the comet approaches the Sun within a distance of about 2 AU. Surface sublimation of nucleus material ejects both gas and meteoroids. Since meteoroids are ejected in random directions from the nucleus with relative velocities about few meters to few hundred meters per second, the meteoroids gradually separate from the comet and form a dust stream. Both comets and meteoroids move around the Sun in elliptical orbits. The orbit of a comet is generally perturbed strongly by the gravitational field of planets. Moreover, non-gravitational force induced by the sublimating gas will modify the orbital period of a comet, further complicating the orbital evolution. Thus, the orbit of dust streams generated by a comet at different perihelion passages is slightly different. Some streams are named 'dust trails.'

Dust trails were detected by the Infra-Red Astronomical Satellite (IRAS) [61]. Reach et al. (2000) observed the dust trail of comet 2P/Encke with the Infrared Space Observatory ISOCAM [82]. While Ishiguro et al. (2007) detected dust trails and its structures of short-period comets from ground-based observations in the visible wavelength [46]. The central core of a dust trail has a size of a few 10^4 km and this determines the duration of a meteor shower of several ten minutes. Nowadays, trail locations can be precisely calculated by numerical integrations, allowing predictions of meteor showers [8]. Several models of dust trail position and density were successfully fitted to the observed rate of Leonid meteor showers. Based on this prediction which has been in development since the Leonid meteor storm of 1998, the time of maximum can be predicted to 10 min accuracy or less [70]. Massive numerical integration for particles ejected from a comet, through a cometary model, allows for the computation of ephemeris of meteors showers and the spatial density of meteors streams more precisely [99]. During the Leonid shower in 1998, Nakamura et al. (2000) detected a meteoric cloud scattered by small meteoroids in the dust trail along the orbit of comet 55P/Tempel–Tuttle [74]. It was observed as a local enhancement, 2–3% of the background zodiacal light that corresponds to the number density of 1.2×10^{-10} m^{-3} assuming typical particle size of 10 μm and albedo of 0.1.

Fig. 5.1. The 1999 Leonid meteor storm observed by Intensified High-Definition TV (HDTV) camera (NHK and H. Yano) [1]. This composite of meteor images (60 s), the first and greatest TV recording of a meteor storm in history, was recorded from the Leonid Multi-Instrument Aircraft Campaign (Leonid MAC [50]) at the peak of the 1999 Leonid meteor shower. FOV is 60°×34°. An ideal hourly rate, Zenith Hourly Rate (ZHR), of the shower was estimated to ~ 4,000 hr^{-1} at 02:02 UT on November 18, 1999. During 1998–2002 Leonid meteor showers, Leonid MAC observing campaigns have been accomplished (principal investigator: Peter Jenniskens, SETI Institute). They were sponsored by NASA and the U.S. Air Force, with additional support from NSF, ESA, ISAS, NAOJ, and many other institutes, universities, and organizations. This airborn-observing campaigns made dramatic progress in the meteor science, such as wide wavelength observations from ultraviolet to mid-infrared with spectrographs

During atmospheric entry, the meteoroids are heated, vaporized, and even partially ionized in the upper atmosphere. An incoming meteoroid begins to feel drag from the atmosphere of the Earth at an altitude of about 120 km, depending on the entry velocity. When a meteoroid reaches around 120 km altitude, the density of the Earth's atmosphere is only one ten-millionth of its density at sea level. Yet the air is dense enough so that the collisions between the meteoroid and the molecules in the atmosphere cause the meteoroid to begin to undergo rapid heating. Ablation begins and leads to rapid loss of mass, with a consequent development of a tail. This ablation process is responsible for the observed meteor phenomenon depending on the meteoroid's size, density, composition, speed, and entry angle.

5.2 Classifications

Meteoroids are generally regarded as solid bodies larger than dust particles and smaller than asteroids and comets. From the observational point of view, we can classify meteoroids according to observation or collection methods, that are limited to certain sizes. Meteoroids are classified into three groups: 'micrometeorites,' 'meteors', and 'meteorites' by detection methods. Figure 5.2 indicates this classification of meteoroids by its size when compared with other larger or smaller bodies in our solar system.

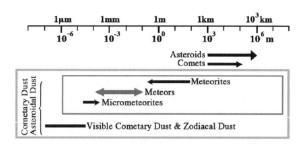

Fig. 5.2. Detection methods of meteoroids by their size: collected micrometeorites and observed meteors and dusts

• **Micrometeorite, IDP.** The smallest collected samples are called 'micrometeorites' or, more commonly, 'interplanetary dust particles (IDPs).' These particles in the mass range of 10^{-12}–10^{-6} g (corresponding to 0.6–60 μm in radius for a typical density of 1 g cm^{-3}) are sufficiently gently decelerated in the upper atmosphere before reaching evaporation temperature. The concept of micrometeorites was introduced by Whipple in 1950 [105]. Micrometeorites cannot be observed by optical or radar methods since they do not produce a significant ionization and luminosity. They are collected from stratosphere, polar ice cores, deep sea, and in-land sediments and also suffer from selection effects in terms of size, shape, magnetism, time resolution, etc. As yet none of these methods enables the determination of micrometeorites' orbits.

- **Meteor**. It is a luminous phenomenon that is generated when a meteoroid enters into the Earth's atmosphere and can be observed by visual, photographic, or radar methods. These meteors form between approximately 130 and 70 km altitude. The smallest meteoroid size which is able to produce a meteor is roughly estimated to 0.01 mm, depending on the velocity. Typical meteors are associated with meteoroid sizes between 0.01 mm and 20 cm in diameter, which corresponds to the mass range of 10^{-9}–10^4 g. In fact, there is no cut-off toward large sizes. Most meteors, especially known meteor showers have cometary origin.

- **Meteorite**. The mass of meteorites that have actually been found ranges from 10^{-2} to 10^8 g. These are their final masses, after they have passed through the atmosphere. The ability to penetrate into the atmosphere depends strongly on the meteoroid velocity. Especially the mass loss due to severe ablation causes a practical upper velocity limit of about 30 km s^{-1} for the occurrence of a meteorite fall. A meteoroid larger than ~20 cm (for a velocity of 15 km s^{-1}) is able to survive the ablation in the atmosphere because there is not enough time to ablate the entire meteoroid mass, before the body slows down to a critical ablation limit of ~3 km s^{-1}. Meteorites are best studied with regard to their physics, chemistry, and mineralogy, but accurate orbits are known for a very small number of meteorites. Meteorites are usually thought to result from fragmentation of asteroids.

We can distinguish the following three phases of the meteoroid entry: 'sputtering (preheating),' 'ablation', and 'dark flight.' Figure 5.3 indicates these three regimes of a meteoroid motion in the atmosphere. Numerical simulations of the entry process, ablation and non-ablation states, are shown in Fig. 5.4. It shows the importance of the ablation process for producing the observed luminosity.

- **Sputtering, preheating**. At altitude around 400–120 km when the meteoroid approaches the Earth, the surface temperature of the meteoroid rises but due to low atmospheric density, ablation according to standard theory is not significant above 150 km. The mean free path (MFP) of atmospheric species is about 1.5 m (MFP~ few tens of centimeters at 100 km, approximately few centimeters at 90 km). Nevertheless, high altitude meteors become luminous at altitudes above ~130 km, where the standard ablation theory is not applicable. A high altitude meteor above 160 km was first detected by means of multi-station TV and photographic observation of a bright Leonid meteor [42]. While the typical light curve (brightness changing with time) of a meteor shows a smooth increase or decrease below 130 km due to ablation process, it fluctuates above 130 km. A comet-like diffuse appearance was also reported [87].

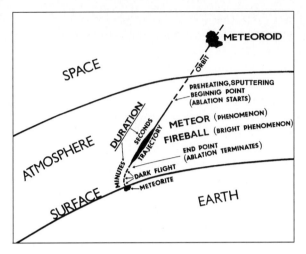

Fig. 5.3. Basic terminology for meteors (Ceplecha et al. 1998) [35]. A phenomena of a meteorite fall is indicated as three regimes, sputtering/preheating (400–120 km), ablation (120–20 km), and dark flight (<20 km)

Also, radar echoes from Leonid meteors were recorded at altitudes up to 400 km. Sputtering was proposed as a mechanism to explain this high altitude luminosity [23]. Monte-Carlo simulations suggest that fast particles sputtered from the meteoroid surface release 10–20% of the incoming energy in the atmosphere [81, 100]. The sputtering model can be applied for observational light curves of high altitude meteors in high-speed meteor showers, such as Leonids and Perseids [60].

- **Ablation**. Thermal ablation begins at altitudes between 130 and 120 km and this altitude is not related to the meteoroid mass. Below this altitude, hydrodynamic approach is useful for describing the meteor luminosity. Thermal ablation occurs because of momentum transfer in collisions between atmospheric gases and the meteoroid. When the meteoroids' temperature rises close to about 2,000 K evaporation starts, which depends on melting point of the containing material. Majority of the meteoroids' kinetic energy is consumed during ablation. As the meteoroid penetrates into the Earth's atmosphere more, deceleration is more obvious. Typical meteor ablation occurs at approximately 130 and 70 km altitude.

- **Dark flight**. When a meteoroid (or fragments) decelerates below the speed at which ablation takes place, the dark flight starts without emitting light. The terminal velocity observed for many bright fireballs (e.g., [35]) is typically 3–4 km s^{-1}. If there is no wind, the remnant of the meteoroid follows a free fall. However, the location of fragments on the ground should be predicted based on a dark flight computation considering atmospheric wind effects (e.g., [16, 20, 25, 83]). The impact velocities on the Earth' surface are in the order of several 10 m s^{-1} for smaller meteorites, ~10 g, and several 100 m s^{-1} for larger meteorites, ~10 kg.

Additional phenomena connected with a meteor are 'wakes,' 'trains,' and 'dust clouds.' These phenomena are not well understood, but may provide important clues about chemical reactions in the meteor wake. They are described below.

- **Meteor wake**. Meteor wake is the radiation emitted just behind the meteoroid, whereas the radiation surrounding the body is sometimes called head radiation. The wake is caused by atmospheric atoms and molecules penetrating the skin of a dense plasma of ablated material in front of the meteoroid. The collisions hasten the meteoric ablation materials and the heated air past the meteoroid, where they expand into the meteor wake. Typical dimensions of the wake are several hundred meters to several kilometers behind the body and a typical duration is in the order of several tenths of seconds. There are two types of wakes depending on the origin of the radiation: (a) gaseous and (b) particulate. Spectral records of wakes show lines of the same elements as in the head radiation, but the excitation energy is significantly lower [35]. The most striking aspect of the images of wakes were observed by freezing meteoroids' motion in high-frame rate of 1,000 frames/s video images with an intensified camera [90, 91] (see Fig. 5.5). This high-frame rate image shows a bowshock-like structure and a spherical luminosity before the shock, which looks like a cometary coma and plasma tail. The ablated vapor cloud extends the size of the wake largely and thus increases the collision cross section of the meteoroid dramatically. The observation suggests that the vapor volume is large enough to induce a strong shock. Jet-like features have been observed [93] (see Fig. 5.6) but the details of the ejection mechanism are not clear.
- **Short duration meteor train and afterglow**. It is evident that short-lasting trains called 'short-duration trains,' which last few seconds at the most, emit a forbidden line of [O I] at 557.7 nm known as the aurora green line. The forbidden green line has an evidently different origin than the other lines. First entry-speed meteors (more than 60 km s^{-1}), such as Leonids, Perseids, and Orionids, are significant. On the other hand, 'afterglow,' which also lasts few seconds, is dominated by atomic lines of low excitation. The afterglow emission is distinguishable from the forbidden line by spectroscopy. The luminosity of afterglow is explained by atomic recombination [19].
- **Persistent meteor train**. Bright fireballs sometimes leave a self-luminous long-lasting plasma at altitudes of about 80–90 km that is called 'persistent meteor trains' or 'persistent trains' (see Fig. 5.7). The persistent train lasts long after the disappearance of its parent meteor. Occasionally such trains may last for hours. After a rapid decay in intensity, it is generally believed that the luminosity of persistent trains is fueled by reactions involving O_3 and atomic O, efficiently catalyzed by metals from the freshly ablated meteoroids. Rapid temperature decay in the afterglow spectrum was estimated [18]. Atomic recombination followed by electron downward cascade is a likely mechanism to produce the short-lasting phase of the train [19].

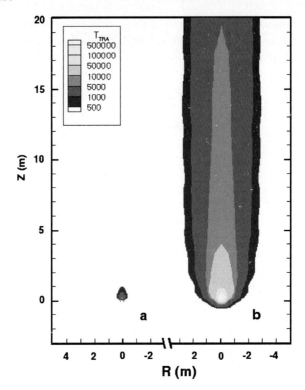

Fig. 5.4. The computations of gas flow field, including meteoroid ablation and translational temperature field of a −1 magnitude Leonid meteor at 95 km altitude and a velocity of 72 km s^{-1}[13, 51]. In the case of non-ablation (a), multiple collisions quickly stop the accelerated air atoms and molecules. Whereas the model considering meteor ablation of Mg I atoms (b) results in increasing the temperature around the meteoroid. The figure shows that a large region of the flow field is affected by meteoroid ablation that produces an extended wake at high temperature, around 5,000 K, in a state of thermal equilibrium. These findings are in qualitative agreement with high-speed imaging [90, 91] and spectroscopic observations of the Leonid meteoroids

Recent spectroscopic observation (Fig. 5.8) in the extremely wide wavelength range between 300 and 930 nm identified some molecules, O_2 at 865 nm and OH at 309 nm, as well as atoms, Na, Mg, Fe, Ca, Si, etc. [3]. O_2 rotational temperature of 250 K at altitude of 88.0 ± 0.5 km and a final exothermic temperature of 130 K at about 15 s after the fireball were estimated. The long-lasting phase of the persistent train after the final exothermic temperature is probably explained by the chemiluminescence mechanism.

• **Meteoric dust cloud**. Dust clouds (smoke trails) left by the meteorites' fragmentation at lower altitudes around 20–30 km are frequently observed. The meteoric dust cloud is visible only when it is illuminated by sunlight.

Note that these trains and dust clouds are different from 'Noctilucent clouds' which are known as polar mesospheric clouds at altitudes of around 85 km observed in the summer months at high latitude region.

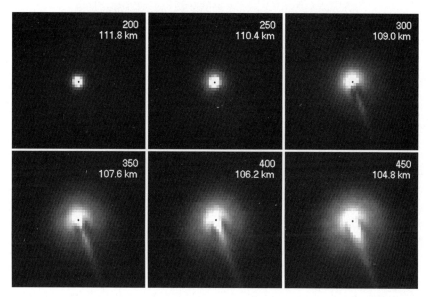

Fig. 5.5. High-frame rate meteor images. Each is a $0.94° \times 0.94°$ section from an original $6.4° \times 6.4°$ image, which shows the development of the meteor morphology. The images were recorded at 1,000 frames/s. The frame number within the sequence is shown in the *upper right corner*. The calculated position of the meteoroid is shown by a *dot* [53, 90, 91]

5.3 Observational Techniques

Various techniques can be used to study meteors: visual and near-infrared observations (300–900 nm) using visible and IR photography, video technique, image-intensifiers, and spectrographs, radio observations (HF: 3–30 MHz, VHF: 30–300 MHz, and UHF: 300–3000 MHz), infrasound detection (< 20 Hz), and seismogram (shockwave recording) methods.

The generated ion train along the meteor path reflects radio waves in the decameter, 10–100 m wavelength range. The ion trains of bright meteors are visible even to the naked-eye. For optical observations, video detection has the advantage that it allows to monitor meteor phenomena and the ablation process in real time. These days, the charge coupled device (CCD) invented in 1970 is the dominant detector for TV meteor astronomy. Though the spatial resolution of TV, 720×480 active pixels, is much lower than that of a

Fig. 5.6. High spatial resolution, narrow band image measurements of the Mg I emission at 518 nm have been used to clearly identify jet-like features in the 1999 Leonid meteor head [93]. The unusual jet-like filaments suggest that the parent meteoroids are spinning and as the whirling fragments are knocked away by the impinging air molecules, or by grain–grain collisions in the fragment ensemble, they ablate quickly generating an extended area of structured luminosity up to about 1–2 km from the meteoroid center. The jet-like features were found to be present in up to ∼8% of whole data of the 1999 Leonids

photographic plate, TV observations can record fainter meteors with higher temporal resolution than photographic techniques.

Atmospheric trajectory and extrapolated interplanetary orbit can be calculated from time synchronized meteor data with velocity information obtained from triangulation stations. Triangulation of meteors by two or more observers spaced few tens to around 100 km apart determines a very pronounced parallax. Photographic observations using a fast rotating shutter, chopper, in front of the lens allow to measure the velocity of meteoroids. For highly sensitive CCD cameras, the video rate recording (generally in 0.03–0.04-s intervals) acts as a chopper and also allows to determine meteoroids' velocity.

Meteor forward scattering is a well-known classical method to detect meteor echoes using radio telescopes (antennas). Signals from distant transmitters are reflected by the meteor trails generated behind the meteoroids. Radar technique has enhanced mass sensitivity when compared with optical cameras and video recording. For example, Kyoto University's newly developed ultra-multi-channel receiving system of the middle and upper atmosphere (MU) radar (46.5 MHz) with 25 channel interferometers can detect 40,000–50,000 meteoroids' echoes a day with a very precise determination of

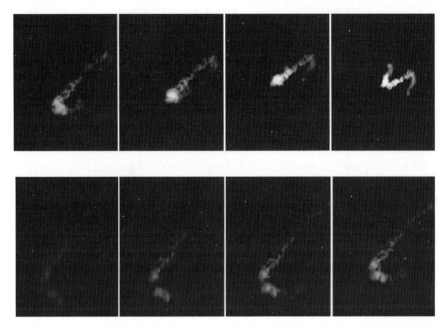

Fig. 5.7. Time sequence photographs, *top right to bottom left*, of a persistent train observed on November 18, 1998. The photographs were taken by Masayuki Toda at Mt Fuji, Japan with a photographic camera Nikon F4s with Nikkor ED f=200 mm/F2.0 lens. Each exposure time was 4 s and the interval 0.1 s. The bluish part in the first picture comes from Mg I (383 nm) emission. The train was blown away by the upper atmospheric wind at the rate of about several tens to 100 m s^{-1}

the location of meteor trails. The emitted radio signal is scattered by electrons in the plasma (meteor trail) created in the vicinity of the meteoroid due to its interaction with the atmosphere. The scattering by a meteor trail typically lasts for 0.1–1 s. The plasma immediately surrounding a meteoroid upon its entry into the atmosphere can also scatter the radar signal. Echoes due to such a scatter process are called meteor head echoes. Head echoes were first detected in 1949 [68, 69]. Much recent progress on head echo observations has been achieved through the use of high-power large-aperture (HPLA) radars, such as Arecibo (430 MHz) (e.g., Mathews et al., 1997 [65]), EISCAT (930 and 224 MHz) (e.g., Pellinen-Wannberg & Wannberg, 1994 [79]), Jicamarca (50 MHz) (e.g., Chapin & Kudeki, 1994 [36]), Millstone Hill (440 MHz) (e.g., Evans, 1966 [41]), ALTAIR (422 and 160 MHz) (e.g., Close et al., 2004 [37]) and the MU Radar (e.g., Sato, Nakamura & Nishimura [86]). With the head echoes, the speed of the meteoroid can be more directly measured through the analyses of Doppler frequency shifts or range-time intensity (RTI) plots. Such studies contributed much to the understanding of meteoroids inherent properties, specifically regarding meteoroid decelerations and velocities. Using the MU Radar, Sato et al. (2000) developed a Doppler pulse compression scheme

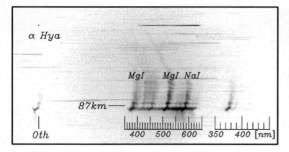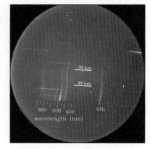

Fig. 5.8. *Left*: Photographic spectrum of 1998 Leonid meteor's persistent train taken by Hideyuki Murayama. This is the same train as shown in Fig. 5.7. Spectrum is obtained with an exposure time of 10 s. The dispersion direction is from left to right and the fireball moved from top to bottom. At the earlier stage, the strength of the line at Mg I (383 nm) is stronger than that of Mg I (518 nm) because of its higher excitation potential. The continuum between 550 and 600 nm originated from orange arc emission of FeO. *Right*: TV spectrum of a meteor persistent train observed after a Leonid fireball taken by Satoru Sugimoto. The spectrum is obtained by averaging of 30 video frames (duration of 1 s), providing a circular field of view of 26° in diameter. The dispersion direction is from right to left and the fireball moved from top to bottom. The strong atoms are Na I at 589 and 818 nm, Mg I at 458 and 518 nm, and many Fe lines, while molecules are FeO around 586 nm and O_2 around 865 nm

that can enhance the S/N ratio of the radar echoes with very large Doppler shifts. It also allows to determine the range of the head echoes with a resolution of 200 m [86]. Doppler velocities, echo powers, and ranges are determined from range-frequency spectra derived with the Doppler pulse compression. The direction of the meteor head can also be determined with interferometry: in this case the three-dimensional location of meteor head can be measured for each single transmission pulse.

Until today, quantitative meteor spectroscopy is the only tool for deriving elemental compositions of millimeter–centimeter-sized dust grains, because in-situ data from space missions are limited and typically measure smaller dusts. Spectroscopic observations of evaporated gas of meteoroids reveal not only chemical composition of dusts but also emission processes caused by hypervelocity impacts in the atmosphere. Millman et al. (1971) first observed meteor spectra by television techniques [72]. Most early spectroscopic studies were concerned with line identification. J. Borovička et al. [14, 15, 17, 21], S. Abe et al. [1, 4, 6], J.M. Trigo-Rodríguez et al. [94], T. Kasuga et al. [58], and P. Jenniskens [56] have carried out spectroscopic studies including quantitative analysis of meteors' spectra.

Typically, the intensified video data are recorded in an 8-bit ($2^8 = 256$) analog video system in PAL (mostly in European region) or NTSC (mostly in US and Japan regions) format. Abe et al. (2000) reported on the first imaging and spectroscopic observations using Intensified High-Definition TV

(II-HDTV) [1]. Earlier, II-HDTV developed by NHK (Nippon Hoso Kyokai) was used to monitor the prominent activity of the 1998 Giacobinid meteor shower, associated with comet 21P/Giacobini–Zinner [103]. This technique increases the number of TV lines from about 576 (NTSC) to 1150, and has a 10-bit ($2^{10} = 1024$) dynamic range, four times higher than previous video systems. For a given field of view, the system is more sensitive than conventional intensified CCD cameras. Meteors as faint as +8 magnitude and stars of +10 magnitude can routinely be observed even with a wide $37° \times 21°$ field lens. The development of ultraviolet visible II-HDTV (with spectral range of 250–1,000 nm) allowed to identify new molecules, N_2^+ and OH [4] (see Figs. 5.11, 5.17). The resulting spectral resolution of II-HDTV was about 1.5 nm in FWHM (full width at half maximum). While using a cooled CCD camera, Jennikens et al. (2007) first studied fainter lines with high resolution, ~ 0.13 nm [56]. The limiting magnitude for meteors was about +5.7, while for spectra it was about +4 magnitude with a field of view of 4.7°.

5.4 Meteoroids to Zodiacal Clouds

A young meteoroid stream of dust, pebbles, and large chunks will become widely dispersed in space by planetary perturbations, collisions, and solar radiation forces (details are shown below). The initial stage of meteoroid stream dispersion is caused by planetary perturbations that can change their orbits significantly. This dispersion time scale of the meteoroid stream is a few hundreds up to 1,000 years [39]. On longer time scale, >10,000 years, the evolution of dust particles is dominated by a mixture of delicately complex forces such as the Poynting–Robertson and Yarkovsky forces. Most millimeter–centimeter-sized meteoroids originated from Jupiter-family comets (JFCs) with high eccentricities are spread by the Poynting–Robertson force over a period of 10^8–10^9 years [29, 73].

A small particle in orbit around the Sun is moving relative to the solar radiation field. From the perspective of a dust particle, solar radiation pressure force is not perfectly radial on the moving particle but has a small component opposite to the particle motion (ahead of the Sun). This phenomenon is called 'aberration.' The 'Poynting–Robertson effect' (non-radial radiation pressure due to aberration of sunlight) acts as drag force resulting in a gradual decrease in eccentricity and semimajor axis. A similar but weaker effect (the ion drag), about 30% of the Poynting–Robertson drag force, arises from the solar wind. The 'Yarkovsky effect' (non-isotropic thermal emission from an illuminated rotating object) causing a change in semimajor axis depends on the sense of rotation. Retrograde rotators place the hot region in the leading hemisphere. The diurnal variant decreases (increases) the semi major axis for retrograde (prograde) rotating objects. The Poynting–Robertson effect dominates the orbital evolution of millimeter–meter-sized particles rather than the Yarkovsky effect; however in general, these grains are affected by collisions.

Whereas meter-sized bodies are influenced importantly by the Yarkovsky effect which can lead to significant changes in the orbital distribution of the bodies and spin period plays an important role [29].

In interplanetary space, meteoroids breakup by mutual high-speed ($v >$ 1 km s^{-1}) collisions. It has been considered that collisions among dust particles are also an important process in maintaining the evolution of the zodiacal clouds [47]. Meteoroids are modified or destroyed, and lots of new fragment particles are generated by the impacts. One of the evidences for meteoroids' collision was indicated by anomalous meteoroids' orbits. These were observed using multiple-station photographic cameras equipped with rotating shutters during 1998–1999 Leonid outbursts [95]. Though these anomalous meteoroids still belonged to the Leonid dust trail because of the similarity of geocentric radiant to those exhibited by the Leonids, they had lesser geocentric velocities corresponding to shorter semimajor axes and smaller eccentricities. The collision with interplanetary meteoroids is the most probable explanation of the origin of these anomalous Leonid meteoroids. These results suggested that collisions of meteoroids with zodiacal dust are common. The maximum mass for zodiacal dust particles is estimated to be 10^{-6} g because masses greater than this are quickly destroyed by mutual collisions [44], while the masses of meteoroids producing visible Leonid meteors are in the order of 10^{-3} g. The average heliocentric velocities of zodiacal dust and Leonid members are 15 and 41 km s^{-1}, respectively. In fact, it is still impossible to make a laboratory experiment to check such a hypervelocity impact. For the meteoroid flux at 1 AU, the collisional time scale is shorter than the Poynting–Robertson orbital decay time scale for particles with masses $m > 10^{-5}$ g [43]. Although some attempts have been made to explain the flux distribution of interplanetary dust at the near-Earth space (e.g., Mukai (1989) [73], Ishimoto (2000) [47], and reviewed by Mann (2004) [64]), the results have not always been consistent with observations for the entire mass range.

5.5 Origin of Meteoroids

Approximately 230 potential showers were observed; however, the sources of minor meteor showers are still unclear. Of particular interest is whether these weak showers originate from faint comets, dormant comets or asteroids.

5.5.1 Interstellar Origin

A part of meteor echoes showed surprisingly fast speed, more than 72 km s^{-1}, and such fast meteoroids can be explained by interstellar dusts from outside of our solar system. Fewer than one out of ten thousand radar meteors have been identified as interstellar meteoroids that pass through the solar system on a hyperbolic orbit. According to AMOR radar observations in New Zealand, these particles have been found to arrive from southern ecliptic latitudes with enhanced fluxes from discrete sources [10, 92]. The estimated radius of inter-stellar meteoroids observed by radar is about 20 μm, while clearly identified interstellar dust particles detected by *Ulysses* spacecraft at distances between 1 and 5.4 AU from the Sun are much smaller, with sizes ranging from 0.1 to above 1 μm with a maximum of about 0.3 μm. Interstellar meteors were also detected with the Jicamara and MU radars. The observed velocity distributions in the fast interstellar component measured at Jicamarca in Peru is well described by a Gaussian function with an average velocity of ~63±6.6 km s^{-1} [49] and agrees very well with distributions reported using the MU radar in Japan [86].

5.5.2 Cometary Origin

Some meteor showers are directly linked to a comet, e.g., short-period comet 2P/Encke, several Jupiter family comets (21P/Giacobini–Zinner, 45P/Honda–Mrkos–Pajdusakova), Halley-type comets (109P/Swift–Tuttle, 55P/Tempel–Tuttle), and the long-period comets (C/1861 I Thatcher, C/1911N$_1$ Kiess).

On the otherhand, for example, the Geminid meteor shower follows an orbit that is clearly associated with the near-Earth asteroid (3200) Phaethon. The recently discovered Apollo asteroid 2005 UD is the most likely candidate for being a large member of the Phaethon–Geminid stream Complex (PGC) [76]. Asteroid 2003 EH$_1$ with high orbital inclination angle, about 71°, was identified as the parent of Quadrantids which is one of the most intense annual meteor showers [54]. Parent bodies of the Geminids (3200 Phaethon and 2005 UD), the Quadrantids (2003 EH1), the Daytime Arietids (Marsden group of sungrazers), The Andromedids (fragments of 3D/Biela), and Phoenicids (2003 WY25) [55, 104] are probably the dormant or extinct cometary nuclei, which originated from breakup of comets.

5.5.3 Asteroidal Origin

Asteroidal particles have been proposed as a major source of the IDPs. A recent theory and observation of asteroidal dust bands and the young asteroid families suggested that the zodiacal cloud is not dominated by asteroidal particles [75]. Considerable fraction of mass-loss ejected from short-period comets were estimated by their dust trails, which suggested that short-period comets could be considered as a source of the IDPs supply in the interplanetary space [46].

Magnitude, altitude, entry angle to the atmosphere, velocity, and deceleration history of a fireball enable to investigate not only orbital elements but also mineral materials using Ceplecha's Classification [33]. One attempt of a survey of asteroidal meteors among the multi-station optical meteor database of ~15,000 orbits succeeded in the discovery of pebbles, five meteoroid candidates, from the near-Earth asteroid (25143) Itokawa [77]. Using the Ceplecha's classification, the classified mineral of most probable candidate originated from the asteroid Itokawa indicated ordinary chondrite, which is similar to Itokawa's surface composition of a LL chondrite analog. Note that Japanese spacecraft 'Hayabusa' explored the Earth-crossing asteroid Itokawa in 2005, which brought us most precise scientific data, e.g., mass, bulk density, porosity, morphology (irregular shape, craters, and boulders), mineralogy, space weathering, that was never ever seen. S-type asteroid Itokawa is a chondritic ruble pile object and is considered as a source of ordinary meteorites.

Most of meteors have been treated as cometary origin, while meteorites are usually thought to be associated with asteroids. Orbital and mineralogical (including reflectance spectrum) links have already been noted that imply an asteroidal origin for most meteorites. In fact, collected meteorites are biased and unrepresentative of the meteoroid population in the near-Earth space. Generalizations about their parent bodies and dynamic evolution of meteorites are incomplete.

Accurate orbital determinations on meteorites are potentially of great value in establishing where the members of particular meteorite classes come from. Precise triangulation measurements during the meteorite entry make it possible to calculate the orbit that the meteorite was following before the encounter with the Earth.

5.6 Meteorites' Orbits and Meteor Observation Networks

There are over 33,000 meteorites in collections worldwide. Japanese expeditions in Antarctic discovered a lot of Antarctic meteorites and Japan is now the most holding country (more than 16,000) [108]. In order to determine the orbit of meteorites, dedicated photographic fireball networks have started since the fall of 1963. However, only nine meteorites have been identified with their interplanetary orbits until 2007 and none of them is associated with known a small solar system bodies. Among them, seven meteorites are classified as ordinary chondrites, Tagish Lake and Neuschwanstein were found to be a unique carbonaceous and enstatite chondrite, respectively. Information on these nine meteorites are summarized in Table 5.1 and their interplanetary orbits are displayed in Fig. 5.9.

5.6.1 Classical Meteor Networks

- The Czechoslovak (Czech) network in the former Czechoslovakia (1963–present) was established by Ceplecha [31]. After the addition of more stations in other European countries, the European Fireball Network (EN) including Czech network have been operating until today by Spurný, which consists of 50 stations spaced by about 100 km, one all-sky camera (fish-eye lens) at each station, and covering about a million square kilometer.

Table 5.1. Orbital elements of fall meteorites

Meteorite	Year of fall (network)*	Recovered mass (kg)	Type	Orbital elements			Ref.
				a (AU)	e (°)	i (°)	
Příbram	1959 [EN]	5.8	H5	2.401±0.002	0.6711±0.0003	10.482±0.004	[31]
Lost City	1970 [PN]	17	H5	1.66±0.01	0.417±0.001	12.0±0.1	[67]
Innisfree	1977 [MORP]	4.58	L5	1.872±0.001	0.4732±0.0001	12.27±0.01	[45]
Peekskill	1992	12.57	H6	1.49±0.03	0.41±0.01	4.9±0.2	[24]
Tagish Lake	2000	5–10	C	2.1±0.2	0.57±0.05	1.4±0.9	[26]
Morávka	2000 [EN]	1.4	H5–6	1.85±0.07	0.47±0.02	32.2±0.5	[20]
Neuschwanstein	2002 [EN]	6.2	EL6	2.40±0.02	0.670±0.002	11.41±0.03	[89]
Park Forest	2003	18	L5	2.53±0.19	0.680±0.023	3.2±0.3	[28]
Villalbeto de la Peña	2004	~5	L6	2.3±0.2	0.63±0.04	0.0±0.2	[63, 97]

* EN: European fireball Network, PN: Prairie Network, MORP: Meteorite Observation, and Recovery Project

- The Prairie Network (PN: 1963–1973), with 16 stations spaced about by 250 km, four cameras with 90° field of view, and covering about a million square kilometer in the US, was built up by Fred Whipple.
- The Meteorite Observation and Recovery Project (MORP: 1971–1985) in Canada was led by Ian Halliday.
- Fainter meteors were systematically photographed from 1952 to 1954 using Super-Schmidt cameras [48, 66]. Trajectory and orbital data on more than 2000 faint meteors are available from this project.

More details on the classical fireball networks and references to original chapters can be found in [32].

On April 7, 1959, the first recovered meteorites to have been photographically tracked during atmospheric entry was the Příbram in the former

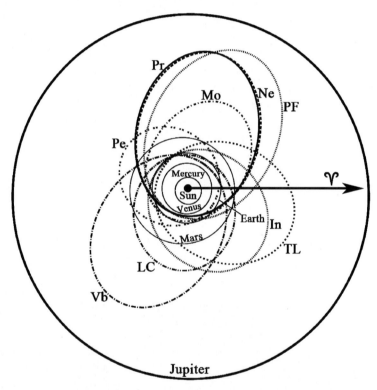

Fig. 5.9. Orbits determined from triangulation observations coverage for nine recovered chondrite falls: Pr–Příbram (H5, Apr. 7, 1959); LC–Lost City (H5, Jan. 3, 1970); In–Innisfree (L5, Feb. 5, 1977); Pe–Peekskill (H6, Oct. 9, 1992); TL–Tagish Lake (C, Jan. 18, 2000); Mo–Morávka (H5-6, May 6, 2000); Ne–Neuschwanstein (EL6, Apr. 6, 2002); PF–Peak Forest (L5, Mar. 26, 2003); Vb–Villalbeto de la Peña (L6, Jan. 4, 2004). Příbram and Neuschwanstein had identical orbits, but are of different chondritic types. The orbits are projected onto the ecliptic plane. Vernal equinox is to the right

Czechoslovakia [31]. The orbit of the Příbram meteorite is very close to the Hilda asteroidal family beyond the outer edge of the main Asteroid Belt. On April 6, 2002, 43 years after the Příbram, a new fall had happened at western Austria and southern Bavaria near the sublime castle of Neuschwanstein and fortunately was photographed by seven cameras in Germany and Czech. The orbital elements of the Neuschwanstein are in agreement with the Příbram meteorite with a high degree of accuracy, which implies that the two recovered samples are members of a stream of similar objects [89]. Surprisingly, the Neuschwanstein was classified as a rare (1.5% of all falls) enstatite chondrite (EL6) [12], while the Příbram was classified as a H5 ordinary chondrite. The cosmic-ray exposure ages of Příbram and Neuschwanstein differ significantly, 12 and 48 Myr, respectively. Both these ages, which are believed to represent the time span from the release of the meteoroid from the parent body to its encounter with the Earth, are significantly longer than the typical survival times of meteoric streams. However, the most mysterious question why a completely different classifications of the two meteorites originated from the same parent is still subject to debate.

5.6.2 Recent Meteor Networks

In recent years, several meteor photographic networks are operating together with serious amateur astronomers, especially in Europe and Japan, and their results are regularly published (WGN, the *Journal of the International Meteor Organization* (IMO); Radiant, the *Journal of the Dutch Meteor Society* (DMS); *Memoirs of the Nippon Meteor Socicty*, Japan (NMS)). Several other meteor networks are in operation, e.g., North American Meteor Network (NAMN), Polish Fireball Network (PFN), French Meteor Observing Network (REFORME), Japan Fireball Network (JN) and Tokyo Meteor Network (TMN). In parallel to that, new generation meteor networks begin to operate in the world. In the following, the major recent-established networks are listed.

- **Czech, UK, & Australia group**. Three autonomous fireball network cameras using photographic plates, covering approximately 0.3–0.4×10^6 km^2, have now been established in the desert regions of Australia on December 2005 initiated by Spurný. Meteorite fall locations will be determined within 1 km^2 accuracy for later recovery by field survey.
- **Spanish group**. The SPanish Meteor Network (SPMN) uses newly developed innovative low-scan all-sky high-resolution CCD cameras that reach $+2/+3$ meteor limiting magnitude [98]. The astrometric accuracy of the camera is ~ 1.5 arcmin, providing better resolution than many commercial video cameras, and close to the astrometric resolution of conventional photography on large-format EN cameras [88].
- **Japanese group**. Observations with the automatic detection software has a great advantage for meteor observations. In order to extend the

number of meteor orbits and spectra, a commercial CCD video camera (e.g., WATEC with $f = 6\,\mathrm{mm/F0.8}$ reaching $+3/+4$ meteor limiting magnitude) with an automatic software (UFOCapture developed by a Japanese software engineer Mr. SonotaCo) is used by amateur astronomers. One rare occasion observed from several stations using this system was the Earth-grazing meteoroid: a large meteoroid that entered into Earth's atmosphere at a shallow angle, passing through the atmosphere for very long distance and returning to space after the perigee passage. Interestingly, this observed Earth-grazing fireball was the third case in history, first case with a spectrum, and was identified as a carbonaceous meteoroid with an Apollo asteroid-type orbit [7].

5.7 Meteoroid Impacts on Other Planets

The fall of Comet Shoemaker-Levy 9 (S-L 9) into the Jupiter atmosphere in July 1994 showed us a powerful collision with no direct analog observed in nature. Impacts of cometary and asteroidal fragments and meteoroids are also observable on other planets and on the moon. Theoretical models based on the number of planet-approaching cometary orbits predicted ample opportunities for observing activity at Mars and Venus [38]. The ratio of planet-approaching Jupiter family comets (JFCs) at Mars, Earth, and Venus is 4:2:1 indicating that JFC-related meteor showers would be more frequent at Mars than at the Earth. On the other the relative numbers of planet-approaching Halley-type comets (HTCs) imply that the respective levels of annual meteor activity at those three planets are similar.

During 1999 Leonids maximum, more than 10 lunar impact flashes were successfully observed on the night side of a 10 day-old Moon from US, Mexico, and Japan [40, 78, 106]. After the discovery of impact flashes caused by Perseids (a non-Leonid meteor shower) [107], lunar impacts are recognized as common phenomena. However, the details of impact flash, the luminous efficiency and the flash materials are still uncertain. The flash magnitude, observed from Earth, typically ranges between 7 and 10. Current meteoroid models indicate that the moon is struck by a meteoroid with a mass greater than 1 kg over 260 times per year. Clearly more observations are needed to establish the rate of large meteoroids impacting the moon.

5.8 Meteoroids Influx on the Earth

The flux of small bodies onto the Earth is poorly constrained. Ceplecha et al. (1998) compiled existing data and applied improved observation analysis to determine total mass influx of all interplanetary bodies to the entire Earth's surface (atmosphere). The data are given in Fig. 5.10. From this, the flux in the mass range of 10^{-21}–10^{15} kg is estimated to be 1.3×10^8 kg [34, 35].

In Fig. 5.10, there are three mass ranges with populations: ($10^{-9.5}$ kg $< m < 10^{-6}$ kg), (10^{-4} kg $< m < 10^{-2.5}$ kg) and (10^5 kg $< m < 10^9$ kg). According to a study by Grün et al. (1985) [43], the first region corresponds to the situation where the losses of particles caused by Poynting–Robertson effect outweigh the collisional gains of particles from bigger bodies. The second region at $m = 10^{-3}$ kg corresponds to the mass range where meteor showers are the most dominant phenomena. Perhaps, the third region corresponds to a majority of small cometary fragments that are observed as dormant or extinct cometary bodies.

According to the data in Fig. 5.10, the influx per year per Earth's surface is 0.8×10^6 kg for the mass range 10^{-7}–10^2 kg, 20×10^6 kg for the mass range 10^2–10^9 kg, 100×10^6 kg for the mass range 10^9–10^{15} and it is 4×10^6 for particles with mass smaller than 10^{-7} kg.

In contrast, Love and Brownlee (1993) studied the flux of meteoroids from the analysis of hypervelocity impact craters on the space-facing end of the Long Duration Exposure Facility satellite (LDEF). They determined the mass influx of particles, in the mass range 10^{-12}–10^{-7} kg $(40\pm20) \times 10^6$ kg per year on a continual basis and about 5–10% of the mass of small particles survives atmospheric entry [62]. This points out the differences in flux between particles crossing the Earth's orbit, particles being detected during entry, and particles reaching the Earth's surface.

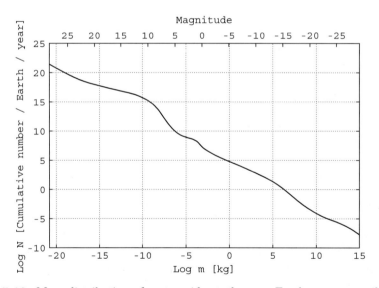

Fig. 5.10. Mass distribution of meteoroids at the near-Earth space compiled by Ceplecha et al. (1998) [35] with considering corresponding visible magnitude of meteor. Cumulative numbers N (all bodies with masses larger than the given mass m) of interplanetary bodies coming to the entire Earth's surface per year is plotted against logarithm of the mass m. Visible magnitude was estimated by a mass-magnitude relationship

Many authors have estimated the influx of near-Earth space in the limited size range. Nevertheless, results are biased by observational techniques and conditions. Based on observations from US military satellites in geostationary orbits, Brown et al. (2002b), for instance, found that the flux of objects in the 1–10 m size range has the same power-law distribution as bodies with diameters larger than 50 m [27].

5.9 Spectrum and Composition of Meteoroids

Spectroscopic observations of meteors reveal not only the chemical composition of (the cometary) meteoroids, but also emission processes of hypervelocity impacts in the atmosphere, which are difficult to reproduce in laboratory experiments at present. If the sizes of entering meteoroids are between 0.05 and 0.5 mm, the body is heated throughout. In the case of a meteoroid larger than 0.5 mm, only a surface layer down to a few tenths of a millimeter is heated. When the surface temperature reaches about 2,000 K, occurring at heights around 100 km, the meteoroid material starts to sublimate from the surface and is surrounded by evaporated vapors. Excited states of atoms of these vapors are gradually de-excited by radiation. Meteor luminosity consists mostly of radiation of discrete emission spectral lines belonging for the most part to metals and mainly to iron. More than 90% of the meteor luminosity originates from radiation of single-excited atoms of meteoroid material with excitation temperatures of 3,000–5,000 K. Figure 5.11 shows a fireball meteor spectrum observed with an intensified TV camera system. The typical first meteor spectrum shown in Fig. 5.12, 72 km s^{-1} for Leonids, indicates that atomic lines, such as Fe I, Mg I, Na I, Ca I originating from the meteoroid emits in the shorter wavelength, while atmospheric emissions such as O I, N I, and N$_2$ are seen in the longer wavelength region.

A typical temperature of 'main(warm) component,' which contains most of the spectral lines, e.g., Fe I, Mg I, Ca I, and Na I, is $T \sim 4,500$ K. The 'second(hot) component' is excited at $T \sim 10,000$ K and consists of a few ionized elements, such as Ca II and Mg II. Note that the rotational temperatures are typically less than the translational temperatures, but they seem to be under equilibrium state in the wake.

The line intensities depend not only on abundances of the respective elements in the meteoroids but also on the physical conditions during the ablation. Line intensities observed in the meteors are strongly affected by their excitation temperature in the thermal equilibrium state. The excitation temperature depends on the entry speed relative to the atmosphere. We assume that the most of the observed meteor emission originates from the nearly equilibrated gas. Figure 5.13 shows the observed Mg/Na line intensity ratio in meteor spectrum as a function of meteor speed. Precise temperature estimations are required for determining elemental abundances from the observed spectra.

The ternary diagram, Fig. 5.14, compiled by Borovička et al. (2005) shows relative intensities of the Mg, Na, and Fe multiplets [21]. The theoretical values for chondritic composition are plotted in the same figure, which are marked with the corresponding temperatures. Elemental abundances of meteoroids derived from fireball spectra when compared with CI abundances and comet 1P/Halley's dust are shown in Fig. 5.15 [22]. The chemical composition of cometary grains does not largely differ from chondritic material. On the other hand, Halley-type comets seem to be richer in Na, Si, and Mg and poorer in Fe, Cr, and Mn in comparison with chondrites.

The meteoroids coming from active comets are chemically relatively homogeneous. Larger diversity is found among sporadic meteoroids of cometary origin and sizes of several millimeters. Part of these are probably fragments of cometary crust irradiated by the solar radiation. The latter are depleted in volatiles (Na) and are mechanically significantly stronger than normal cometary material. The loss of volatiles and general compaction also occurs in the vicinity of the Sun. Small meteoroids with perihelia within 0.2 AU are chemically and physically altered. This process can be at least partly responsible for the high density of Geminid meteoroids which parent, asteroid (3200) Phaethon, approaches the Sun as close as 0.14 AU.

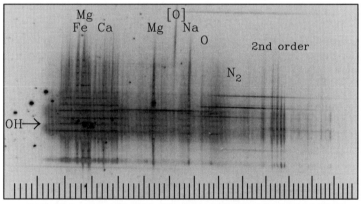

wavelength (nm)

Fig. 5.11. Spectrum of a Leonid fireball observed by UV sensitive II-HDTV system on November 18, 2001. This image (the field of view is $23° \times 13°$) is composed of 15 consecutive frames taken during the total duration of 0.5 s. The meteor moved from top to bottom and the dispersion direction is from left to right and parts of the second and third order spectra are on the right. A part of radiation comes from forbidden lines of neutral oxygen at 557.7 nm. The analyzed spectrum of this fireball is shown in Fig. 5.17

5.10 Differential Ablation, Density, and Strength of Meteoroids

A model of meteor differential ablation predicted that volatile atoms, such as Na, vaporize earlier than Mg, Fe, and Si, while refractory Ca vaporizes later on and not fully [71]. The model explained much higher abundance of Na than Ca in atmospheric metal layers and this fact was confirmed by LIDAR (Light Detection and Ranging) observations of Leonids and other meteor showers [101]. LIDAR is an optical remote sensing technology that measures properties of scattered light to find range and other information of a distant target. Detection of differential ablation by optical spectroscopy was first reported during the 1998 Leonids [17] and was confirmed during the 1999 Leonids [1]. Figure 5.16 represents atomic light curves of the 1999 Leonids compared with the 1999 Taurids. In Leonids, the Na line often starts and ends earlier than the Mg line. However, the effect of early release of Na varies from meteor to meteor. In the Taurids' light curve, in contrast, Na emission closely follows

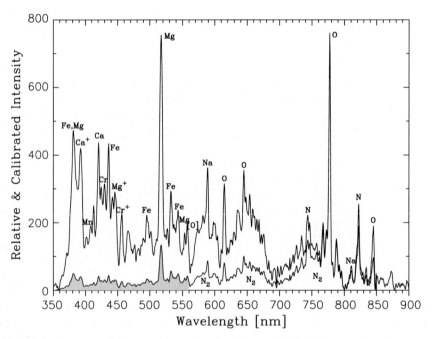

Fig. 5.12. Spectrum of a Leonid meteor observed by II-HDTV system on November 18, 1999 [1]. *Upper thin line* is the observed spectrum and *lower bold line* is the spectrum after sensitivity calibration. Meteor emission originates from a mixture of atoms and molecules ablated from the meteoroid itself (*blue rectangle*), Fe, Mg, Na, and Ca as well as the surrounding air plasma, O, N, and N_2. The atmospheric emissions are particularly prominent for high-speed meteor showers, such as Leonids, Perseids, and Orionids

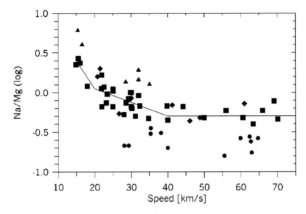

Fig. 5.13. Observed Mg/Na line intensity ratio in meteors as a function of meteor speed [21]. The approximate fit (*solid line*) is drawn through the meteors classified as having normal Mg and Na abundances (marked by *large symbols*). The Na/Mg line ratio is speed independent for speeds larger than 40 km s^{-1} but increases for lower speeds, especially below 20 km s^{-1}. The Na/Mg line ratio for high speeds correspond nearly to the temperature of 5,000 K, while at 15 km s^{-1} the temperature is nearly 4,000 K. The temperature curve in Fig. 5.14 can be therefore treated as a speed curve. Any meteor with chondritic composition should lie near this curve

Mg emission as shown in Fig. 5.16. The effect of sodium early release is strong in the Leonids (comet 55P/Tempel–Tuttle), Orionids (comet 1P/Halley), and Leo Minorids. On the other hand, it is weaker in Perseids (comet 109P/Swift–Tuttle) and very weak (or no) sodium preferential ablation was observed in Geminids (asteroid (3200) Phaethon), Taurids (comet 2P/Encke), and the Ursa Minorids (comet 8P/Tuttle). It seems reasonable to suppose that dust-ball-meteoroid grains originating from active comets, comet-asteroid transition objects (CATs), and very young distinct comets show differential ablation. The first high-quality spectra of Quadrantid meteor shower were obtained in Miyazaki prefecture, Japan (Koji Maeda), and these also show the clear differential ablation of Na [5]. Na emission appeared earlier than that of other metal atoms, such as Mg and Fe. Of particular interest to Quadrantids is that the parent body was discovered as an asteroid, 2003 EH$_1$ [54]. According to signatures of initial velocity, beginning and terminal heights of 44 Quadrantids with their spectral light curves supposed that Quadrantid meteoroids have partially lost their volatile component, but are not depleted to the same extent as Geminid meteoroids [59]. In consideration of the orbital history of 2003 EH$_1$, these results lead us to the conclusion that the parent body is a dormant comet. However, the effect of early sodium release is not universal and seems to be influenced by the structure and disruption altitude of the meteoroid. Detailed correlation of spectral data with altitude information will improve understanding of differential ablation of meteoroids in the future.

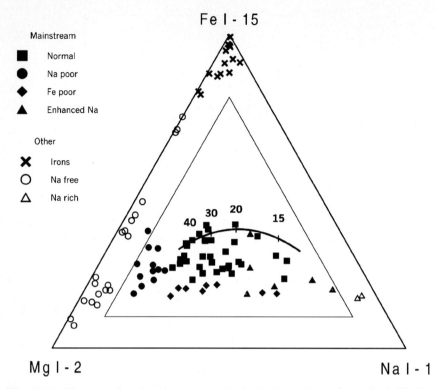

Fig. 5.14. Diagram showing the measured relative intensities of the Mg I (2), Na I (1), and Fe I (15) multiplets in 96 meteor spectra [21]. This diagram forms the basis of spectral classification of faint meteors. Different classes are marked by different symbols. The *solid curve* shows the expected range for chondritic composition as a function of meteor speed. The speeds (in km s^{-1}) are marked with numbers. For speeds larger than 40 km s^{-1}, the line ratios should not change substantially. There are three classes of them: irons, Na-free meteoroids, and Na-rich meteoroids. Among 97 meteoroids, 14 are classified as irons. Na-free meteoroids are those showing no Na line but not classified as irons are 21 meteoroids, while two Na-rich meteoroids were obtained. The meteoroids which occupy middle parts of the ternary of the diagram will be collectively called mainstream meteoroids, because they form the majority of the meteoroids and their spectra are closer to the expected chondritic spectra. Detailed classification of these meteoroids with their orbit information when compared with Tisserand parameters relative to Jupiter are published in J. Borovička et al. [21]. Surprisingly, typical asteroidal-chondritic orbits with low inclinations and aphelia inside the asteroid belt are occupied mostly by iron meteoroids

The dynamic pressure of the atmosphere at the point of disruption, which is approximately equal to the bulk crushing (mechanical) strength of the meteoroid, can be estimated by a simple equation. Though this value is still based on the assumptions of several parameters (shape, heat transfer coefficient, and density), estimated densities and strengths are summarized in

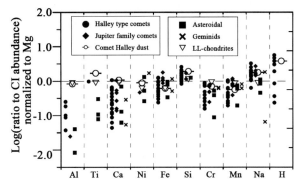

Fig. 5.15. Elemental abundances of meteoroids derived from fireball spectra categorized as Halley, Jupiter comets, asteroidal, or Geminids by their orbits compared with CI abundances (compiled by J. Borovička [22]). The fireballs are divided according to their origin, cometary or asteroidal. The volatility of the atoms increases from left to right. The in situ measured abundances of the dust of comet 1P/Halley indicated as 'Comet Halley dust' (Jessberger et al. 1988 [57]) and of LL ordinary chondrites (Wasson & Kallemeyn 1988 [102]) are plotted for comparison. Though chemical composition of cometary meteoroids does not largely differ from chondritic material, it seems that chemically homogeneous and diversity are found among active cometary meteoroids and sporadic meteoroids coming from cometary origin, respectively

Table 5.2. Trans-Jovian cometary fireballs are those that are in retrograde heliocentric orbits, have aphelia beyond 5 AU, and are associated with known meteor streams. The strength of the trans-Jovian fireballs were estimated to be 10^3–10^6 Pa (1 Pa = 10 g cm^{-1} s^{-2}). While the most fragile meteorites that survive to the Earth's surface have the strength of $\sim 10^6$ Pa.

5.11 Organics and Water in Meteoroids

Cometary meteoroids are a potential source of organic matter. It is important to understand how meteoroids supply the Earth with space matter including organics and waters. Meteors represent a unique chemical pathway toward prebiotic compounds on the early Earth and a significant fraction of organic matter is expected to survive. Searches for organics related with CHON have been made in the recent years in the spectra of meteors. Hydrogen can be seen in the high temperature component of the spectra of fast fireballs. The derived H/Mg abundance varies widely from less than in CI chondrites to somewhat more than in Halley dust (Fig. 5.15). The hydrogen may come either from water embedded in partly hydrated minerals [84] or from organic material. Atomic carbon has been positively identified in a UV Leonid spectrum from spectrographic imagers onboard the Midcourse Space Experiment (MSX) satellite [30]. Since ozone in the stratosphere strongly absorbs below

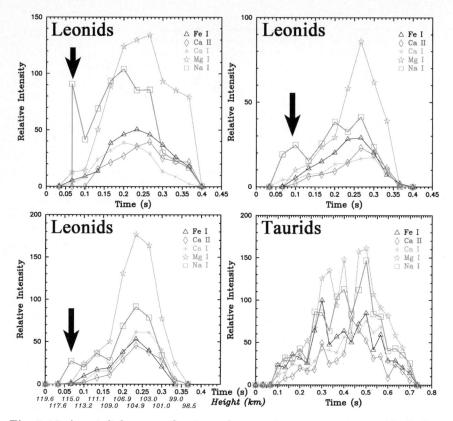

Fig. 5.16. Atomic light curve of active and non-active cometary meteoroids. Sodium releases earlier than other atoms, so-called differential ablation, are clearly seen in the Leonids (cometary meteoroids), but in the Taurids (more asteroidal meteoroids). Height (km) is indicated for a light curve (*lower left*), which was observed by triangulation method

290 nm, preventing the UV light from reaching the Earth's surface, observations from the low-Earth-orbit (LEO) provide a golden opportunity to search for new atoms and molecules including organics in the UV region. The search for the main band of CN at 388 nm in Leonid spectra was unsuccessful because of its overlap with numerous Fe lines. Russell et al. (2000) detected CO, CO_2, H_2O, and probably CH_4 in the mid-infrared spectrum of a meteor train several minutes after the fireball passage [85]. Pellinen-Wannberg et al. (2004b) reported the detection of water in an Leonid meteor but their evidence is indirect [80]. The ultraviolet band of OH at 309 nm was tentatively detected by Jenniskens et al. (2002) [52] and Abe et al. (2002, 2005) [2, 4]. Figures 5.11 and 5.17 show the UV spectrum of a Leonid fireball. UV spectra of a meteor persistent train during 2001 Leonid meteor shower also indicated possible OH emission around 309 nm [3]. If OH originates from the meteoroids, the most

Table 5.2. Bulk densities and strengths of meteoroids. S_{breaks} and S_{max} are estimated strengths at the meteor breaks and maximum brightness point, respectively

Shower	Parent body	Bulk density (kg m^{-3})		Strength* (MPa)	
		i	ii	S_{breaks}	S_{max}
Geminids	3200 Phaethon	2,900±600	1,940±700	0.022	0.077
δ-Aquarids	Marsden and Kracht group	2,400±600		0.012	0.016
Quadrantids	2003EH$_1$	1,900±200	790±1000	∼0.02	0.031
Taurids	2P/Encke	1,500±200	420±200	0.034	0.073
Perseids	109P/Swift–Tuttle	1,300±200	600±100	0.012	0.024
α-Capricornids			450±200	0.022	0.013
Leonids	55P/Tempel–Tutlle	400±100		0.006	0.007
Sporadic		2,200±300		0.010	0.021

1 Pa = 10 g cm^{-1} s^{-2} = 10 dyn cm^{-2} = 10^{-5} bar
i Babadzhanov (2002) [9]
ii Bellot Rubio et al. (2002) [11]
* Trigo-Rodríguez (2006) [96]

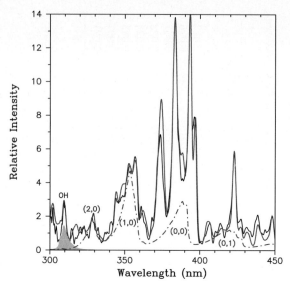

Fig. 5.17. An ultraviolet visible spectrum of a Leonid fireball between 300 and 450 nm, obtained from Fig. 5.11. It is generated by entry of a cometary meteoroid originating from 55P/Tempel–Tuttle and was investigated precisely. The observed spectrum (*black line*) is compared with a synthetic spectrum considering atoms and molecules of $N_2^+(1-)$ and OH(A-X) (*red line*) assuming local thermal equilibrium. The synthetic spectrum consists of Mg I at 383.8 nm and Ca II at 393.4 and 396.8 nm, Ca I at 422.7 nm and Mg II at 448.1 nm with numerous iron lines at the temperature of 4,500 K. The *dash-dot blue line* indicates $N_2^+(1-)$ at 10,000 K temperature. N_2^+ (0,1), (0,0), (1,0), and (2,0) bands corresponding to wavelengths of 427.8, 391.4, 353.4, and 329.3 nm, respectively. In general, meteor spectra consist of two components at different temperatures [15]. The gray filled area near 309 nm indicates OH A-X bands, which were tentatively identified. The most likely scenario of the induced $N_2^+(1-)$ in the meteoroid will result in the effect of large dimensions of high temperature regions just ahead and behind the meteoroid caused by large meteoroids' vapor cloud. Strong $N_2^+(1-)$ was observed by the spectrum of Stardust (cometary dust sample return) reentry capsule in January 2006. Hayabusa (asteroidal material sample return) reentry directly from the interplanetary space in June 2010 will be a good opportunity for artificial fireball spectroscopy tests in the future. On the other hand, the most likely mechanism for emitting OH A-X band in the meteor is caused by the dissociation of water or mineral water in the meteoroid.

likely mechanism for emitting in the OH band in the meteor and in the train is the dissociation of water or mineral water in the meteoroid. However, further observations and explanations of the emission of organics and water are needed for future confirmation.

5.12 Association Between Comets and Asteroids Through Meteoroids

The orbits of minor meteor showers have large uncertainties, owing to the observational biases, e.g., the small number of measured orbits and observational limiting magnitude. Moreover, questions to address are the composition, size and spatial distribution of dusts in the interplanetary space contributed form different sources (asteroids, comets, planets, and interstellar dust). Studying problems should be alleviated over the next decade by means of new generation observation techniques. The first integrated asteroid detection project, the Panoramatic Survey Telescope And Rapid Response System (Pan-STARRS), to repeatedly survey covering three quarters of the entire sky will discover a very large number of new near-Earth objects, ~50,000/3 years. Meteor and meteorite associations must be identified by Pan-STARRS survey. Quantitative understanding of the connection between small bodies and meteoroids will generate new insights of our solar system (see Fig. 5.18).

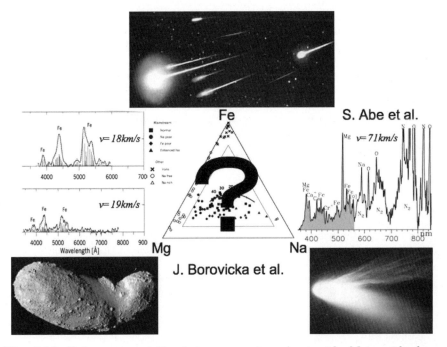

Fig. 5.18. Unknown association between comets and asteroids. Meteoroids, fragments of small bodies are key material to investigate the relationship between comets and asteroids

Acknowledgments

I would like to particularly thank Edmond Murad for many helpful comments. I would also like to thank Ingrid Mann, Tadashi Mukai, and Jiří Borovička for their constructive comments and kind advises. This research is supported by the 21st Century COE Program 'Origin and Evolution of Planetary Systems' under the Ministry of Education, Culture, Sports, Science, and Technology (MEXT) and also supported by the Grant-in-Aid for Young Scientists (B), Japan Society for the Promotion of Science (JSPS).

References

1. S. Abe, H. Yano, N. Ebizuka, and J.-I. Watanabe: *First results of high-definition TV spectroscopic observations of the 1999 Leonid meteor shower*, Earth, Moon, Planets **82/83**, 369–378 (2000)
2. S. Abe, H. Yano, N. Ebizuka, T. Kasuga, J.-I. Watanabe, M. Sugimoto, N. Fujino, T. Fuse, and R. Ogasawara: *First results of OH emission from meteor and afterglow: search for organics in cometary meteoroids*, Proc. 2002 Asteroids, Comets, Meteors, In: ESA SP **500**, Warmbein (Ed.), 213–216 (2002)
3. S. Abe, N. Ebizuka, H. Murayama, K. Ohtsuka, S. Sugimoto, M. Yamamoto, H. Yano, J.-I. Watanabe, and J. Borovička: *Video and photographic spectroscopy of 1998 and 2001 Leonid persistent trains from 300 to 930 nm*, Earth, Moon, Planets **95** (1–4), 265–277 (2004)
4. S. Abe, N. Ebizuka, H. Yano, J.-I. Watanabe, and J. Borovička: *Detection of N2+ first negative system in a bright Leonid fireball*, Astrophys. J. **618**, 141–144 (2005)
5. S. Abe, J. Borovička, K. Maeda, N. Ebizuka, and J.-I. Watanabe: *First results of quadrantid meteor spectrum*, Proc. Lunar Planetary Sci. Conf. **36**, 1536 (2005)
6. S. Abe, N. Ebizuka, H. Yano, J.-I. Watanabe, and J. Borovička: *Search for OH(A-X) and detection of N_2^+ (B-X) in ultraviolet meteor spectrum*, Advances Space Res. **39**, 538–543, (2007a)
7. S. Abe, J. Borovička, P. Koten, and Japanese Meteor Network: *Earth-grazing fireball: Caboneceous meteor from Apollo type asteroid*, Astron. Astrophys., (2007b) inpress
8. D. J. Asher: *The dynamical structure of meteor streams and meteor shower predictions*, Proc. Dynamics of Populations of Planetary Systems, In: IAU Colloquium **197**, Z. Knežević, and A. Milani (Eds.), 375–382 (2005)
9. P. B. Babadzhanov: *Fragmentation and densities of meteoroids*, Astron. Astrophyss **384**, 317–321 (2002)
10. W. J. Baggaley, D. P. Galligan: *Mapping the interstellar dust flow into the solar system*, Proc. Meteoroids 2001, In: ESA SP **495**, B. Warmbein (Ed.), 663–666 (2001)
11. L. R. Bellot Rubio, M. J. Martínez González, L. Ruiz Herrera, J. Licandro, D. Martínez-Delgado, P. Rodríguez-Gil, and M. Serra-Ricart: *Modeling the photometric and dynamical behavior of Super-Schmidt meteors in the Earth's atmosphere*, Astron. Astrophyss **389**, 680–691 (2002)

12. A. Bischoff, and J. Zipfel: *Mineralogy of the Neuschwanstein (EL6) chondrite - first results*, Proc. Lunar Planetary Sci. Conf. **34**, 1212 (2003)
13. I. D. Boyd: *Computation of atmospheric entry flow about a Leonid meteoroid*, Earth, Moon, Planets **82/83**, 93–108 (2000)
14. J. Borovička: *Line identifications in a fireball spectrum*, Astron. Astrophys. Suppl. Ser. **103**, 83–96 (1994a)
15. J. Borovička: *Two components in meteor spectra*, Planet. Space Sci. **42**, 145–150 (1994b)
16. J. Borovička, O. P. Popova, I. V. Nemtchinov, P. Spurný, and Z. Ceplecha: *Bolides produced by impacts of large meteoroids into the Earth's atmosphere: comparison of theory with observations*, Astron. Astrophys. **334**, 713–728 (1998)
17. J. Borovička, R. Štork, and J. Boček: *First results from video spectroscopy of 1998 Leonid meteors*, Meteoritics Planet. Sci. **34**, 987–994 (1999)
18. J. Borovička and P. Jenniskens: *Time resolved spectroscopy of a Leonid fireball afterglow*, Earth, Moon, Planets **82/83**, 399–428 (2000)
19. J. Borovička and P. Koten: *Three phases in the evolution of Leonid meteor trains*, Proc. 2002 International Science Symposium on the Leonid Meteor Storms, In: ISAS SP **15**, H. Yano, S. Abe, and M. Yoshikawa (Eds.), 165–173 (2003)
20. J. Borovička, P. Spurný, P. Kalenda, and E. Tagliaferri: *The Morávka meteorite fall: 1. Description of the events and determination of the fireball trajectory and orbit from video records*, Meteoritics Planet. Sci. **38**, 975–987 (2003)
21. J. Borovička, P. Koten, P. Spurný, J. Boček, and R. Štork: *A survey of meteor spectra and orbits: evidence for three populations of Na-free meteoroids*, Icarus **174**, 15–30 (2005)
22. J. Borovička: *Physical and chemical properties of meteoroids as deduced from observations*, Proc. 2005 Asteroids, Comets, Meteors, In: IAU Symposium **229**, D. Lazzaro, S. Ferraz-Mello, and J. A. Fernández (Eds.), 249–271 (2005)
23. N. Brosch, L. S. Schijvarg, M. Podolak, and M. R. Rosenkrantz: *Meteor observations from Israel*, Proc. Meteoroids 2001, In: ESA SP **495**, B. Warmbein (Ed.), 165–173 (2001)
24. P. G. Brown, Z. Ceplecha, R. L. Hawkes, G. Wetherill, M. Beech, and K. Mossman: *The orbit and atmospheric trajectory of the Peekskill meteorite from video records*, Nature **367**, 624–626 (1994)
25. P. Brown, A. R. Hildebrand, D. W. E. Green, D. Page, C. Jacobs, D. ReVelle, E. Tagliaferri, J. Wacker, and B. Wetmiller: *The fall of the St-Robert meteorite*, Meteoritics Planet. Sci. **31**, 502–517 (1996)
26. P. Brown, D. ReVelle, E. Tagliaferri, and A. Hildebrand: *An entry model for the Tagish Lake fireball using seismic, satellite, and infrasound records*, Meteoritics Planet. Sci. **37**, 661–675 (2002a)
27. P. Brown, R. Spalding, D. ReVelle, E. Tagliaferri, and S. Worden: *The flux of small near-Earth objects colliding with the Earth*, Nature **420**, 294–296 (2002b)
28. P. Brown, D. Pack, W. N. Edwards, D. ReVelle, B. Yoo, R. Spalding, and E. Tagliaferri: *The orbit, atmospheric dynamics, and initial mass of the Park Forest meteorite*, Meteoritics Planet. Sci. **39**, 1781–1796 (2004)
29. J. A. Burns, P. L. Lamy, S. Soter: *Radiation forces on small particles in the solar system*, Icarus **40**, 1–48 (1979)
30. J. F. Carbary, D. Morrison, G. J. Romick, J.-H. Yee: *Leonid meteor spectrum from 110 to 860 nm*, Icarus **161** (2), 223–234 (2003)

31. Z. Ceplecha: *Multiple fall of Příbram meteorites photographed: Double-station photographs of the fireball and their relations to the found meteorites*, Bull. Astron. Inst. Czech **12**, 21–47 (1961)

32. Z. Ceplecha: *Photographic Fireball Networks*, In: Asteroids, Comets, Meteors II, C.-I. Lagerkvist, B. A. Lindblad, H. Lundsted, and H. Rickman (Eds.), 575–582 (1986)

33. Z. Ceplecha: *Earth's influx of different populations of sporadic meteoroids from photographic and television data*, Bull. Astron. Inst. Czsl. **39**, 221–236 (1988)

34. Z. Ceplecha: *Luminous efficiency based on photographic observations of the Lost City fireball and implications for the influx of interplanetary bodies onto Earth*, Astron. Astrophys. **311**, 329–332 (1996)

35. Z. Ceplecha, J. Borovička, W. G. Elford, D. O. Revelle, R. L. Hawkes, V. Porubčan, and M. Šimek: *Meteor phenomena and bodies*, Space Sci Rev **84** (3/4), 327–471 (1998)

36. E. Chapin and E. Kudeki: *Radar interferometric imaging studies of longduration meteor echoes observed at Jicamarca*, J. Geophys. Res. **99** 8937–8949 (1994)

37. S. Close, M. Oppenheim, S. Hunt, and A. Coster: *A technique for calculating meteor plasma density and meteoroid mass from radar head echo scattering*, Icarus **168**, 43–52 (2004)

38. A. A. Christou: *Predicting Martian and Venusian meteor shower activity*, Earth, Moon, Planets **95** (1–4), 425–431 (2004)

39. G. Cremonse, M. Fulle, F. Marzari, and V. Vanzani: *Orbital evolution of meteoroids from short period comets*, Astron. Astrophys. **324**, 770–777 (1997)

40. D. W. Dunham, B. Cudnik, D. M. Palmer, P. V. Sada, J. Melosh, M. Frankenberger, R. Beech, L. Pellerin, R. Venable, D. Asher, R. Sterner, B. Gotwols, D. Wun, D. Stockbauer: *The first confirmed videorecordings of lunar meteor impacts*, Proc. Lunar Planet. Sci. XXXI. **1547**, (2000)

41. J. Evans: *Radar observations of meteor deceleration*, J. Geophys. Res. **71**, 171–188 (1996)

42. Y. Fujiwara, M. Ueda , Y. Shiba, M. Sugimoto, M. Kinoshita, C. Shimoda, and T. Nakamura: *Meteor luminosity at 160 km altitude from TV observations for bright Leonid meteors*, Geophys. Res. Lett. **25**, 285–288 (1998)

43. E. Grün, H. A. Zook, H. Fechtig, R. H. Giese: *Collisional balance of the meteoritic complex*, Icarus **62**, 244–272 (1985)

44. E. Grün, V. Dikarev, H. Krüger, and M. Landgraf: *Space dust measurements*, In: Meteors in the Earth's Atmosphere, E. Murad and I. P. Williams (Eds.), 35–75 (2002)

45. I. Halliday, A. T. Blackwell, and A. A. Griffin: *The Innisfree meteorite and the Canadian camera network*, J. Royal Astron. Soc. Can. **72**, 15–39 (1978)

46. M. Ishiguro, Y. Sarugaku, M. Ueno, N. Miura, F. Usui, M.-Y. Chunm, and S. M. Kwon: *Dark red debris from three short-period comets: 2P/Encke, 22P/Kopff, and 65P/Gunn*, Icarus **189** (1), 169–183 (2007)

47. H. Ishimoto: *Modeling the number density distribution of interplanetary dust on the ecliptic plane within 5AU of the sun*, Astron. Astrophys. **362**, 1158–1173 (2000)

48. L. G. Jacchia, and F. L. Whipple: *Precision orbits of 413 photographic meteors*, Smithsonian Contr. Astrophys. **4**, 97–129 (1961)

49. D. Janches and J. L. Chau: *Observed diurnal and seasonal behavior of the micrometeor flux using the Arecibo and Jicamarca radars*, J. Atmospheric, and Solar-Terrestrial Phys., **67**, 1196–1210 (2005)
50. P. Jenniskens, S. J. Butow, and M. Fonda: *The 1999 Leonid Multi-Instrument Aircraft Campaign – an early review*, Earth, Moon, Planets, **82/83**, 1–26 (2000)
51. P. Jenniskens, M. A. Wilson, D. Packan, C. O. Laux, C. H. Krger, I. D. Boyd, O. P. Popova, and M. Fonda: *Meteors: A delivery mechanism of organic matter to the early Earth*, Earth, Moon, and Planets, **82/83**, 57–70 (2000)
52. P. Jenniskens, E. Tedesco, J. Murthy, C. O. Laux, and S. Price: *Spaceborne ultraviolet 251-384 nm spectroscopy of a meteor during the 1997 Leonid shower*, Meteoritics Planet. Sci. **37**, 1071–1078 (2002)
53. P. Jenniskens and H. C. Stenbaek-Nielsen: *Meteor wake in high frame-rate images-implications for the chemistry of ablated organic compounds*, Astrobiology **4** (1), 95–108 (2004a)
54. P. Jenniskens: *2003 EH₁ is the Quadrantid shower parent comet*, Astron. J. **127**, 3018–3022 (2004b)
55. P. Jenniskens and E. Lyytinen: *Meteor showers from the debris of broken comets: D/1819 W1 (Blanpain), 2003 WY25, and the Phoenicids*, Astron. J. **130**, 1286–1290 (2005)
56. P. Jenniskens: *Quantitative meteor spectroscopy: Elemental abundances*, Advances Space Res **39**, 491–512 (2007)
57. E. K. Jessberger, A. Christoforidis, and A. Kissel: *Aspects of the major element composition of Halley's dust*, Nature **322**, 691–695 (1988)
58. T. Kasuga, T. Yamamoto, J.-I. Watanabe, N. Ebizuka, H. Kawakita, and H. Yano: *Metallic abundances of the 2002 Leonid meteor deduced from the high-definition TV spectra*, Astron. Astrophys. **435**, 341–351 (2005)
59. P. Koten, J. Borovička, P. Spurný, S. Evans, R. Štork, and A. Elliott: *Double station and spectroscopic observations of the Quadrantid meteor shower and the implications for its parent body*, Mon. Not. R. Astron. Soc. **366** (4), 1367–1372 (2006)
60. P. Koten, P. Spurný, J. Borovička, S. Evans, A. Elliott, H. Betlem, R. Štork, and K. Jobse: *The beginning heights and light curves of high-altitude meteors*, Meteoritics Planet. Sci. **41** (9), 1305–1320 (2006)
61. L. Kresák: *Cometary dust trails and meteor storms*, Astron. Astrophys. **279** (2), 646–660 (1993)
62. S. G Love and D. E. Brownlee: *A direct measurement of the terrestrial mass accretion rate of cosmic dust*, Science **262** (5133), 550–553 (1993)
63. J. Llorca, J. M. Trigo-Rodríguez, J. L. Ortiz, J. A. Docobo, J. Garciá-Guinea, A. J. Castro-Tirado, A. E. Rubin, O. Eugster, W. Edwards, M. Laubenstein, and I. Casanova: *The Villalbeto de la Peña meteorite fall: I. Fireball energy, meteorite recovery, strewn field, and petrography*, Meteoritics Planet. Sci. **40**, 795–804 (2005)
64. I. Mann, H. Kimura, D. A. Biesecker, B. T. Tsurutani, E. Grün, R. B. McKibben, J.-C. Liou, R. M. MacQueen, T. Mukai, M. Guhathakurta, and P. Lamy: *Dust near the sun*, Space Sci. Rev. **110** (3), 269–305 (2004)
65. J. D. Mathews, D. D. Meisel, K. P. Hunter, V. S. Getman, Q. Zhou: *Very high resolution studies of micrometeors using the Arecibo 430 MHz radar*, Icarus **126**, 157–169 (1997)
66. R. E. McCrosky and A. Posen: *Orbital elements of photographic meteors*, Smithsonian Contr. Astrophys. **4**, 15–84 (1961)

67. R. E. McCrosky, A. Posen, G. Schwartz, and C.-Y. Shao: *The Lost City mete-orite: Its recovery and a comparison with fireballs*, J. Geophys. Res. **76**, 4090–4108 (1971)
68. D. W. R. McKinley and P. M. Millman: *A phenomenological theory of radar echoes from meteors*, Proc. I.R.E., **37**, 364–375 (1949)
69. D. W. R. McKinley: *Meteor Science and Engineering* (McGraw-Hill, New York, 1961)
70. R. H. McNaught, and D. J. Asher: *Leonid dust trails and meteor storms*, WGN, J. IMO, **27** (2), 85–102 (1999)
71. W. J. McNeil, S. T. Lai, and E. Murad: *Models of thermospheric sodium, calcium and magnesium at the magnetic equator*, Advances Space Res. **21** (6), 863–866 (1998)
72. P. E. Millman, A. F. Cook, and C. L. Hemenway: *Spectroscopy of Perseid meteors with an image orthicon*, Canadian J. Phys. **49**, 1365–1373 (1971)
73. T. Mukai: *Cometary dust and interplanetary particles*, Proc. the International School of Physics 'Enrico Fermi' Course CI, In: Evolution of Interstellar Dust and Related Topics, A. Bonetti, J. M. Greenberg, and S. Aiello (Eds.), 397 (1989)
74. R. Nakamura, Y. Fujii, M. Ishiguro, K. Morishige, S. Yokogawa, P. Jenniskens, and T. Mukai: *The discovery of a faint glow of scattered sunlight from the dust trail of the Leonid parent comet 55P/Tempel–Tuttle*, Astrophys. J. **540** (2), 1172–1176 (2000)
75. D. Nesvorný, D. Vokrouhlický, W. F. Bottke, and M. V. Sykes: *Physical properties of asteroid dust bands and their sources*, Icarus **181**, 107–144 (2006)
76. K. Ohtsuka, T. Sekiguchi, D. Kinoshita, J.-I. Watanabe, T. Ito, H. Arakida, and T. Kasuga: *Apollo asteroid 2005 UD: split nucleus of (3200) Phaethon?*, Astron. Astrophys. **450**, 25–28 (2006)
77. K. Ohtsuka, S. Abe, M. Abe, H. Yano, and J.-I. Watanabe: *Are There Meteors Originated from Near Earth Asteroid (25143) Itokawa?*, Proc. 2nd Hayabusa Symposium, (2007) inpress
78. J. L. Ortiz, P. V. Sada, L. R. Bellot Rubio, F. J. Aceituno, J. Aceituno, P. J. Gutierrez, and U. Thiele: *Optical detection of meteoroidal impacts on the Moon*, Nature **405**, 921–923 (2000)
79. A. Pellinen-Wannberg and G. Wannberg: *Meteor observations with the European incoherent scatter UHF radar*, J. Geophys. Res. textbf99, 11379–11390 (1994)
80. A. Pellinen-Wannberg, E. Murad, B. Gustavsson, U. Brändström, C.-F. Enell, C. Roth, I. P. Williams, and A. Steen: *Optical observations of water in Leonid meteor trails*, Geophys. Ress Lett. **31** (3), L03812 (2004)
81. O. P. Popova, A. S. Strelkov, and S. N. Sidneva: *Sputtering of fast meteoroids' surface*, Advances Space Res. **39**, 567–573 (2007)
82. W. T. Reach, M. V. Sykes, D. Lien, and J. K. Davies: *The formation of Encke meteoroids and dust trail*, Icarus **148** (1), 80–94 (2000)
83. D. ReVelle, P. Brown, and P. Spurný: *Entry dynamics and acoustics/infrasonic/seismic analysis for the Neuschwanstein meteorite fall*, Meteoritics Planet. Sci. **39**, 1605–1626 (2004)
84. F. J. M. Rietmeijer: *Interplanetary dust and carbonaceous meteorites: Constraints on porosity, mineralogy and chemistry of meteors from rubble-pile planetesimals*, Earth, Moon, Planets **95** (1–4), 321–338 (2004)

85. R. W. Russell, G. S. Rossano, M. A. Chatelain, D. K. Lynch, T. K. Tessensohn, E. Abendroth, D. Kim, and P. Jenniskens: *Mid-infrared spectroscopy of persistent Leonid trains*, Earth, Moon, Planets **82/83**, 439–456 (2000)
86. T. Sato, T. Nakamura, and K. Nishimura: *Orbit determination of meteors using the MU radar*, IEICE Trans. Commun. **E83-B** (9), 1990–1995 (2000)
87. P. Spurný, H. Betlem, K. Jobse, P. Koten, and J. Van't Leven: *New type of radiation of bright Leonid meteors above 130 km*, Meteoritics Planet. Sci. **35**, 1109–1115 (2000)
88. P. Spurný and J. Borovička: *The autonomous all-sky photographic camera for meteor observation*, Proc. 2002 Asteroids, Comets, Meteors, In ESA SP **500** B. Warmbein (Ed.), 257–259 (2002)
89. P. Spurný, J. Oberst, and D. Heinlein: *Photographic observations of Neuschwanstein, a second meteorite from the orbit of the Příbram chondrite*, Nature **423**, 151–153 (2003)
90. H. C. Stenbaek-Nielsen and P. Jenniskens: *Leonid at 1000 frames per second*, Proc. 2002 International Science Symposium on the Leonid Meteor Storms, In: ISAS SP **15**, H. Yano, S. Abe, and M. Yoshikawa (Eds.), 207–214 (2003)
91. H. C. Stenbaek-Nielsen and P. Jenniskens: *A 'shocking' Leonid meteor at 1000 fps*, Advances Space Res. **33**, 1459–1465 (2004)
92. A. D. Taylor, W. J. Baggaley, and D. I. Steel: *Discovery of interstellar dust entering the Earth's atmosphere*, Nature **380**, 323–325 (1996)
93. M. J. Taylor, L. C. Gardner, I. S. Murray, P. Jenniskens: *Jet-like structures and wake in Mg I (518 nm) images of 1999 Leonid storm meteors*, Earth, Moon, and Planets **82/83**, 379–389 (2000)
94. J. M. Trigo-Rodríguez, J. Llorca, J. Borovička, and J. Fabregat: *Chemical abundances determined from meteor spectra: I. Ratios of the main chemical elements*, Meteoritics Planet. Sci. **38**, 1283–1294 (2003)
95. J. M. Trigo-Rodríguez, H. Betlem, and E. Lyytinen: *Leonid meteoroid orbits perturbed by collisions with interplanetary dust*, Astron. J. **621**, 1146–1152 (2005)
96. J. M. Trigo-Rodríguez and J. Llorca: *The strength of cometary meteoroids: clues to the structure and evolution of comets*, Mon. Not. R. Astron. Soc. **372**, 655–660 (2006a)
97. J. M. Trigo-Rodríguez, J. Borovička, P. Spurný, J. L. Ortiz, J. A. Docobo, A. J. Castro-Tirado, and J. Llorca: *The Villalbeto de la Peña meteorite fall: II. Determination of atmospheric trajectory and orbit*, Meteoritics Planet. Sci. **41** (4), 505–517 (2006b)
98. J. M. Trigo-Rodríguez, J. M. Madiedo, A. J. Castro-Tirado, J. L. Ortiz, P. S. Gural, J. Llorca, J. Fabregat, S. Vítek, P. Pujols, and B. Troughton: *Spanish meteor network: 2006 all-sky and video monitoring highlights*, Proc. Lunar Planetary Sci. Conf. **38**, 1584 (2007)
99. J. Vaubaillon, F. Colas, and L. Jorda: *A new method to predict meteor showers. I. Description of the model*, Astron. Astrophys. **439** (2), 751–760 (2005)
100. D. Vinković: *Thermalization of sputtered particles as the source of diffuse radiation from high altitude meteors*, Advances Space Res. **39**, 574–582, (2007)
101. U. von Zahn, M. Gerding, J. Höffner, W. J. McNeil, and E. Murad: *Iron, calcium, and potassium atom densities in the trails of Leonids and other meteors: Strong evidence for differential ablation*, Meteoritics Planet. Sci. **34** (6), 1017–1027 (1999)

102. J. T. Wasson and G. W. Kallemeyn: *Compositions of chondrites*, Phil. Trans. R. Soc. London A **325**, 535–544 (1988)

103. J.-I. Watanabe, S. Abe, M. Takanashi, T. Hashimoto, O. Iiyama, Y. Ishibashi, K. Morishige, and S. Yokogaw: *HD TV observation of the strong activity of the Giacobinid meteor shower in 1998*, Geophys. Res. Lett. **26**, 1117–1120 (1998)

104. J.-I. Watanabe, M. Sato, and T. Kasuga: *Phoenicids in 1956 revised*, Publ. Astron. Soc. Japan **57**, 45–49 (2005)

105. F. L. Whipple: *A Comet model. I. The acceleration of comet Encke*, Astrophys. J. **111**, 375–394 (1950)

106. M. Yanagisawa and N. Kisaichi: *Lightcurves of 1999 Leonid impact flashes on the Moon*, Icarus **159**, 31–38 (2002)

107. M. Yanagisawa, K. Ohnishi, Y. Takamura, H. Masuda, Y. Sakai, M. Ida, M. Adachi, and M. Ishida: *The first confirmed Perseid lunar impact flash*, Icarus **182** (2), 489–495 (2006)

108. K. Yanai: *Yamato-74 meteorites collection, Antarctica, from November to December 1974*, Mem. Natl. Inst. Polar Res. **8**, 1–37 (1978)

6

Optical Properties of Dust

A. Li

Department of Physics and Astronomy, University of Missouri,
Columbia, MO 65211, USA
LiA@missouri.edu

Abstract Except in a few cases, cosmic dust can be studied in situ or in terrestrial laboratories, essentially all of our information concerning the nature of cosmic dust depends on its interaction with electromagnetic radiation. This chapter presents the theoretical basis for describing the optical properties of dust—how it absorbs and scatters starlight and reradiates the absorbed energy at longer wavelengths.

6.1 Introduction

Dust is everywhere in the Universe: it is a ubiquitous feature of the cosmos, impinging directly or indirectly on most fields of modern astronomy. It occurs in a wide variety of astrophysical regions, ranging from the local environment of the Earth to distant galaxies and quasars: from meteorites originated in the asteroid belt, the most pristine solar system objects—comets, and stratospherically collected interplanetary dust particles (IDPs) of either cometary or asteroidal origin to external galaxies (both normal and active, nearby and distant) and circumnuclear tori around active galactic nuclei; from circumstellar envelopes around evolved stars (cool red giants, AGB stars) and Wolf-Rayet stars, planetary nebulae, nova and supernova ejecta, and supernova remnants to interstellar clouds, and star-forming regions; from the terrestrial zodiacal cloud to protoplanetary disks around young stellar objects and debris disks around main-sequence stars, etc.

Dust plays an increasingly important role in astrophysics. It has a dramatic effect on the Universe by affecting the physical conditions and processes taking place within the Universe and shaping the appearance of dusty objects (e.g., cometary comae, reflection nebulae, dust disks, and galaxies) (i) as an absorber, scatterer, polarizer, and emitter of electromagnetic radiation; (ii) as a revealer of heavily obscured objects (e.g., IR sources) of which we might otherwise be unaware; (iii) as a driver for the mass loss of evolved stars; (iv) as a sink of heavy elements which if otherwise in the gas phase, would profoundly affect the interstellar gas chemistry; (v) as an efficient catalyst for

Li, A.: *Optical Properties of Dust.* Lect. Notes Phys. **758**, 167–188 (2009)
DOI 10.1007/978-3-540-76935-4_6   © Springer-Verlag Berlin Heidelberg 2009

the formation of H_2 and other simple molecules as well as complex organic molecules (and as a protector by shielding them from photodissociating ultraviolet [UV] photons) in the interstellar medium (ISM); (vi) as an efficient agent for heating the interstellar gas by providing photoelectrons; (vii) as an important coolant in dense regions by radiating infrared (IR) photons (which is particularly important for the process of star formation in dense clouds by removing the gravitational energy of collapsing clouds and allowing star formation to take place); (viii) as an active participant in interstellar gas dynamics by communicating radiation pressure from starlight to the gas, and providing coupling of the magnetic field to the gas in regions of low fractional ionization; (ix) as a building block in the formation of stars and planetary bodies; and finally, (x) as a diagnosis of the physical conditions (e.g., gas density, temperature, radiation intensity, electron density, magnetic field) of the regions where dust is seen.

The dust in the space between stars—interstellar dust—is the most extensively studied cosmic dust type, with circumstellar dust ("stardust"), cometary dust and IDPs coming second. Interstellar dust is an important constituent of the Milky Way and external galaxies. The presence of dust in galaxies limits our ability to interpret the local and distant Universe because dust extinction dims and reddens the galaxy light in the UV-optical-near-IR windows, where the vast majority of the astronomical data have been obtained. In order to infer the stellar content of a galaxy, or the history of star formation in the Universe, it is essential to correct for the effects of interstellar extinction. Dust absorbs starlight and reradiates at longer wavelengths. Nearly half of the bolometric luminosity of the local Universe is reprocessed by dust into the mid- and far-IR.

Stardust, cometary dust, and IDPs are directly or indirectly related to interstellar dust: stardust, condensed in the cool atmospheres of evolved stars or supernova ejecta and subsequently injected into the ISM, is considered as a major source of interstellar dust, although the bulk of interstellar dust is not really stardust but must have recondensed in the ISM [11]. Comets, because of their cold formation and cold storage, are considered as the most primitive objects in the solar system and best preserve the composition of the presolar molecular cloud among all solar system bodies. Greenberg [20] argued that comets are made of unaltered pristine interstellar materials with only the most volatile components partially evaporated, although it has also been proposed that cometary materials have been subjected to evaporation, recondensation, and other reprocessing in the protosolar nebula, and therefore have lost all the records of the presolar molecular cloud out of which they have formed. Genuine presolar grains have been identified in IDPs and primitive meteorites based on their isotopic anomalies [8], indicating that stardust can survive journeys from its birth in stellar outflows and supernova explosions, through the ISM, the formation of the solar system, and its ultimate incorporation into asteroids and comets.

Except in a few cases, cosmic dust can be studied in situ (e.g., cometary dust [32], dust in the local interstellar cloud entering our solar system [22]) or in terrestrial laboratories (e.g., IDPs and meteorites [8], cometary dust [7]), our knowledge about cosmic dust is mainly derived from its interaction with electromagnetic radiation: extinction (scattering and absorption), polarization, and emission. Dust reveals its presence and physical and chemical properties, and provides clues about the environment where it is found by scattering and absorbing starlight (or photons from other objects) and reradiating the absorbed energy at longer wavelengths.

This chapter deals with the optical properties of dust, i.e., how light is absorbed, scattered, and reradiated by cosmic dust. The subject of light scattering by small particles is a vast, fast-developing field. In this chapter, I restrict myself to astrophysically-relevant topics. In Sect. 6.2, I present a brief summary of the underlying physics of light scattering. The basic scattering terms are defined in Sect. 6.3. In Sect. 6.4, I discuss the physical basis of the dielectric functions of dust materials. In Sect. 6.5 and Sect. 6.6, I summarize the analytic and numerical solutions for calculating the absorption and scattering parameters of dust, respectively.

6.2 Scattering of Light by Dust: A Conceptual Overview

When a dust grain, composed of discrete electric charges, is illuminated by an electromagnetic wave, the electric field of the incident electromagnetic wave will set the electric charges in the dust into oscillatory motion. These accelerated electric charges radiate electromagnetic energy in all directions, at the same frequency as that of the incident wave. This process, known as "scattering" (to be more precise, elastic scattering), removes energy from the incident beam of electromagnetic radiation. Absorption also arises as the excited charges transform part of the incident electromagnetic energy into thermal energy. The combined effect of absorption and scattering, known as "extinction," is the total energy loss of the incident wave [6].

The scattering of light by dust depends on the size, shape, and chemical composition of the dust and the direction at which the light is scattered. This can be qualitatively understood, as schematically shown in Fig. 6.1, by conceptually subdividing the dust into many small regions, in each of which a dipole moment will be induced when illuminated by an incident electromagnetic wave. These dipoles oscillate at the frequency of the incident wave and therefore scatter secondary radiation in all directions. The total scattered field of a given direction (e.g., P_1, P_2) is the sum of the scattered wavelets, with their phase differences taken into account. Since these phase relations change for a different scattering direction, the scattered field varies with scattering

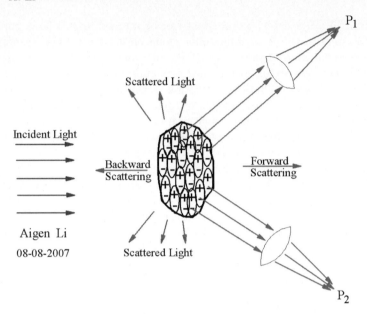

Fig. 6.1. A conceptual illustration of the scattering of light by dust. The dust is conceptually subdivided into many small regions. Upon illuminated by an incident electromagnetic wave, in each small region a dipole moment (of which the amplitude and phase depend on the composition of the dust) will be induced. These dipoles oscillate at the frequency of the incident wave and scatter secondary radiation in all directions. The total scattered field of a given direction (e.g., P_1, P_2) is the sum of the scattered wavelets, where due account is taken of their phase differences. Since the phase relations among the scattered wavelets change with the scattering direction and the size and shape of the dust, the scattering of light by dust depends on the size, shape, and chemical composition of the dust and the direction at which the light is scattered

direction.[1] The phase relations among the scattered wavelets also change with the size and shape of the dust. On the other hand, the amplitude and phase of the induced dipole moment for a given frequency depend on the material of which the dust is composed. Therefore, the scattered field is sensitive to the size, shape, and chemical composition of the dust [6].

6.3 Scattering of Light by Dust: Definitions

As discussed in Sect. 6.2, when light impinges on a grain it is either scattered or absorbed. Let $I_o(\lambda)$ be the intensity of the incident light at wavelength λ, the

---

[1] An exception to this is the dust in the Rayleigh regime (i.e., with its size being much smaller than the wavelength) which scatters light nearly isotropically, with little variation with direction since for dust so small all the secondary wavelets are approximately in phase.

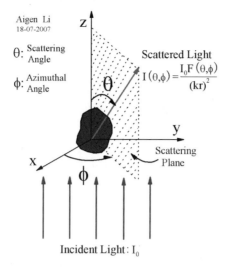

Fig. 6.2. Schematic scattering geometry of a dust grain in an incident radiation field of intensity I_0 which scatters radiation of intensity $I(\theta,\phi)$ into a scattering angle θ ($\theta = 0°$: forward scattering; $\theta = 180°$: backward scattering), an azimuthal angle ϕ, and a distance r from the dust. In a Cartesian coordinate system, the incident direction defines the $+z$ axis. The scattering and the incident directions define the scattering plane. In the far-field region (i.e., $kr \gg 1$), $I = I_0 \, F(\theta,\phi)/k^2 r^2$, where $k = 2\pi/\lambda$ is the wave number in vacuum

intensity of light scattered into a direction defined by θ and ϕ (see Fig. 6.2) is

$$I(\lambda) = \frac{I_0(\lambda) \, F(\theta, \phi)}{k^2 \, r^2} \tag{6.1}$$

where $0° \leq \theta \leq 180°$ is the *scattering angle* (the angle from the incident direction), $0° \leq \phi \leq 360°$ is the azimuthal angle which uniquely determines the *scattering plane* defined by the incident direction and the scattering directions (see Fig. 6.2),[2] $F(\theta, \phi)$ is the (dimensionless) angular scattering function, $r \gg \lambda/2\pi$ is the distance from the scatterer, and $k = 2\pi/\lambda$ is the wave number in vacuum. The *scattering cross-section* C_{sca}, defined as the area on which the incident wave falls with the same amount of energy as that scattered in all directions by the dust, may be obtained by integrating the angular scattering distribution $F(\theta, \phi)/k^2$ over all solid angles

$$C_{\mathrm{sca}}(\lambda) = \frac{1}{k^2} \int_0^{2\pi} \int_0^\pi F(\theta, \phi) \sin\theta \, d\theta \, d\phi \quad , \tag{6.2}$$

---

[2] When the scattering is along the incident direction ($\theta = 0°$, i.e., "forward scattering") or the scattering is on the opposite direction of the incident direction ($\theta = 180°$, i.e., "backward scattering"), any plane containing the z axis is a suitable scattering plane.

where $F(\theta, \phi)/k^2$ (with a dimension of area),[3] after normalized by C_{sca}, is known as the *phase function* or *scattering diagram*

$$p(\theta, \phi) \equiv \frac{F(\theta, \phi)/k^2}{C_{sca}} \ . \tag{6.3}$$

The *asymmetry parameter* (or *asymmetry factor*) g is defined as the average cosine of the scattering angle θ

$$g \equiv \langle \cos \theta \rangle = \int_{4\pi} p(\theta, \phi) \cos \theta \, d\Omega = \frac{1}{k^2 C_{sca}} \int_0^{2\pi} \int_0^{\pi} F(\theta, \phi) \cos \theta \sin \theta \, d\theta \, d\phi \ . \tag{6.4}$$

The *asymmetry parameter* g, specifying the degree of scattering in the forward direction ($\theta = 0°$), varies from -1 (i.e., all radiation is backward scattered like a "mirror") to 1 (for pure forward scattering). If a grain scatters more light toward the forward direction, $g > 0$; $g < 0$ if the scattering is directed more toward the back direction; $g = 0$ if it scatters light isotropically (e.g., small grains in the Rayleigh regime) or if the scattering is symmetric with respect to $\theta = 90°$ (i.e., the scattered radiation is azimuthal independent and symmetric with respect to the plane perpendicular to the incident radiation).

In radiative transfer modeling of dusty regions, astronomers often use the empirical Henyey–Greenstein phase function to represent the anisotropic scattering properties of dust [26]

$$H(\theta) \equiv \frac{1}{4\pi} \frac{1 - g^2}{(1 + g^2 - 2g \cos \theta)^{3/2}} \ . \tag{6.5}$$

Draine (2003) proposed a more general analytic form for the phase function

$$H_\eta(\theta) \equiv \frac{1}{4\pi} \frac{3 (1 - g^2)}{3 + \eta (1 + 2g^2)} \frac{1 + \eta \cos^2 \theta}{(1 + g^2 - 2g \cos \theta)^{3/2}} \tag{6.6}$$

where η is an adjustable parameter. For $\eta = 0$ (6.6) reduces to the Henyey–Greenstein phase function. For $g = 0$ and $\eta = 1$ (6.6) reduces to the phase function for Rayleigh scattering [12].

As discussed in Sect. 6.2, both scattering and absorption (the sum of which is called *extinction*) remove energy from the incident beam. The *extinction cross-section*, defined as

$$C_{ext} = C_{sca} + C_{abs} = \frac{\text{total energy scattered and absorbed per unit time}}{\text{incident energy per unit area per unit time}} \ , \tag{6.7}$$

---

[3] Also called the "differential scattering cross-section" $dC_{sca}/d\Omega \equiv F(\theta, \phi)/k^2$, it specifies the angular distribution of the scattered light (i.e., the amount of light [for unit incident irradiance] scattered into a unit solid angle about a given direction).

is determined from the *optical theorem* which relates C_{ext} to the real part of the complex scattering amplitude $S(\theta, \phi)^4$ in the forward direction alone [28]

$$C_{\text{ext}} = -\frac{4\pi}{k^2} \operatorname{Re}\{S(\theta = 0^\circ)\} \ . \tag{6.8}$$

The *absorption cross-section* C_{abs} is the area on which the incident wave falls with the same amount of energy as that absorbed inside the dust; C_{ext}, having a dimension of area, is the "effective" blocking area to the incident radiation (for grains much larger than the wavelength of the incident radiation, C_{ext} is about twice the geometrical blocking area). For a grain (of size a and complex index of refraction m) in the Rayleigh limit (i.e., $2\pi a/\lambda \ll 1$, $2\pi a|m|/\lambda \ll 1$), the absorption cross-section C_{abs} is much larger than the scattering cross-section C_{sca} and therefore $C_{\text{ext}} \approx C_{\text{abs}}$. Non-absorbing dust has $C_{\text{ext}} = C_{\text{sca}}$.

The *albedo* of a grain is defined as $\alpha \equiv C_{\text{sca}}/C_{\text{ext}}$. For grains in the Rayleigh limit, $\alpha \approx 0$ since $C_{\text{sca}} \ll C_{\text{abs}}$. For non-absorbing dust, $\alpha = 1$.

In addition to energy, light carries momentum of which the direction is that of propagation and the amount is $h\nu/c$ (where h is the Planck constant, c is the speed of light, and ν is the frequency of the light). Therefore, upon illuminated by an incident beam of light, dust will acquire momentum and a force called *radiation pressure* will be exerted on it in the direction of propagation of the incident light. The radiation pressure force is proportional to the net loss of the forward component of the momentum of the incident beam. While the momentum of the *absorbed* light (which is in the forward direction) will be transferred to the dust, the forward component of the momentum of the *scattered* light will not be removed from the incident beam. Therefore, the radiation pressure force exerted on the dust is

$$F_{\text{pr}} = I_o C_{\text{pr}}/c \ , \quad C_{\text{pr}} = C_{\text{abs}} + (1 - g)\, C_{\text{sca}}, \tag{6.9}$$

where C_{pr} is the radiation pressure cross-section and I_o is the intensity (irradiance) of the incident light.

In literature, one often encounters Q_{ext}, Q_{sca}, Q_{abs}, and Q_{pr}—the extinction, scattering, absorption, and radiation pressure efficiencies. They are defined as the extinction, scattering, absorption, and radiation pressure cross-sections divided by the geometrical cross-section of the dust C_{geo},

$$Q_{\text{ext}} = \frac{C_{\text{ext}}}{C_{\text{geo}}} \ ; \quad Q_{\text{sca}} = \frac{C_{\text{sca}}}{C_{\text{geo}}} \ ; \quad Q_{\text{abs}} = \frac{C_{\text{abs}}}{C_{\text{geo}}} \ ; \quad Q_{\text{pr}} = \frac{C_{\text{pr}}}{C_{\text{geo}}} \ . \tag{6.10}$$

For spherical grains of radii a, $C_{\text{geo}} = \pi a^2$. For non-spherical grains, there is no uniformity in choosing C_{geo}. A reasonable choice is the geometrical cross-section of an "equal volume sphere" $C_{\text{geo}} \equiv \pi\, (3V/4\pi)^{2/3} \approx 1.21\, V^{2/3}$, where V is the volume of the non-spherical dust.

---

[4] The angular scattering function $F(\theta, \phi)$ is just the absolute square of the complex scattering amplitude $S(\theta, \phi)$: $F(\theta, \phi) = |S(\theta, \phi)|^2$.

6.4 Scattering of Light by Dust: Dielectric Functions

The light scattering properties of dust are usually evaluated based on its optical properties (i.e., *dielectric functions* or *indices of refraction*) and geometry (i.e., size and shape) by solving the Maxwell equations

$$\nabla \times \mathbf{E} + \frac{1}{c}\frac{\partial \mathbf{B}}{\partial t} = 0; \nabla \times \mathbf{H} - \frac{1}{c}\frac{\partial \mathbf{D}}{\partial t} = \frac{4\pi}{c}\mathbf{J}; \nabla \cdot \mathbf{D} = 4\pi\rho; \nabla \cdot \mathbf{B} = 0; \quad (6.11)$$

where $\mathbf{E}$ is the electric field, $\mathbf{B}$ is the magnetic induction, $\mathbf{H}$ is the magnetic field, $\mathbf{D}$ is the electric displacement, $\mathbf{J}$ is the electric current density, and ρ is the electric charge density. They are supplemented with the *constitutive relations* (or "material relations")

$$\mathbf{J} = \sigma\mathbf{E}; \ \mathbf{D} = \varepsilon\mathbf{E} = (1 + 4\pi\chi)\mathbf{E}; \mathbf{B} = \mu\mathbf{H}; \quad (6.12)$$

where σ is the *electric conductivity*, $\varepsilon = 1 + 4\pi\chi$ is the *dielectric function* (or *permittivity* or **dielectric permeability**), χ is the *electric susceptibility*, and μ is the *magnetic permeability*. For time-harmonic fields $\mathbf{E}, \mathbf{H} \propto \exp(-\mathrm{i}\omega t)$, the Maxwell equations are reduced to the Helmholtz wave equations

$$\nabla^2\mathbf{E} + \tilde{k}^2\mathbf{E} = 0; \ \nabla^2\mathbf{H} + \tilde{k}^2\mathbf{H} = 0; \quad (6.13)$$

where $\tilde{k} = m\omega/c$ is the complex wave number, $m = \sqrt{\mu(\varepsilon + \mathrm{i}4\pi\sigma/\omega)}$ is the complex refractive index, and $\omega = 2\pi c/\lambda$ is the circular frequency. These equations should be considered for the field outside the dust (which is the superposition of the incident field and the scattered field) and the field inside the dust, together with the boundary conditions (i.e., any tangential and normal components of $\mathbf{E}$ are continuous across the dust boundary). For non-magnetic dust ($\mu = 1$), the complex refractive index is $m = \sqrt{\varepsilon + \mathrm{i}4\pi\sigma/\omega}$. For non-magnetic, non-conducting dust, $m = \sqrt{\varepsilon}$. For highly-conducting dust, $4\pi\sigma/\omega \gg 1$, therefore, both the real part and the imaginary part of the index of refraction are approximately $\sqrt{2\pi\sigma/\omega}$ at long wavelengths.

The complex refractive indices m or dielectric functions ε of dust are often called *optical constants*, although they are not constant but vary with wavelengths. They are of great importance in studying the absorption and scattering of light by dust. They are written in the form of $m = m' + \mathrm{i}m''$ and $\varepsilon = \varepsilon' + \mathrm{i}\varepsilon''$. The sign of the imaginary part of m or ε is opposite to that of the time-dependent term of the harmonically variable fields [i.e., m'', $\varepsilon'' > 0$ for $\mathbf{E}, \mathbf{H} \propto \exp(-\mathrm{i}\omega t)$; m'', $\varepsilon'' < 0$ for $\mathbf{E}, \mathbf{H} \propto \exp(\mathrm{i}\omega t)$]. The imaginary part of the index of refraction characterizes the attenuation of the wave ($4\pi m''/\lambda$ is called the *absorption coefficient*), while the real part determines the *phase velocity* (c/m') of the wave in the medium: for an electric field propagating in an absorbing medium of $m = m' + \mathrm{i}m''$, say, in the x direction, $\mathbf{E} \propto \exp(-\omega m''x/c)\exp[-\mathrm{i}\omega(t - m'x/c)]$.

The physical basis of the dielectric function ε can readily be understood in terms of the classical *Lorentz harmonic oscillator* model (for insulators) and the *Drude* model (for free-electron metals). In the Lorentz oscillator model, the *bound* electrons and ions of a dust grain are treated as simple harmonic oscillators subject to the driving force of an applied electromagnetic field. The applied field distorts the charge distribution and therefore produces an *induced dipole moment*. To estimate the induced moments, we consider a (polarizable) grain as a collection of identical, independent, isotropic, harmonic oscillators with mass m and charge q; each oscillator is under the action of three forces: (i) a restoring force $-K\mathbf{x}$, where K is the force constant (i.e., stiffness) of the "spring" to which the bound charges are attached and $\mathbf{x}$ is the displacement of the bound charges from their equilibrium; (ii) a damping force $-b\dot{\mathbf{x}}$, where b is the damping constant; and (iii) a driving force $q\mathbf{E}$ produced by the (local) electric field $\mathbf{E}$. The equation of motion of the oscillators is

$$m\ddot{\mathbf{x}} + b\dot{\mathbf{x}} + K\mathbf{x} = q\mathbf{E}. \tag{6.14}$$

For time harmonic electric fields $\mathbf{E} \propto \exp\left(-i\omega t\right)$, we solve for the displacement

$$\mathbf{x} = \frac{(q/m)\,\mathbf{E}}{\omega_o{}^2 - \omega^2 - i\gamma\omega} \ , \ \ \omega_o{}^2 = K/m \ , \ \ \gamma = b/m \ , \tag{6.15}$$

where ω_o is the frequency of oscillation about equilibrium (i.e., the restoring force is $-m\omega_o{}^2\mathbf{x}$). The induced dipole moment $\mathbf{p}$ of an oscillator is $\mathbf{p} = q\mathbf{x}$. Let n be the number of oscillators per unit volume. The *polarization* (i.e., dipole moment per unit volume) $\mathbf{P} = n\mathbf{p} = nq\mathbf{x}$ is

$$\mathbf{P} = \frac{1}{4\pi}\frac{\omega_{\mathrm{p}}{}^2}{\omega_o{}^2 - \omega^2 - i\gamma\omega}\mathbf{E} \ , \ \ \omega_{\mathrm{p}}{}^2 = 4\pi nq^2/m, \tag{6.16}$$

where ω_{p} is the plasma frequency. Since $\mathbf{P} = \chi\mathbf{E} = (\varepsilon - 1)/4\pi\,\mathbf{E}$, the dielectric function for a one-oscillator model around a resonance frequency ω_o is

$$\varepsilon = 1 + 4\pi\chi = 1 + \frac{\omega_{\mathrm{p}}{}^2}{\omega_o{}^2 - \omega^2 - i\gamma\omega} \tag{6.17}$$

$$\varepsilon' = 1 + 4\pi\chi' = 1 + \frac{\omega_{\mathrm{p}}{}^2\left(\omega_o{}^2 - \omega^2\right)}{\left(\omega_o{}^2 - \omega^2\right)^2 + \gamma^2\omega^2} \tag{6.18}$$

$$\varepsilon'' = 4\pi\chi'' = \frac{\omega_{\mathrm{p}}{}^2\gamma\omega}{\left(\omega_o{}^2 - \omega^2\right)^2 + \gamma^2\omega^2}. \tag{6.19}$$

We see from (6.19) $\varepsilon'' \geq 0$ for all frequencies. This is true for materials close to thermodynamic equilibrium, except those with population inversions. If there are many oscillators of different frequencies, the dielectric function for a multiple-oscillator model is

$$\varepsilon = 1 + \sum_j \frac{\omega_{\mathrm{p},j}^2}{\omega_j^2 - \omega^2 - i\gamma_j\omega} \tag{6.20}$$

where $\omega_{p,j}$, ω_j, and γ_j are, respectively, the plasma frequency, the resonant frequency, and the damping constant of the jth oscillator.

The optical properties associated with *free electrons* are described by the *Drude* model. The free electrons experience no restoring forces when driven by the electric field of a light wave and do not have natural resonant frequencies (i.e., $\omega_o = 0$). The Drude model for metals is obtained directly from the Lorentz model for insulators simply by setting the restoring force in (6.14) equal to zero. The dielectric function for free electrons is

$$\varepsilon = 1 - \frac{\omega_{p,e}^2}{\omega^2 + i\gamma_e\omega} \quad , \quad \omega_{p,e}^2 = 4\pi n_e e^2/m_e \tag{6.21}$$

$$\varepsilon' = 1 - \frac{\omega_{p,e}^2}{\omega^2 + \gamma_e^2} \tag{6.22}$$

$$\varepsilon'' = \frac{\omega_{p,e}^2 \gamma_e}{\omega\left(\omega^2 + \gamma_e^2\right)} \tag{6.23}$$

where $\omega_{p,e}$ is the plasma frequency, n_e is the density of free electrons, and m_e is the effective mass of an electron. The damping constant $\gamma_e = 1/\tau_e$ is the reciprocal of the mean free time between collisions (τ_e) which are often determined by electron–phonon scattering (i.e., interaction of the electrons with lattice vibrations).[5] For dust materials (e.g., graphite) containing both bound charges and free electrons, the dielectric function is

$$\varepsilon = 1 - \frac{\omega_{p,e}^2}{\omega^2 + i\gamma_e\omega} + \sum_j \frac{\omega_{p,j}^2}{\omega_j^2 - \omega^2 - i\gamma_j\omega}. \tag{6.24}$$

The optical response of free electrons in metals can also be understood in terms of the electric current density $\mathbf{J}$ and conductivity $\sigma(\omega)$. The free electrons in metals move between molecules. In the absence of an external field, they move in a random manner and hence they do not give rise to a net current flow. When an external field is applied, the free electrons acquire an additional velocity and their motion becomes more orderly which gives rise

---

[5] For nano-sized metallic dust of size a (which is smaller than the mean free path of conduction electrons in the bulk metal), τ_e and γ_e are increased because of additional collisions with the boundary of the dust: $\gamma_e = \gamma_{\text{bulk}} + v_F/(\varsigma a)$, where γ_{bulk} is the bulk metal damping constant, v_F is the electron velocity at the Fermi surface, and ς is a dimensionless constant of order unity which depends on the character of the scattering at the boundary (ςa is the effective mean free path for collisions with the boundary): $\varsigma = 1$ for classic isotropic scattering, $\varsigma = 4/3$ for classic diffusive scattering, $\varsigma = 1.16$ or 1.33 for scattering based on the quantum particle-in-a-box model (see [9] and references therein). Since $\omega^2 \gg \gamma_e^2$ in metals near the plasma frequency, ε'' can be written as $\varepsilon'' = \varepsilon''_{\text{bulk}} + v_F\omega_p^2/(\varsigma a\omega^3)$. This, known as the "electron mean free path limitation effect," indicates that for a metallic grain ε'' increases as the grain becomes smaller.

to an induced current flow. The current density $\mathbf{J} = -n_e e \dot{\mathbf{x}}$ is obtained by solving the equation of motion of the free electrons $m_e \ddot{\mathbf{x}} = -e\mathbf{E} - m_e \dot{\mathbf{x}}/\tau_e$

$$\dot{\mathbf{x}} = \frac{-\tau_e e}{m_e} \frac{1}{1 - i\omega\tau_e} \mathbf{E} \ . \tag{6.25}$$

Since $\mathbf{J} = \sigma \mathbf{E}$, the a.c. conductivity is $\sigma = \sigma_o / (1 - i\omega\tau_e)$, where $\sigma_o = n_e \tau_e e^2 / m_e$ is the d.c. conductivity. The dielectric function for a free-electron metal is therefore

$$\varepsilon = 1 + \frac{i\,4\pi\sigma}{\omega} = 1 - \frac{\omega_{p,e}{}^2}{\omega^2 + i\omega/\tau_e}. \tag{6.26}$$

The real and imaginary parts of the dielectric function are not independent. They are related through the *Kramers–Kronig* (or *dispersion*) relations

$$\varepsilon'(\omega) = 1 + \frac{2}{\pi} P \int_0^\infty \frac{x\varepsilon''(x)}{x^2 - \omega^2} \, dx \ , \quad \varepsilon''(\omega) = \frac{-2\omega}{\pi} P \int_0^\infty \frac{\varepsilon'(x)}{x^2 - \omega^2} \, dx, \tag{6.27}$$

where P is the Cauchy Principal value of the integral

$$P \int_0^\infty \frac{x\varepsilon''(x)}{x^2 - \omega^2} \, dx = \lim_{a \to 0} \left[\int_0^{\omega - a} \frac{x\varepsilon''(x)}{x^2 - \omega^2} \, dx + \int_{\omega + a}^\infty \frac{x\varepsilon''(x)}{x^2 - \omega^2} \, dx \right]. \tag{6.28}$$

The real and imaginary parts of the index of refraction are also connected through the Kramers–Kronig relation. This also holds for the real and imaginary parts of the electric susceptibility.

$$m'(\omega) = 1 + \frac{2}{\pi} P \int_0^\infty \frac{x m''(x)}{x^2 - \omega^2} \, dx \ , \quad m''(\omega) = \frac{-2\omega}{\pi} P \int_0^\infty \frac{m'(x)}{x^2 - \omega^2} \, dx \ , \tag{6.29}$$

$$\chi'(\omega) = \frac{2}{\pi} P \int_0^\infty \frac{x\chi''(x)}{x^2 - \omega^2} \, dx \ , \quad \chi''(\omega) = \frac{-2\omega}{\pi} P \int_0^\infty \frac{\chi'(x)}{x^2 - \omega^2} \, dx \ . \tag{6.30}$$

The Kramers–Kronig relation can be used to relate the extinction cross-section integrated over the entire wavelength range to the dust volume V

$$\int_0^\infty C_{\text{ext}}(\lambda) \, d\lambda = 3\pi^2 FV, \tag{6.31}$$

where F, a dimensionless factor, is the orientationally averaged polarizability relative to the polarizability of an equal-volume conducting sphere, depending only on the grain shape and the static (zero-frequency) dielectric constant ϵ_o of the grain material [64]. This has also been used to place constraints on interstellar grain models based on the interstellar depletions [46, 50] and the carrier of the mysterious $21\,\mu m$ emission feature seen in 12 protoplanetary nebulae [44].

For illustration, we show in Fig. 6.3 the refractive indices of glassy SiO_2 and neutral and singly charged silicon nanoparticles (SNPs).[6] The optical

---

[6] SNPs were proposed as the carrier of the "extended red emission" (ERE), a broad, featureless emission band between ~ 5400 and $9000\,\text{Å}$ seen in a wide variety of dusty environments [72, 41] (but see [48]). SNPs were also suggested to be present in the inner corona of the Sun [24] (but see [54, 55]).

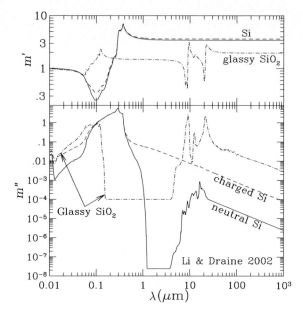

Fig. 6.3. Refractive indices m' (*upper panel*), m'' (*lower panel*) of neutral silicon nanoparticles (Si; *solid lines*), singly, positively charged ($Z = +1$) silicon nanoparticles of size $a = 10\,\text{Å}$ (*dashed lines*), and SiO$_2$ glass (*dot-dashed lines*). Compared to neutral Si, in charged Si free electrons (if negatively charged) or holes (if positively charged) contribute to the dielectric function. Crystalline Si is IR inactive since its lattice vibrations have no dipole moment; the bands at 6.91, 7.03, 7.68, 8.9, 11.2, 13.5, 14.5, 16.4, and 17.9 μm are due to multi-phonon processes. Taken from [48]

properties of SNPs depend on whether any free electrons or holes are present. The contribution of free electrons or holes to the dielectric function of SNPs is approximated by $\delta\varepsilon \approx -\omega_p{}^2\tau^2 / \left(\omega^2\tau^2 + \mathrm{i}\omega\tau\right)$, $\omega_p{}^2 = 3|Z|\,e^2/a^3 m_{\text{eff}}$, where e is the proton charge, Ze is the grain charge, a is the grain radius, $\tau \approx a/v_F$ is the collision time (we take $v_F \approx 10^8\,\text{cm}\,\text{s}^{-1}$), and m_{eff} is the effective mass of a free electron or hole. In Fig. 6.3 the charged SNPs are taken to contain only one hole (i.e., $Z = +1$). We take $m_{\text{eff}} \approx 0.2\,m_e$.

6.5 Scattering of Light by Dust: Analytic Solutions

For dust with sizes much smaller than the wavelength of the incident radiation, analytic solutions to the light scattering problem exist for certain shapes. Let a be the characteristic length of the dust and $x \equiv 2\pi a/\lambda$ be the dimensionless *size parameter*. Under the condition of $x \ll 1$ and $|mx| \ll 1$ (i.e., in the "Rayleigh" regime), $C_{\text{abs}} = 4\pi k\,\text{Im}\{\boldsymbol{\alpha}\}$, $C_{\text{sca}} = (8\pi/3)\,k^4\,|\boldsymbol{\alpha}|^2$, where $\boldsymbol{\alpha}$ is the complex electric *polarizability* of the dust. Apparently, $C_{\text{sca}} \ll C_{\text{abs}}$ and

$C_{\text{ext}} \approx C_{\text{abs}}$. In general, $\boldsymbol{\alpha}$ is a diagonalized tensor;[7] for homogeneous spheres composed of an isotropic material, it is independent of direction

$$\alpha_{jj} = \frac{3V}{4\pi} \frac{\varepsilon - 1}{\varepsilon + 2}, \tag{6.32}$$

where V is the dust volume.[8] For a homogeneous, isotropic ellipsoid, the polarizability for electric field vector parallel to its principal axis j is

$$\alpha_{jj} = \frac{V}{4\pi} \frac{\varepsilon - 1}{(\varepsilon - 1)L_j + 1}, \tag{6.35}$$

where L_j is the "depolarization factor" along principal axis j (see [13]). The electric polarizability $\boldsymbol{\alpha}$ is also known for concentric core-mantle spheres [69], confocal core-mantle ellipsoids [13, 19], and multi-layered ellipsoids of *equal eccentricity* [18]. For a thin *conducting* cylindrical rod with length $2l$ and radius $r_{\text{a}} \ll l$, the polarizability along the axis of the rod is [38]

$$\alpha_{jj} \approx \frac{l^3}{3 \log\left(4l/r_{\text{a}}\right) - 7}. \tag{6.36}$$

In astronomical modeling, the most commonly invoked grain shapes are spheres and spheroids (oblates or prolates).[9] In the Rayleigh regime, their absorption and scattering properties are readily obtained from (6.32, 6.35). For both dielectric and conducting spheres (as long as $x \ll 1$ and $|mx| \ll 1$)

$$C_{\text{abs}}/V = \frac{9\omega}{c} \frac{\varepsilon''}{\left(\varepsilon' + 2\right)^2 + \varepsilon''^2} \gg C_{\text{sca}}/V = \frac{3}{2\pi} \left(\frac{\omega}{c}\right)^4 \left|\frac{\varepsilon - 1}{\varepsilon + 2}\right|^2. \tag{6.37}$$

---

[7] $\boldsymbol{\alpha}$ can be diagonalized by appropriate choice of Cartesian coordinate system. It describes the linear response of a dust grain to applied electric field $\mathbf{E}$: $\mathbf{p} = \boldsymbol{\alpha}\mathbf{E}$, where $\mathbf{p}$ is the induced electric dipole moment.

[8] For a dielectric sphere with dielectric function given in (6.17), in the Rayleigh regime the absorption cross-section is

$$C_{\text{abs}}(\omega)/V = \frac{9}{c} \frac{\gamma\omega_{\text{p}}^2 \omega^2}{\left(3\omega^2 - \omega_{\text{p}}^2 - 3\omega_{\text{o}}^2\right)^2 + 9\gamma^2\omega^2}. \tag{6.33}$$

Similarly, for a metallic sphere with dielectric function given in (6.21),

$$C_{\text{abs}}(\omega)/V = \frac{9}{c} \frac{\gamma_{\text{e}}\omega_{\text{p,e}}^2 \omega^2}{\left(3\omega^2 - \omega_{\text{p,e}}^2\right)^2 + 9\gamma_{\text{e}}^2\omega^2}. \tag{6.34}$$

It is seen that the frequency-dependent absorption cross-section for both dielectric and metallic spheres is a Drude function. This is also true for ellipsoids.

[9] Spheroids are a special class of ellipsoids. Let r_{a}, r_b, and r_c be the semi-axes of an ellipsoid. For spheroids, $r_b = r_c$. Prolates with $r_{\text{a}} > r_b$ are generated by rotating an ellipse (of semi-major axis r_{a} and semi-minor axis r_b) about its major axis; oblates with $r_{\text{a}} < r_b$ are generated by rotating an ellipse (of semi-minor axis r_{a} and semi-major axis r_b) about its minor axis.

At long wavelengths, for dielectric dust $\varepsilon'' \propto \omega$ while ε' approaches a constant much larger than ε'' (see 6.18, 6.19), we see $C_{\text{abs}} \propto \omega\varepsilon'' \propto \omega^2$; for metallic dust, $\varepsilon'' \propto 1/\omega$ while ε' approaches a constant much smaller than ε'' (see 6.22, 6.23), we see $C_{\text{abs}} \propto \omega/\varepsilon'' \propto \omega^2$; therefore, for both dielectric and metallic dust $C_{\text{abs}} \propto \lambda^{-2}$ at long wavelengths![10]

It is also seen from (6.37) that for spherical dust in the Rayleigh regime the albedo $\alpha \approx 0$, and the radiation cross-section $C_{\text{pr}} \approx C_{\text{abs}}$. This has an interesting implication. Let $\beta_{\text{pr}}(a)$ be the ratio of the radiation pressure force to the gravity of a spherical grain of radius a in the solar system or in debris disks illuminated by stars (of radius $R_\star$ and mass $M_\star$) with a stellar flux of $F_\lambda^\star$ at the top of the atmosphere [53],

$$\beta_{\text{pr}}(a) = \frac{3\,R_\star^2 \int F_\lambda^\star \left[C_{\text{abs}}(a,\lambda) + (1-g)\,C_{\text{sca}}(a,\lambda)\right] d\lambda}{16\pi\,c\,GM_\star a^3 \rho_{\text{dust}}} \qquad (6.38)$$

where G is the gravitational constant and ρ_{dust} is the mass density of the dust. For grains in the Rayleigh regime ($g \approx 0$; $C_{\text{sca}} \ll C_{\text{abs}}$; $C_{\text{abs}} \propto a^3$), we see $\beta_{\text{pr}} \propto C_{\text{abs}}/a^3$ is independent of the grain size a (see Fig. 6.4)!

Spheroids are often invoked to model the interstellar polarization. In the Rayleigh approximation, their absorption cross-sections for light polarized parallel ($\parallel$) or perpendicular ($\perp$) to the grain symmetry axis are[11]

$$C_{\text{abs}}^{\parallel,\perp}/V = \frac{\omega}{c}\text{Im}\left\{\frac{\varepsilon - 1}{(\varepsilon - 1)L_{\parallel,\perp} + 1}\right\} , \qquad (6.39)$$

where the depolarization factors parallel ($L_\parallel$) or perpendicular ($L_\perp$) to the grain symmetry axis are not independent, but related to each other through $L_\parallel + 2\,L_\perp = 1$, with

$$L_\parallel = \frac{1 - \xi_e^2}{\xi_e^2}\left[\frac{1}{2\xi_e}\ln\left(\frac{1 + \xi_e}{1 - \xi_e}\right) - 1\right] , \quad \xi_e = \sqrt{1 - (r_b/r_a)^2} \qquad (6.40)$$

---

[10] However, various astronomical data suggest a *flatter* wavelength-dependence (i.e., $C_{\text{abs}} \propto \lambda^{-\beta}$ with $\beta < 2$): $\beta < 2$ in the far-IR/submillimeter wavelength range has been reported for interstellar molecular clouds, circumstellar disks around young stars, and circumstellar envelopes around evolved stars. Laboratory measurements have also found $\beta < 2$ for certain cosmic dust analogs. In literature, the *flatter* ($\beta < 2$) long-wavelength opacity law is commonly attributed to grain growth by coagulation of small dust into large fluffy aggregates (see [47] and references therein). However, as shown in (6.31), the Kramers–Kronig relation requires that β should be larger than 1 for $\lambda \to \infty$ since F is a finite number and the integration in the left hand side of (6.31) should be convergent although we cannot rule out $\beta < 1$ over a finite range of wavelengths.

[11] For grains spinning around the principal axis of the largest moment of inertia, the polarization cross-sections are $C_{\text{pol}} = (C_{\text{abs}}^{\parallel} - C_{\text{abs}}^{\perp})/2$ for prolates and $C_{\text{pol}} = (C_{\text{abs}}^{\perp} - C_{\text{abs}}^{\parallel})$ for oblates; the absorption cross-sections for randomly oriented spheroids are $C_{\text{abs}} = (C_{\text{abs}}^{\parallel} + 2C_{\text{abs}}^{\perp})/3$ [42].

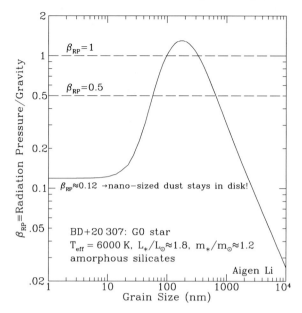

Fig. 6.4. $\beta_{\rm pr}$—Ratio of radiation pressure to gravity for compact silicate grains in the debris disk around the Sun-like star BD+20 307 (G0, age ~ 300 Myr). We note for nano-sized grains $\beta_{\rm pr} \approx 0.12$, independent of grain size; for grains larger than $\sim 0.3 \, \mu$m, $\beta_{\rm pr}$ is inverse proportional to grain size. Song et al. [68] attributed the dust in this disk to recent extreme collisions between asteroids. Taken from [51]

for prolates $(r_{\rm a} > r_b)$ where $\xi_{\rm e}$ is the eccentricity, and

$$L_{\parallel} = \frac{1 + \xi_{\rm e}^2}{\xi_{\rm e}^2} \left(1 - \frac{1}{\xi_{\rm e}} \arctan \xi_{\rm e} \right), \quad \xi_{\rm e} = \sqrt{(r_b/r_{\rm a})^2 - 1} \qquad (6.41)$$

for oblates $(r_{\rm a} < r_b)$. For spheres $L_{\parallel} = L_{\perp} = 1/3$ and $\xi_{\rm e} = 0$. For extremely elongated prolates or "needles" $(r_{\rm a} \gg r_b)$, it is apparent $C_{\rm abs}^{\perp} \ll C_{\rm abs}^{\parallel}$, we thus obtain

$$C_{\rm abs}/V \approx \frac{\omega}{3c} \frac{\varepsilon''}{\left[L_{\parallel}(\varepsilon' - 1) + 1 \right]^2 + \left(L_{\parallel} \varepsilon'' \right)^2} \qquad (6.42)$$

where $L_{\parallel} \approx (r_b/r_{\rm a})^2 \ln(r_{\rm a}/r_b)$. For dielectric needles, $C_{\rm abs} \propto \omega \varepsilon'' \propto \lambda^{-2}$ at long wavelengths since $L_{\parallel}(\varepsilon' - 1) + 1 \gg L_{\parallel} \varepsilon''$ (see [45]); for *metallic* needles, for a given value of ε'' one can always find a sufficiently long needle with $L_{\parallel} \varepsilon'' < 1$ and $L_{\parallel}(\varepsilon' - 1) \ll 1$ so that $C_{\rm abs} \propto \omega \varepsilon'' \propto \sigma$ which can be very large (see [45]). Because of their unique optical properties, metallic needles with high electrical conductivities (e.g., iron needles, graphite whiskers) are resorted to explain a wide variety of astrophysical phenomena: (1) as a source of starlight opacity to create a non-cosmological microwave background by the thermalization of starlight in a steady-state cosmology [27]; (2) as a source of the gray opacity needed to explain the observed redshift-magnitude relation of Type Ia supernovae without invoking a positive cosmological constant [1]; (3) as the source for the submillimeter excess observed in the Cas A supernova remnant [16]; and (4) as an explanation for the flat 3–8 μm extinction observed for line of sight toward the Galactic Center and in the Galactic plane [15]. However, caution should be taken in using (6.42) (i.e., the Rayleigh approximation) since the Rayleigh criterion $2\pi r_a|m|/\lambda \ll 1$ is often not satisfied for highly conducting needles (see [45]).[12]

In astronomical spectroscopy modeling, the continuous distribution of ellipsoid (CDE) shapes has been widely used to approximate the spectra of irregular dust grains by averaging over all ellipsoidal shape parameters [6]. In the Rayleigh limit, this approach, assuming that all ellipsoidal shapes are equally probable, has a simple expression for the average cross-section

$$\langle C_{abs}/V \rangle = \frac{\omega}{c}\text{Im}\left\{\frac{2\varepsilon}{\varepsilon-1}\text{Log}\,\varepsilon\right\} \ , \ \text{Log}\,\varepsilon \equiv \ln\sqrt{\varepsilon'^2+\varepsilon''^2} + i\arctan\left(\varepsilon''/\varepsilon'\right)$$

(6.43)

where $\text{Log}\,\varepsilon$ is the principal value of the logarithm of ε. The CDE approach, resulting in a significantly broadened spectral band (but with its maximum reduced), seems to fit the experimental absorption spectra of solids better than Mie theory. Although the CDE may indeed represent a distribution of shape factors caused either by highly irregular dust shapes or by clustering of spherical grains into irregular agglomerates, one should caution that the shape distribution of cosmic dust does not seem likely to resemble the CDE, which assumes that extreme shapes like needles and disks are equally probable. A more reasonable shape distribution function would be like $dP/dL_\parallel = 12L_\parallel$ $(1-L_\parallel)^2$ which peaks at spheres ($L^\parallel = 1/3$). This function is symmetric about spheres with respect to eccentricity e and drops to zero for the extreme cases: infinitely thin needles ($e \to 1$, $L_\parallel \to 0$) or infinitely flattened pancakes ($e \to \infty$, $L_\parallel \to 1$). Averaging over the shape distribution, the resultant absorption cross-section is $C_{abs} = \int_0^1 dL_\parallel dP/dL_\parallel C_{abs}(L_\parallel)$, where $C_{abs}(L_\parallel)$ is the absorption cross-section of a particular shape $L_\parallel$ [48, 49, 62]. Alternatively, Fabian et al. [17] proposed a quadratic weighting for the shape distribution, "with near-spherical shapes being most probable."

When a dust grain is very large compared with the wavelength, the electromagnetic radiation may be treated by geometric optics: $Q_{ext} \equiv C_{ext}/C_{geo} \to 2$

---

[12] The "antenna theory" has been applied for conducting needle-like dust to estimate its absorption cross-sections [75]. Let it be represented by a circular cylinder of radius r_a and length l ($r_a \ll l$). Let ρ_R be its resistivity. The absorption cross-section is given by $C_{abs} = (4\pi/3c)(\pi r_a^2 l/\rho_R)$, with a long-wavelength cutoff of $\lambda_o = \rho_R c(l/r_a)^2/\ln(l/r_a)^2$, and a short-wavelength cutoff of $\lambda_{min} \approx (2\pi c m_e)/(\rho_R n_e e^2)$, where m_e, e, and n_e are, respectively, the mass, charge, and number density of the charge-carrying electrons.

if $x \equiv 2\pi a/\lambda \gg 1$ and $|m - 1| x \gg 1$.[13] For these grains ($g \approx 1$; $C_{\mathrm{abs}} \approx C_{\mathrm{geo}}$), the ratio of the radiation pressure to gravity $\beta_{\mathrm{pr}} \propto C_{\mathrm{abs}}/a^3 \propto 1/a$. This is demonstrated in Fig. 6.4. For dust with $x \gg 1$ and $|m - 1| x \ll 1$, one can use the "anomalous diffraction" theory [69]. For dust with $|m - 1| x \ll 1$ and $|m - 1| \ll 1$, one can use the Rayleigh–Gans approximation[14] to obtain the absorption and scattering cross-sections [6, 39, 69]:

$$Q_{\mathrm{abs}} \approx \frac{8}{3}\mathrm{Im}\left\{mx\right\}, \quad Q_{\mathrm{sca}} \approx \frac{32\,|m - 1|^2 x^4}{27 + 16x^2}. \tag{6.44}$$

It is important to note that the Rayleigh–Gans approximation is invalid for modeling the X-ray scattering by interstellar dust at energies below 1 keV. This approximation systematically and substantially overestimates the intensity of the X-ray halo below 1 keV [66].

6.6 Scattering of Light by Dust: Numerical Techniques

While simple analytic expressions exist for the scattering and absorption properties of dust grains which are either very small or very large compared to the wavelength of the incident radiation (see Sect. 6.5), however, in many astrophysical applications we are concerned with grains which are neither very small nor very large compared to the wavelength. Moreover, cosmic dust would, in general, be expected to have non-spherical, irregular shapes.

Our ability to compute scattering and absorption cross-sections for non-spherical particles is extremely limited. So far, exact solutions of scattering

---

[13] At a first glance, $Q_{\mathrm{ext}} \to 2$ appears to contradict "common sense" by implying that a large grain removes twice the energy that is incident on it! This actually can be readily understood in terms of basic optics principles: (1) on one hand, all rays impinging on the dust are either scattered or absorbed. This gives rise to a contribution of C_{geo} to the extinction cross-section. (2) On the other hand, all the rays in the field which do not hit the dust give rise to a diffraction pattern that is, by Babinet's principle, identical to the diffraction through a hole of area C_{geo}. If the detection excludes this diffracted light, then an additional contribution of C_{geo} is made to the total extinction cross-section [6].

[14] The conditions for the Rayleigh–Gans approximation to be valid are $|m - 1| \ll 1$ and $|m - 1| x \ll 1$. The former ensures that the reflection from the surface of the dust is negligible (i.e., the impinging light enters the dust instead of being reflected); the latter ensures that the phase of the incident wave is not shifted inside the dust. For sufficiently small scattering angles, it is therefore possible for the waves scattered throughout the dust to add coherently. The intensity (I) of the scattered waves is proportional to the number (N) of scattering sites squared: $I \propto N^2 \propto \rho^2 a^6$ (where ρ is the mass density of the dust). This is why the X-ray halos (usually within $\sim 1°$ surrounding a distant X-ray point source; [63]) created by the small-angle scattering of X-rays by interstellar dust are often used to probe the size (particularly the large size end; [14, 67, 73]), morphology (compact or porous; [58, 66]), composition, and spatial distribution of dust.

problems exist only for bare or layered spherical grains ("Mie theory;" [6]), infinite cylinders [52], and spheroids [3, 4, 70]. The "T-matrix" (transition matrix) method, originally developed by Barber & Yeh [5] and substantially extended by Mishchenko et al. [59], is able to treat axisymmetric (spheroidal or finite cylindrical) grains with sizes comparable to the wavelength. The discrete dipole approximation (DDA), originally developed by Purcell & Pennypacker [65] and greatly improved by Draine [10], is a powerful technique for irregular heterogeneous grains with sizes as large as several times the wavelength. The VIEF (volume integration of electric fields) method developed by Hage & Greenberg [25], based on an integral representation of Maxwell's equations, is physically similar to the DDA method. The microwave analog methods originally developed by Greenberg et al. [21] provide an effective experimental approach to complex particles [23].

Although interstellar grains are obviously non-spherical as evidenced by the observed polarization of starlight, the assumption of spherical shapes (together with the Bruggeman or the Maxwell–Garnett effective medium theories for inhomogeneous grains; [6]) is usually sufficient in modeling the interstellar absorption, scattering, and IR (continuum) emission. For IR polarization modeling, the dipole approximation for spheroidal grains is proven to be successful in many cases.

The DDA method is highly recommended for studies of inhomogeneous grains and irregular grains, such as cometary, interplanetary, and protoplanetary dust particles which are expected to have a porous aggregate structure. Extensive investigations using the DDA method have been performed for the scattering, absorption, thermal IR emission, and radiation pressure properties of fluffy aggregated dust (e.g., see [2, 23, 29, 30, 31, 33, 34, 35, 36, 37, 40, 43, 56, 57, 60, 61, 71, 74, 76, 77]).

Acknowledgements

I thank Drs I. Mann and T. Mukai for their very helpful comments and support. Partial support by a Chandra Theory program and HST Theory Programs is gratefully acknowledged.

References

1. A.N. Aguirre: *Intergalactic dust and observations of type IA supernovae*, Astrophys. J. **512**, L19 (1999)
2. A.C. Andersen, H. Mutschke, T. Posch, M. Min and A. Tamanai: *Infrared extinction by homogeneous particle aggregates of SiC, FeO and SiO₂: comparison of different theoretical approaches*, J. Quant. Spectrosc. Radiat. Transfer **100**, 4 (2006)
3. S. Asano and M. Sato: *Light scattering by randomly oriented spheroidal particles*, Appl. Opt. **19**, 962 (1980)

4. S. Asano and G. Yamamoto: *Light scattering by a spheroidal particle*, Appl. Opt. **14**, 29 (1975)
5. P. Barber and C. Yeh: *Scattering of electromagnetic waves by arbitrarily shaped dielectric bodies*, Appl. Opt. **14**, 2864 (1975)
6. C.F. Bohren and D. R. Huffman: *Absorption and Scattering of Light by Small Particles*, Wiley, New York, (1983)
7. D.E. Brownlee, et al.: *Comet 81P/Wild 2 under a microscope*, Science **314**, 1711 (2006)
8. D.D. Clayton and L.R. Nittler: *Astrophysics with presolar stardust*, Ann. Rev. Astro. Astrophys. **42**, 39 (2004)
9. E.A. Coronado and G.C. Schatz: *Surface plasmon broadening for arbitrary shape nanoparticles: a geometrical probability approach*, J. Chem. Phys. **119**, 3926 (2003)
10. B.T. Draine: *The discrete-dipole approximation and its application to interstellar graphite grains*, Astrophys. J. **333**, 848 (1988)
11. B.T. Draine: *Evolution of Interstellar Dust*, In ASP Conf. Ser. 12, The Evolution of the Interstellar Medium, L. Blitz (Ed.) (ASP, San Francisco, 1990), 193
12. B.T. Draine: *Scattering by interstellar dust grains. I. optical and ultraviolet*, Astrophys. J. **598**, 1017 (2003)
13. B.T. Draine and H.M. Lee: *Optical properties of interstellar graphite and silicate grains*, Astrophys. J. **285**, 89 (1984)
14. B.T. Draine and J.C. Tan: *The scattered X-ray halo around nova cygni 1992: testing a model for interstellar dust*, Astrophys. J. **594**, 347 (2003)
15. E. Dwek: *Galactic center extinction: evidence of metallic needles in the general interstellar medium*, Astrophys. J. **611**, L109 (2004a)
16. E. Dwek: *The detection of cold dust in cassiopeia A: evidence for the formation of metallic needles in the ejecta*, Astrophys. J. **607**, 848 (2004b)
17. D. Fabian, Th. Henning, C. Jäger, H. Mutschke, J. Dorschner and O. Wehrhan: *Steps toward interstellar silicate mineralogy. VI. Dependence of crystalline olivine IR spectra on iron content and particle shape*, Astron. Astrophys. **378**, 228 (2001)
18. V.G. Farafonov: *Light scattering by multilayer nonconfocal ellipsoids in the rayleigh approximation*, Opt. Spectrosc. **90**, 574 (2001)
19. D.P. Gilra: *Collective excitations and dust particles in space*, In Scientific Results from the Orbiting Astronomical Observatory (OAO-2), **310**, 295 (1972)
20. J.M. Greenberg: *What Are Comets Made of?* In: Comets, L.L. Wilkening (Ed.) (Univ. of Arizona Press, Tuscon, 1982), 131
21. J.M. Greenberg, N.E. Pedersen and J.C. Pedersen: *Microwave analog to the scattering of light by nonspherical particles*, J. Appl. Phys. **32**, 233 (1961)
22. E. Grün, et al.: *Kuiper prize lecture: Dust astronomy*, Icarus, **174**, 1 (2005)
23. B.Å.S. Gustafson and L. Kolokolova: *A systematic study of light scattering by aggregate particles using the microwave analog technique: Angular and wavelength dependence of intensity and polarization*, J. Geophys. Res. **104**, 31711 (1999)
24. S.R. Habbal, et al.: *On the detection of the signature of silicon nanoparticle dust grains in coronal holes*, Astrophys. J. **592**, L87 (2003)
25. J.I. Hage and J.M. Greenberg: *A model for the optical properties of porous grains*, Astrophys. J. **361**, 251 (1990)

26. L.G. Henyey and J.L. Greenstein: *Diffuse radiation in the galaxy*, Astrophys. J. **93**, 70 (1941)

27. F. Hoyle and N.C. Wickramasinghe: *Metallic particles in astronomy*, Astrophys. Space Sci., **147**, 245 (1988)

28. J.D. Jackson: *Classical electrodynamics* (3rd ed), (Wiley, New York, 1998)

29. H. Kimura, H. Okamoto and T. Mukai: *Radiation pressure and the poynting-robertson effect for fluffy dust particles*, Icarus **157**, 349 (2002)

30. H. Kimura, L. Kolokolova and I. Mann: *Optical properties of cometary dust: Constraints from numerical studies on light scattering by aggregate particles*, Astron. Astrophys. **407**, L5 (2003)

31. H. Kimura, L. Kolokolova and I. Mann: *Light scattering by cometary dust: Numerically simulated with aggregate particles consisting of identical spheres*, Astron. Astrophys. **449**, 1243 (2006)

32. J. Kissel, et al.: *COSIMA – High resolution time-of-flight secondary ion mass spectrometer for the analysis of cometary dust particles onboard rosetta*, Space Sci. Rev. **128**, 823 (2007)

33. M. Köhler, H. Kimura and I. Mann: *Applicability of the discrete-dipole approximation to light-scattering simulations of large cosmic dust aggregates*, Astron. Astrophys. **448**, 395 (2006)

34. M. Köhler, T. Minato, H. Kimura and I. Mann: *Radiation pressure force acting on cometary aggregates*, Adv. Space Res. **40**, 266 (2007)

35. L. Kolokolova, M.S. Hanner, A.-C. Levasseur-Regourd and B.Å.S. Gustafson: *Physical properties of cometary dust from light scattering and thermal emission*, Comets **II**, 577 (2004)

36. T. Kozasa, J. Blum and T. Mukai: *Optical properties of dust aggregates. I. Wavelength dependence*, Astron. Astrophys. **263**, 423 (1992)

37. T. Kozasa, J. Blum, H. Okamoto and T. Mukai: *Optical properties of dust aggregates. II. Angular dependence of scattered light*, Astron. Astrophys. **276**, 278 (1993)

38. L.D. Landau, E.M. Lifshitz and L.P. Pitaevskii: *Electrodynamics of Continuous Media* (2nd ed.), (Pergamon, Oxford, 2000)

39. A. Laor and B.T. Draine: *Spectroscopic constraints on the properties of dust in active galactic nuclei*, Astrophys. J. **402**, 441 (1993)

40. J. Lasue and A.C. Levasseur-Regourd: *Porous irregular aggregates of submicron sized grains to reproduce cometary dust light scattering observations*, J. Quant. Spectrosc. Radiat. Transfer **100**, 220 (2006)

41. G. Ledoux, et al.: *Silicon as a candidate carrier for ERE*, Astron. Astrophys. **333**, L39 (1998)

42. H.M. Lee and B.T. Draine: *Infrared extinction and polarization due to partially aligned spheroidal grains – models for the dust toward the BN object*, Astrophys. J. **290**, 211 (1985)

43. A.C. Levasseur-Regourd, T. Mukai, J. Lasue and Y. Okada: *Physical properties of cometary and interplanetary dust*, Planet. Space Sci. **55**, 1010 (2007)

44. A. Li: *On TiC nanoparticles as the origin of the 21µm emission feature in post-asymptotic giant branch stars*, Astrophys. J. **599**, L45 (2003a)

45. A. Li: *Cosmic Needles versus cosmic microwave background radiation*, Astrophys. J. **584**, 593 (2003b)

46. A. Li: *Can Fluffy dust alleviate the subsolar interstellar abundance problem?*, Astrophys. J. **622**, 965 (2005a)

47. A. Li: *On the absorption and emission properties of interstellar grains*, In: "The Spectral Energy Distribution of Gas-Rich Galaxies: Confronting Models with Data", C.C. Popescu and R.J. Tuffs, (Ed.) AIP Conf. Ser. **761**, 163 (2005b)
48. A. Li and B.T. Draine: *Are silicon nanoparticles an interstellar dust component?* Astrophys. J. **564**, 803 (2002)
49. A. Li, J.M. Greenberg and G. Zhao: *Modelling the astronomical silicate features – I. On the spectrum subtraction method*, Mon. Not. Roy. Astron. Soc. **334**, 840 (2002)
50. A. Li, K.A. Misselt and Y.J. Wang: *On the unusual depletions toward Sk 155 or what are the small magellanic cloud dust grains made of?*, Astrophys. J. **640**, L151 (2006)
51. A. Li, J. Ortega and J.I. Lunine: *BD+20 307: Attogram dust or extreme asteroidal collisions?*, in preparation (2008)
52. A.C. Lind and J.M. Greenberg: *Electromagnetic scattering by obliquely oriented cylinders*, J. Appl. Phys. **37**, 3195 (1966)
53. I. Mann: *Evolution of dust and small bodies: Physical processes*, in Small Bodies in Planetary Systems, I. Mann, A.M. Nakamura, and T. Mukai, (Ed.) (Springer, Berlin 2007a)
54. I. Mann: *Nanoparticles in the inner solar system*, Planet. Space Sci. **55**, 1000 (2007b)
55. I. Mann and E. Murad: *On the existence of silicon nanodust near the sun*, Astrophys. J. **624**, L125 (2005)
56. I. Mann, H. Okamoto, T. Mukai, H. Kimura and Y. Kitada: *Fractal aggregate analogues for near solar dust properties*, Astron. Astrophys. **291**, 1011 (1994)
57. I. Mann, H. Kimura and L. Kolokolova: *A comprehensive model to describe light scattering properties of cometary dust*, J. Quant. Spectrosc. Radiat. Transfer **89**, 291 (2004)
58. J.S. Mathis, D. Cohen, J.P. Finley and J. Krautter: *The X-ray halo of nova V1974 cygni and the nature of interstellar dust*, Astrophys. J. **449**, 320 (1995)
59. M.I. Mishchenko, L.D. Travis and A. Macke: *Scattering of light by polydisperse, randomly oriented, finite circular cylinders*, Appl. Opt. **35**, 4927 (1996)
60. T. Mukai, H. Ishimoto, T. Kozasa, J. Blum and J.M. Greenberg: *Radiation pressure forces of fluffy porous grains*, Astron. Astrophys. **262**, 315 (1992)
61. Y. Okada, A.M. Nakamura and T. Mukai: *Light scattering by particulate media of irregularly shaped particles: Laboratory measurements and numerical simulations*, J. Quant. Spectrosc. Radiat. Transfer **100**, 295 (2006)
62. V. Ossenkopf, Th. Henning and J.S. Mathis: *Constraints on cosmic silicates*, Astron. Astrophys. **261**, 567 (1992)
63. J.W. Overbeck: *Small-angle scattering of celestial X-rays by interstellar grains*, Astrophys. J. **141**, 864 (1965)
64. E.M. Purcell: *On the absorption and emission of light by interstellar grains*, Astrophys. J. **158**, 433 (1969)
65. E.M. Purcell and C.R. Pennypacker: *Scattering and absorption of light by nonspherical dielectric grains*, Astrophys. J. **186**, 705 (1973)
66. R.K. Smith and E. Dwek: *Soft X-ray scattering and halos from dust*, Astrophys. J. **503**, 831 (1998)
67. R.K. Smith, R.J. Edgar and R.A. Shafer: *The X-ray halo of GX 13+1*, Astrophys. J. **581**, 562 (2002)

68. I. Song, B. Zuckerman, A.J. Weinberger and E.E. Becklin: *Extreme collisions between Planetesimals as the origin of warm dust around a sun-like star*, Nature **436**, 363 (2005)
69. H.C. van de Hulst: *Light Scattering by Small Particles*, (John Wiley & Sons, New York, 1957)
70. N.V. Voshchinnikov and V.G. Farafonov: *Optical properties of spheroidal particles*, Astrophys. Space Sci. **204**, 19 (1993)
71. M. Wilck and I. Mann: *Radiation pressure forces on "Typical" interplanetary dust grains*, Planet. Space Sci. **44**, 493 (1996)
72. A.N. Witt, K.D. Gordon and D.G. Furton: *Silicon nanoparticles: Source of extended red emission?* Astrophys. J. **501**, L111 (1998)
73. A.N. Witt, R.K. Smith and E. Dwek: *X-ray halos and large grains in the diffuse interstellar medium*, Astrophys. J. **550**, L201 (2001)
74. M.J. Wolff, G.C. Clayton and S.J. Gibson: *Modeling composite and fluffy grains. II. Porosity and phase functions*, Astrophys. J. **503**, 815 (1998)
75. E.L. Wright: *Thermalization of starlight by elongated grains – could the microwave background have been produced by stars?* Astrophys. J. **255**, 401 (1982)
76. Z.F. Xing and M.S. Hanner: *Light scattering by aggregate particles*, Astron. Astrophys. **324**, 805 (1999)
77. P.A. Yanamandra-Fisher and M.S. Hanner: *Optical properties of nonspherical particles of size comparable to the wavelength of light: Application to comet dust*, Icarus **138**, 107 (1999)

Evolution of Dust and Small Bodies: Physical Processes

I. Mann

Graduate School of Science, Department of Earth and Planetary Sciences, &
Center for Planetary Science, Kobe 657-8501, Japan,
mann@diamond.kobe-u.ac.jp

Abstract Planetary debris disks are exposed to the brightness of the central star
and for young systems the brightness at wavelengths shorter than the visible is
variable in time. The central star ejects a stellar wind and depending on its parameters compared to those of the local interstellar medium plasma, an astrosphere
may evolve and prevent the low energy part of galactic cosmic rays from entering
the system. The parameters of the astrosphere, stellar radiation, and gravity determine the entry of interstellar medium dust into the system. Both stellar radiation
and stellar wind give rise to a Poynting–Robertson effect, which limits the lifetime
of the dust particles that are in bound orbit about the star. In debris disks with
high dust content lifetimes due to mutual collisions are even shorter. The former
case is denoted as migration-dominated disk, the latter case as collision-dominated
disk. Dust collisions are a potential source of second-generation gas in planetary
debris disks and the gas production mainly results from dust in hyperbolic orbits.
Electrons, protons, and highly charged heavier ions are accelerated in the star or at
plasma structures of the astrosphere. These locally produced energetic particles are
the potentially major particle component causing dust material alteration. Aside
from material alteration due to energetic particles, heating, collision processes, and
alteration in the parent bodies play a role. These different processes of dust material
alteration have to be considered for the interpretation of astronomical observations
in terms of dust evolution. Surface charging and subsequent deflection in magnetic
fields determine the dynamics of dust in the range of 10 s of nanometers and below.
These nano-particles were detected in the solar system, but at present there are no
observational data for planetary debris disks around other stars.

7.1 Introduction

7.1.1 Dust and Small Objects

Dust and small objects in planetary systems are observed in circumstellar
debris disks around stars of spectral types B, A, F, G, K, and M. These systems are characterized by a small amount of circumstellar gas, the presence

Mann, I.: *Evolution of Dust and Small Bodies: Physical Processes.* Lect. Notes Phys. **758**,
189–230 (2009)
DOI 10.1007/978-3-540-76935-4_7

of planetary objects and dust. For stars of spectral class G, K, M also planets were detected, though they are expected to exist around stars of other spectral types as well. After the first detection at Vega planetary debris disks are sometimes denoted as "Vega-like" systems. Because of various processes limiting the lifetime of dust, the dust in planetary debris disks has to be recently produced. It is thought that the disks consist of "first generation" planetary objects and "second generation" debris generated by fragmentation of the planetary objects. The majority of debris or dust production is most likely due to the disintegration of planetesimals caused by catastrophic collisions (like from asteroids in the solar system) or caused by the sublimation of volatile species (like the activity of comets for the case of the solar system). The exact size boundary between first and second generation components may vary within the system as well as from system to system and the size distributions of first and second generation populations may overlap. The small solar system objects (trans-Neptunian objects, asteroids, comets, and meteoroids) and the interplanetary dust cloud also form a planetary debris disk, although the solar system dust density is much lower than for the debris disks that are observed around other stars.

7.1.2 Dust Observations

Information about dust in planetary debris disks is derived from observations of thermal emission brightness, scattered light brightness and in some cases, from polarization. The dominant contribution to the observed brightness of the debris disks is produced by the second generation and for circumstellar disks observers use the term "dust" to denote all the solid particles that make up the disk brightness. For the case of the solar system, the information about dust is complemented by in situ detection from spacecraft, meteor observations, and laboratory study of collected samples (meteorites, interplanetary dust collected in the Earth's atmosphere or isolated from glacial ices, and recently the dust collected during space mission).

7.1.3 Size Distribution

The origin of the small objects from disruption processes as well as the observation of the disks over a broad spectral range suggests that the observed dust disks cover a broad size range. The shape of the size distribution of collision fragments can be estimated from laboratory analog experiments (Nakamura, this issue). The size distribution of fragments ejected during the evaporation from icy objects is not known in detail (see Ishiguro and Ueno, this issue). The forces acting on the released particles further influence the size distribution. Small particles are, for instance, more prone to non-gravitational forces, and ejected from bound orbits.

7.1.4 Environments

Planetary debris disks evolve after removal of the gas component of the protoplanetary disk. The gas content is comparatively low, as is its influence on
the dust dynamics. The gas in the debris disks is usually not detected with
astronomical observations. A presumably second generation gas component is
detected around β Pictoris though and will we discussed later in this chapter.

Similar to the solar system, the prevailing gas component in the inner
region of the planetary debris disks is most likely the stellar wind. In the outer
parts, this is not necessarily the case: interstellar neutral gas enters the system,
gas species are possibly generated by processes within the gas component, and
by the dust component (see discussion below). The gas densities temperatures
and ionization states and the dust densities, around the Sun are listed in
Table 7.1. These parameters may serve as an example for other planetary
systems. Compared to the interstellar medium, the dust is exposed to a denser
environment of electrons, protons, and highly charged ions, as well as to a
stronger stellar magnetic field and stellar brightness. This environment has
some influence on the forces acting on dust particles and the physical processes
that occur.

Table 7.1. Properties of dust and gas around the Sun: the outer solar corona where
the main brightness is generated by the dust particles, the interplanetary medium
near Earth orbit, and the very local interstellar medium beyond the region that is
filled with the Solar wind. Values for the interstellar medium gas are derived from
the helium and hydrogen number densities given for the local interstellar cloud near
the heliosphere [44], basic solar wind parameters are from Allen's Astrophysical
Quantities [1] and dust parameters from in situ measurements of interstellar dust
entering the solar system [36, 52] and studies of dust near the Sun [54], respectively

	Interplanetary medium	Solar F-corona	Very local interstellar medium
Distance from Sun	≈1 AU	≈0.1 AU	>200 AU
M_{gas}/kg m^{-3}	≈1×10^{-19}	3×10^{-18}	1×10^{-20}
M_{dust}/kg m^{-3}	2×10^{-18}	2×10^{-17}	3×10^{-23}
Ionized Gas	≈100%	100%	20%
n_e/m^{-3}	5×10^6	1.6×10^9	4×10^4
T_e/K	2×10^5	8×10^5	7×10^3

7.1.5 Focus of this Chapter

We here discuss physical processes that influence the material composition,
size distribution, and dynamics, of dust and small objects as well as the influence on the gas component that may result from these processes. This chapter

starts by summarizing the present knowledge about the photon fluxes, particle fluxes, and magnetic fields around the host stars (Sect. 2). Based on these parameters, the major physical processes are listed in Sect. (3). Subsequently, we discuss some of the consequences for planetary debris disks: the entry of interstellar matter (Sect. 4), the collisional evolution (Sect. 5), the possible existence of a second generation gas component (Sect. 6), the lifetimes of dust in the debris disks (Sect. 7), the formation of rings as a results of dust sublimation (Sect. 8), and finally the thermal and non-thermal material alteration (Sect. 9).

7.2 The Stellar Environments

7.2.1 Brightness

The photospheric temperatures of the stars surrounded by planetary debris disks range from roughly 3000 to 15,000 K. Correspondingly, the peak emission ranges from UV to near IR. The stellar luminosities range from about 1/10 to several 100 of the solar luminosity. The total main sequence lifetime of stars of the type that may have a debris disk ranges from 10^6 to 10^{11} years. Variations of the total stellar irradiance are of the order of 0.1% for Sun-like stars and a couple of percentage for some other stars. The stellar spectra deviate from that described by a Planck law for the average photospheric temperature (see Fig. 7.1). Deviations from the Planck emission are expected in the UV and X-ray regime and those deviations are variable in time. For the Sun, for instance, the chromospheric emission is correlated with changes in brightness or irradiance [27]. An observational study of other solar-like stars suggests that their chromospheric activity is often higher than that of the Sun [27]. Chromospheric activity is associated with highly variable line emission that is observed to a larger extent for young stars. Space measurements allow today for detailed studies of stellar brightness at wavelength shorter than the visible and also allow, for instance, to study the X-ray and UV emission associated with stellar flares [62]. Solar-like stars during evolution from protostars to the main sequence show enhanced magnetic activity on the surfaces resulting in high X-ray and energetic particle emission [25]. From the analysis of X-ray observations of pre-main-sequence solar-like stars (i.e., with stellar masses between 0.7 and 1.4 solar masses), Feigelson et al. [24] conclude that the X-ray flares are more frequent (by a factor of about 300) and more powerful (by a factor of about 30) than for the Sun.

7.2.2 Particles and Fields Environment

Due to a lack of other observational data, the discussion of the particle and fields environment is largely based on findings about the interplanetary medium of the solar system.

Astrosphere Formation

The astrosphere (in the case of the Sun, the heliosphere) is the region (see Fig. 7.2) around the star that is filled with the stellar wind plasma (as opposed to the interstellar plasma). The stellar wind plasma (for the solar system, the solar wind) has a high temperature and, in the outer part of the astrosphere, low density compared to the surrounding interstellar medium. The interstellar medium includes regions with different densities, temperatures, and states of ionization (see Fig. 7.3). The region in which the Sun is embedded is often called the very local interstellar medium (VLISM). The astropause (heliopause) is the boundary between the astrosphere and the interstellar medium. This boundary arises at the distance from the star where the pressure of the stellar wind equals to that of the ionized component of the interstellar medium. The flow of the neutral component of the interstellar medium is not affected by the astropause and can penetrate into the inner part of the astrosphere. The formation of the astrosphere is closely connected to the appearance of energetic particles (highly ionized electrons and atomic ions) and among them the galactic and anomalous cosmic ray particles.

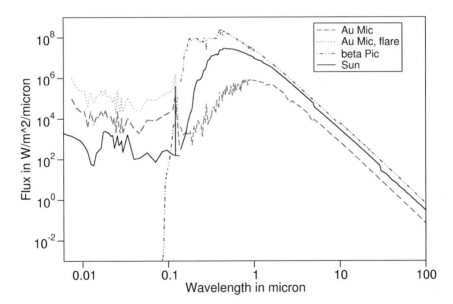

Fig. 7.1. The stellar flux as function of wavelength for Au Mic, β Pic and the Sun. The fluxes are from [5, 31, 32, 75]. For the case of β Pic, no UV data are shown. For AU Mic, the *solid line* denotes the brightness during a quite phase and the *dotted line* illustrates the assumed brightness enhancement during a flare

Structure of Astrospheres

The stars are moving relative to the interstellar medium that they are imbedded in and therefore the astrospheres are asymmetric, with elongated "tails" stretching in the direction opposite to the star motion. This structure is generated by the stellar wind emitted from the star, which acts as an obstacle to the interstellar plasma flow. The Sun is moving with a velocity of about $26\,\mathrm{km\ s^{-1}}$ relative to its local surrounding. The equivalent picture, as seen from the Sun, is that there is a flux of interstellar gas coming towards the Sun: this is sometimes called interstellar wind, and the direction from which it comes defines the "upwind" direction. In the case of the Sun, the velocity vector of the interstellar wind is almost parallel to the ecliptic plane. This is a coincidence and the geometry is different for other astrospheres.

The extension of astrospheres depends on the stellar wind parameters, the local interstellar medium parameters, and the relative velocities between the star and the interstellar medium. For the case of the solar system, the location of the termination shock, where the solar wind radial outward motion is decelerated to subsonic speed by the flow of interstellar gas, was recently determined by the Voyager 1 spacecraft which passed the shock in December 2004 at the distance 94 AU from the Sun. Observations and theory indicate that the location of the termination shock varies with time.

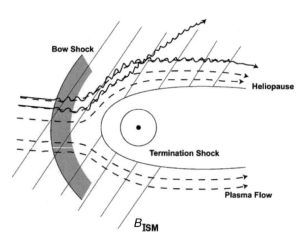

Fig. 7.2. The components of the heliosphere shown for the Sun moving from the right to the left relative to the surrounding interstellar medium plasma. *Thin lines* indicate the direction of the interstellar magnetic field (B_{ISM}), the *dashed lines* the interstellar plasma flow. The *shaded region* behind the bow shock indicates the accumulation of neutral hydrogen ("hydrogen wall") in front of the heliosphere. The motion of two small interstellar dust particles is indicated with *solid lines*: they gyrate and slide along the magnetic field lines carried by the plasma flow (from [55])

Variation of Astrospheres During Main-Sequence Phase of Stars

Since stellar wind varies with the age of the star and since the star passes different regions of the interstellar medium, the astrosphere also varies during the lifetime of the star, including also the main-sequence phase. Therefore, different studies were devoted to estimate the evolution of the heliosphere over the history of the solar system. The stellar winds are typically too low in intensity to be detected by astronomical observations (the exception being strong winds emitted by some hot stars). However, a small number of astrospheres surrounding the stars not too distant from the Sun have been observed. The observations make use of the fact that an enhanced density of neutral hydrogen (the "hydrogen wall") is generated outward from the astropauses. Wood et al. [84] studied the density of neutral hydrogen around selected stars by spectroscopically detecting the absorption of neutral hydrogen distributed along the line of sight in stellar brightness data. In this way, the enhancement in neutral hydrogen can, under certain conditions, be detected around other stars as well as around the solar system. This study allowed estimating the stellar wind parameters of stars (see Table 7.2).

Table 7.2. List of stars for which astrospheres have been detected. R_H denotes the estimated distance range to hydrogen wall in the ISM apex direction

Star	Spectral type	distance (pc)	Mass loss $\dot{M}_\odot$	R_H (AU)
α Cen	G + K	1.35	2	220–400
ε Eri	K	3.22	30	800–1750
61 Cyg A	K	3.48	0.5	20–30
ε Ind	K	3.63	0.5	30–40
36 Oph	K + K	5.99	15	300–600
λ And	G + M	25.8	5	150–200
EV Lac	M	5.05	1	60–100
70 Oph	K + K	5.09	100	1000–1700
ξ Boo	G + K	6.7	5	300–500
61 Vir	G	8.53	0.3	300–450
δ Eri	K	9.04	4	200–300
HD 128987	G	23.6	?	?
DK UMa	G	32.4	0.15	200–400

The Case Without Astrosphere

The shielding effect of the heliosphere results from the solar plasma flow and the magnetic field configuration, which deflects charged particles from entering the heliosphere and creates a boundary between the stellar wind plasma and the interstellar plasma. The charge exchange of solar wind ions with neutrals would plausibly brake and, in the supersonic flow region, heat the wind. In the absence of an astrosphere, the planetary system would be exposed to galactic cosmic rays over a broader range of energies, as well as to the flux of interstellar gas and dust depending on the relative velocity between the central star and the local interstellar medium. In the case of young stars, the formation of magnetospheres is discussed in the literature. Note that the case of the solar magnetic field amounts to about several 10 nT at 1 AU, significantly above the galactic field. Estimates for the galactic magnetic field are typically below 0.40 nT and its structure is probably homogenous on scales of the planetary systems, the heliospheric size or larger.

Particle Environment Beyond or Without Astrospheres

While outside the astrospheres objects would be exposed to the full cosmic ray spectrum, the anomalous cosmic rays would be absent. In the vicinity of an astrosphere, some of the anomalous cosmic rays may be leaking out, but their intensity would be very low. The flux of small interstellar dust would be larger than inside the astrosphere. Outside, the magnetic field would approach the interstellar field, presumably homogenous over large distance scale.

Particle Environment Extrapolated from Solar System Conditions

The high energy spectrum of the galactic and anomalous cosmic rays is rather stable in composition. Its variation over the solar cycle is understood as due to propagation effects. In addition to these rather stable components, transient particle fluxes are produced in the form of energetic particles ejected from the Sun or generated in the interplanetary medium [73]. A variety of acceleration processes produces energetic particles in the heliosphere. These include acceleration by shocks (first-order Fermi acceleration), turbulence (second-order Fermi acceleration), and reconnection processes. The generated particles cover especially the range of intermediate energies ranging from $\approx$ 30 keV/nucleon to $\approx$ 30 MeV/nucleon. The intensity, spectra, and composition of heliospheric particles are highly variable, particularly at solar maximum (see Fig. 7.4). At the lowest, solar wind energies, the typical scale of variations in the velocity, intensity, and composition of the solar wind is a factor of $\approx$ 2, on time scales that range from hours to days to years.

Recently, Mewaldt et al. [59] summarized the particle fluences from $\approx$ 0.3 keV/nucleon to $\approx$ 300 MeV/nucleon in the interplanetary medium based on in situ measurements from space missions covering both solar minimum and solar maximum conditions. They find that the solar wind contributes the majority of particles in the energy range up to $\approx$ 8 keV/nucleon. Several different populations contribute to the intermediate region from $\approx$ 10 keV/nucleon to $\approx$ 5 MeV/nucleon. The fluences in this energy range are time-variable and generated from several different sources. In spite of this, the superimposed spectra of the different populations follow a E^{-2} power-law. In the range of

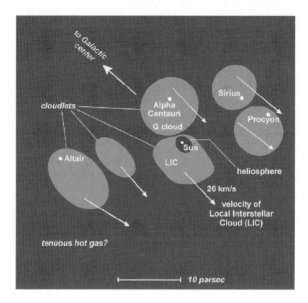

Fig. 7.3. A schematic view of the interstellar medium in the vicinity of the Sun: densities, temperatures, and velocities of the sketched interstellar cloudlets are derived from gas absorption along the line of sight recognized in the spectra of the nearby stars sketched in the figures. The local LIC moves relative to the Sun and the heliosphere with about 26 km s^{-1}. In this case, the results of the astronomical observations agree with in situ measurements of neutral interstellar gas entering the solar system. In this sketch, the heliosphere is enlarged 200 times [44]

5 to 50 MeV/nucleon, intense solar energetic particles make the largest contribution, and galactic cosmic rays at even larger energies.

Magnetic Field in the Solar System

The average magnetic field in the interplanetary medium at moderate latitudes and beyond the inner corona can be described following Parker's model [72] as:

$$B_r = \pm B_0 \left(\frac{r}{r_0}\right)^{-2} \qquad B_\phi = \pm B_0 \left(\frac{r}{r_0}\right)^{-1} \cos\theta \qquad B_\theta \equiv 0 \qquad (7.1)$$

In the solar magnetic coordinate system with $r_0 = 1\,\mathrm{AU}$, and B_0. Except for regions in the vicinity of the Sun or directly above the solar poles, this model serves as a good description of the field. The measured magnetic flux at 1 AU is highly variable and of the order of several to several 10 nT. The polarity structure of the magnetic field depends on the phase of the solar cycle. Near the minimum, the field can be approximated by a model in which northern and southern solar hemisphere correspond to opposite polarity. The dividing

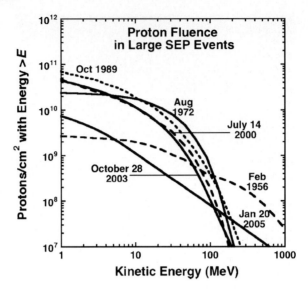

Fig. 7.4. Integral fluence spectra for protons are shown for some of the largest events of solar energetic particles (SEP) of the last 50 years [59]

line does not coincide with the ecliptic, but is in general tilted: this leads to the sector structure of the interplanetary field. The sectors with opposite polarity are separated by a thin (order of 10^5 km at a large distance from the Sun) current sheet which is warped due to the effect of the tilt. Near solar maximum, the field configuration is more complex with the regions of alternating polarity occurring also at high heliolatitudes.

7.3 Physical Processes

7.3.1 Radiation Pressure Force

The infalling stellar photons impose onto the objects the radiation pressure force:

$$F_{\text{RAD}} = \frac{L_*}{4\pi r^2 c} A <Q_{\text{RAD}}> \tag{7.2}$$

where L_* is the stellar luminosity, c is the speed of light, A is the geometrical cross-section of dust, and $<Q_{\text{RAD}}>$ is the average radiation pressure efficiency for the given stellar spectrum $F_*(\lambda)$. For spherically symmetric particles, the force is directed radially outward.

This radiation pressure force counteracts gravitational force acting on a body with mass m:

$$F_{\text{G}} = -G\frac{M_* m}{r^2} \tag{7.3}$$

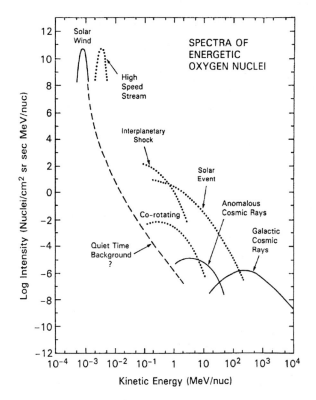

Fig. 7.5. The energy spectrum of particles in the solar system and the approximate intensities: solar wind ions (SW), particles accelerated at interplanetary shocks (ISP), solar energetic particles (SEP), particles accelerated at co-rotating interaction regions (CR), anomalous (ACR), and galactic cosmic ray particles (GCR) [76]

where G is the gravitational constant, M_* is the mass of the star, and r is heliocentric distance of the object from the star.

The ratio β_{RAD} of radiation pressure force to gravity force:

$$\beta_{\mathrm{RAD}} = \frac{F_{\mathrm{RAD}}}{F_{\mathrm{G}}} = \frac{L_* A <Q_{\mathrm{RAD}}>}{4\pi G M_* mc}. \tag{7.4}$$

is independent of the distance r from the star.

The cross section for radiation pressure transfer is a product of the geometrical cross section and the radiation pressure efficiency considered over the spectral range of infalling light:

$$<Q_{\mathrm{RAD}}> = \frac{\int_0^\infty F_*(\lambda)\, Q_{\mathrm{PR}}(m^*, \lambda)\mathrm{d}\lambda}{\int_0^\infty F_*(\lambda)\, \mathrm{d}\lambda}. \tag{7.5}$$

The radiation pressure efficiency is $Q_{\mathrm{RAD}} = Q_{\mathrm{ABS}} + Q_{\mathrm{SCA}}(1-g)$ with the efficiency for absorption Q_{ABS}, the efficiency for scattering Q_{SCA}, and the

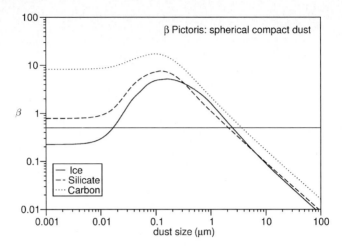

Fig. 7.6. The radiation pressure to gravity ratio calculated for compact spherical dust around β Pictoris. See [41] for detailed discussion

asymmetry parameters g. The latter denotes the ratio of light scattered in forward direction to the light scattered in backward direction.

The cross section for radiation pressure transfer $A < Q_{RAD} >$ is proportional to the geometrical cross-section A when the size of dust is large compared to the wavelength of incident light and in the Rayleigh limit it is proportional to the volume. As a result, $\beta_{RAD} \sim 1/r$ for $s >> \lambda$ and $\beta_{RAD} \sim$ const for particle in the Rayleigh limit ($s << \lambda$). For $s \sim \lambda$, it strongly varies with size depending on the wavelength of light, shape of the particle, and optical constant m^* of its composing material. The maximum of β_{RAD} as function of size lies close to sizes $s \sim \lambda$. The calculated values for compact spherical dust particles orbiting β Pictoris are shown in Fig. 7.6.

7.3.2 Stellar Wind Pressure

Stellar wind particles impacting on the dust exert a force:

$$F_{SW} = \frac{\dot{M}_* v_{SW}}{4\pi r^2} A < Q_{SW} > \tag{7.6}$$

where $\dot{M}_*$ is the mass loss rate of the star, A the geometric cross-section of the dust particle, and $< Q_{SW} >$ the efficiency factor for momentum transfer from stellar wind.

For the large dust particles, the cross section is nearly its geometrical cross section. Minato et al. [60] studied the effect of the passage of the impinging ions through small dust grains and showed that the dependence of the cross-section $A < Q_{SW} >$ on dust size is analogous to that for the electromagnetic radiation force. For dust smaller than the range of impinging ions

$(0.01 - 0.1\,\mu m)$, the cross-section is proportional to the volume of the dust particle $A <Q_{SW}> \propto V$.

In analogy to the radiation pressure force, the value β_{SW} [61] is

$$\beta_{SW} = \frac{F_{SW}}{F_G} = \frac{\dot{M}_* v_{SW} A <Q_{SW}>}{4\pi G M_* m} \tag{7.7}$$

and β_{SW} is independent of the heliocentric distance as long as $v_{SW} = \text{const}$. In the case of the solar system, the solar wind's radial force is negligible compared to the radiation pressure; $F_{SW}/F_{RAD} \sim 10^{-3}$ [67].

7.3.3 Poynting–Robertson Effect

Particles in bound orbit about the star for which radiation pressure force is smaller than the stellar gravity force migrate toward the star due to the azimuthal component of the radiation pressure force and this is called the Poynting–Robertson effect. (Note that the same applies for the stellar wind force.)

The radiation pressure force acting on dust moving with velocity v can be written in the first-order approximation in v/c as [16]

$$\boldsymbol{F}_{RAD} = F_{RAD} \left[\left(1 - \frac{\boldsymbol{v} \cdot \boldsymbol{r}}{c} \right) \frac{\boldsymbol{r}}{r} - \frac{\boldsymbol{v}}{c} \right]. \tag{7.8}$$

The non-radial term in (7.8) is opposed to the velocity vector of the dust, and it reduces the orbital energy and angular momentum of particles. The falling time of dust with circular orbit from heliocentric distance r to the star is given as

$$\tau_{RAD} = \frac{r^2 c}{2 G M_* \, \beta_{RAD}}. \tag{7.9}$$

The falling time scale of $\sim \mu m$-sized dust from 1 AU to the Sun is several thousand years.

Discussing the radiation pressure force by introducing the β_{RAD}-value requires that the stellar brightness spectrum is constant in time. Augereau and Beust [5] recently studied the influence of frequent flares on the dust dynamics around AU Mic and introduced a time-averaged β_{RAD}-value that accounts for the time variation of the stellar spectrum in the UV and EUV.

The stellar wind force in the frame of the moving particles depends on the dust velocity v as:

$$\boldsymbol{F}_{SW} = F_{SW} \left[\left(1 - \frac{\boldsymbol{v} \cdot \boldsymbol{r}}{v_{SW}} \right) \frac{\boldsymbol{r}}{r} - \frac{\boldsymbol{v}}{v_{SW}} \right], \tag{7.10}$$

where F_{SW} is the force on the dust for $v = 0$ and v_{SW} is the bulk velocity of the wind. The non-radial term in (7.10) is referred to as plasma or pseudo Poynting–Robertson drag force.

The lack of clear knowlege about the stellar wind makes it difficult to discuss the wind's forces in debris disks. Assuming, based on the studies of the astrospheres mentioned above, that the mass loss rate $\dot{M}_*$ of Sun-like stars is up to ~100 times stronger than the current solar value, the plasma P–R drag exceeds the (photon) P–R drag.

While in the case of the solar system, the solar wind's radial force is negligible compared to the radiation pressure, the plasma P–R drag is not negligible: The ratio of plasma P–R drags force to (photon) P–R drag force can be written as

$$\frac{F_{\rm SW}}{F_{\rm RAD}} \frac{c}{v_{\rm SW}} \simeq 0.3 \left(\frac{\dot{M}_*}{\dot{M}_\odot}\right) \left(\frac{L_*}{L_\odot}\right)^{-1} \left(\frac{<Q_{\rm SW}>}{<Q_{\rm RAD}>}\right) \tag{7.11}$$

where $L_\odot$ is the solar luminosity and $\dot{M}_\odot$ is the solar mass loss rate. The factor $c/v_{\rm SW}$ results from the difference of the aberration angles for photons or solar wind particles.

7.3.4 Dust Ejection: "β-Meteoroids"

Objects for which radiation pressure force (and/or stellar wind force) exceed gravitational attraction force ($\beta > 1$) cannot reach the vicinity of the star (see entry of interstellar medium dust discussed below). In planetary debris disks large objects ($\beta << 1$) orbit the star and by collisional fragmentation continuously produce smaller particles. Among the newly formed fragments, the small particles are most likely to be in hyperbolic orbits and these particles are often denoted as "β-meteoroids." The dynamics of a released $\beta > 1$ particle is determined by its kinetic energy, orbital angular momentum, and potential energy at the time of the release. Since the gravitational attraction to the star is smaller than for the parent body, the particle has a larger potential energy. If the speed relative to the parent body is small compared to orbital speed, the kinetic energy is identical to that of the parent body. As a result, the particle is less strongly coupled to the star. The orbit of the particle has a higher eccentricity than the orbit of the parent body. The parameters of the new orbit can be derived from considering orbital angular momentum and energy of the particle at the time of the release. A particle released from a parent body in circular orbit is in unbound orbit if $\beta > 0.5$ (if the initial relative velocity to the parent body is small). In the solar system ejection takes place for particles of a fraction of micrometers in size, for the case of β Pictoris for several micrometer in size.

The radial velocity $v_r(r, r_0)$ at r of a species released from a parent body in circular orbit at r_0 is:

$$v_r(r, r_0) = \left(\frac{GM}{r} \left(\frac{r}{r_0} - 1\right) (2\beta + (r_0/r) - 1)\right)^{1/2}, \tag{7.12}$$

where G is the constant of gravitation and M is the mass of the star. The asymptotic radial velocity of a β-meteoroid is $v_\infty(r_0) = ((2\beta - 1)GM/r_0)^{1/2}$. It can be expressed as $v_\infty(r_0) = (2\beta - 1)^{1/2}v_{\mathrm{orb}}(r_0)$, where $v_{\mathrm{orb}}(r_0)$ is the Keplerian velocity at the distance r_0 where the β-meteoroid was created.

For the case of dust in the solar system, most of the β-meteoroids have $0.5 < \beta < 1$ and therefore the ejected fragments do not reach high velocities. This description of the orbit of an released species is the same for dust and gaseous species, the difference lies in the β that in case of the gaseous species is determined by line absorption.

Figure 7.10 shows v_∞ versus mass for dust particles around β Pictoris calculated for different dust compositions and an initial distance r_0=10 AU from the star. In a similar way, the dust particles can be ejected by the stellar wind. This ejecting force also deflects interstellar dust particles approaching the system (see discussion below). The dust ejection processes varies for young systems where UV, X-ray, and stellar wind fluxes are possibly and probably highly variable in time.

7.3.5 Dust Temperature and Dust Sublimation

The equilibrium temperature of a particle is determined from the balance of incoming and outgoing energy [66]: Energy input results from absorption of sunlight and from energy transfer from impinging solar wind particles, the latter being negligible for the case of the present Sun. Energy is lost due to thermal emission of light and due to sublimation, the latter being important only in the range close to sublimation temperature.

Therefore in most cases, the balance of absorbed and emitted light determines the temperature. In the case of a material for which emissivity is constant with wavelength, since emissivity equals absorptivity the temperature can be approximated with the blackbody temperature. When approximating with the Stefan–Boltzmann law both, the in-falling stellar radiation and the thermally emitted radiation of the dust, one obtains the relation:

$$T_{\mathrm{dust}} = T_* \left(\frac{1-A}{4}\right)^{1/4} \left(\frac{R_*}{r}\right)^{1/2} \tag{7.13}$$

for small or for fast rotating particles at distance r from the star, where F_* denotes the brightness of the stellar photosphere, R_* the radius of the star, r, the distance from the star, and A the albedo of the object.

The temperature varies approximately as $r^{-1/2}$ in a given system and for a given distance increases proportionally to the temperature of the stellar photosphere. Assuming $T = 5800$ K for the solar photosphere and $A = 0$ for a blackbody results to the dust temperature in the solar system of 280 K at 1 AU. The temperature of objects deviates from the blackbody temperature since their emissivity varies with wavelength. One can write:

$$T_{\text{dust}} = T_{\odot} \left(\frac{1-A}{4} \right)^{1/2} \left(\frac{R_{\odot}}{r} \right)^{1/2} \left(\frac{C_A}{C_E}^{0.25} \right) \tag{7.14}$$

with C_A being the absorption cross section and C_E the emission cross-section. While according to Kirchhoff's rule at a given wavelength, the efficiencies of absorption and emission are the same $Q_A = Q_E$ for a material, the emission and absorption have their maxima in different spectral regimes with different absorption and emission efficiencies.

Through the absorption and emission behavior, the temperature of small particles strongly depends on the composition, structure, and size. Detailed studies of the temperature are therefore based on specific dust models (see Li, this issue). Temperatures both below and above the blackbody temperature may occur, depending on the particle model as well as on the distance from the star. While, for instance, a model of cometary dust described as a porous silicate with absorbing inclusion leads to temperatures above the blackbody temperature at 1 AU from the Sun, the same model leads to temperatures below that of a blackbody at distances smaller than 0.1 AU [53].

7.3.6 Dust Charging and Related Forces

Surface Charging

The surface charge of the dust is caused by impinging of solar wind particles, photoelectron emission caused by solar radiation, secondary electron emission caused by impact of charged particles with high energy, and emission of thermal ions [68].

The impact of solar wind particles and photons are the predominant mechanisms for dust charging in the interplanetary medium of the solar system, but grain size plays a crucial role for the charging when secondary emission is important [20]. The time scales to reach equilibrium surface charge are short compared to dust lifetimes.

The equilibrium surface charge on dust in the solar system corresponds to an equilibrium potential, U, of a couple of Volts compared to the surrounding medium. The calculated values depend on the material composition and location of the dust particles. In a calculation for specific solar wind conditions and dust material composition at 1 AU, the potential at 1 AU was calculated to be $U = 3.4(+0.02/-0.01)$ V for carbon particles and $U = 3.2(+0.5/-0.05)$ V for silicate particles, where the given variations result from variation of the solar wind conditions during the time span 1965–1996 (see Fig. 7.8). The calculated surface potential is almost constant with distance from the Sun in the region between 1 AU and the termination shock and for particles of sizes above 0.02 µm, it is almost constant with size [36]. The surface charge in this distance range is mainly produced by photoelectron emission and sticking of solar wind electrons. Since both, the electron density and the photon flux decrease with distance from the sun, the surface charge is almost constant with distance.

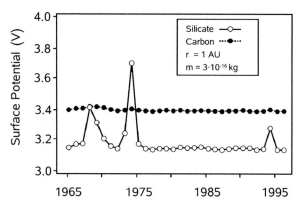

Fig. 7.7. The equilibrium surface potential of compact dust particles calculated for the measured solar wind parameters from 1965 to 1995. The calculations were made for silicate (*open circles*) and carbon (*closed circles*) particles of a mass 3×10^{-16} kg and these results are from [36]

Assuming spherical dust particles, the surface charge is then well approximated with $Q = 4\pi\epsilon_0 aU$, where ϵ_0 is the electric constant and a is the radius of the dust. For particles of 0.1 μm radius, the values given above result in a positive surface charge corresponding to $|Q(a_0 = 0.1\,\mu m)| = 240$ e (elementary charges) for silicate and $|Q(a_0 = 0.1\,\mu m)| = 225$ e for carbon, with size variation $Q(a) = Q(a_0)\, a/a_0$.

Most of the parameters that determine the surface charge are not known for circumstellar systems. To have an idea about possibly occurring dust charges, we calculated surface potentials of dust particles at 1 AU varying the solar wind parameters compared to those for the solar system. We enhanced number density of stellar wind, enhanced the plasma temperature, and enhanced the solar wind speed. All these showed only moderate enhancements of the surface charge (see Fig. 7.8). Aside from the stellar wind parameters the charging of circumstellar dust, through the photo ionization process, depends on the spectrum of the star. Figure 7.9 shows the equilibrium charge as function of distance from the star for the case of β Pictoris and AU Mic compared to the solar system. While beyond a distance of approximately 1 AU, the equilibrium charges slightly differ, close to the star they are determined by the stellar wind paramaters and in the figure are identical, since solar wind conditions were assumed for the calculations.

Lorentz Force

The Lorentz force acting on dust in the solar system reads

$$\boldsymbol{F}_{\mathrm{L}} = q\boldsymbol{V} \times \boldsymbol{B} \qquad (7.15)$$

where q is the electric surface charge, $\boldsymbol{V} = \boldsymbol{v} - \boldsymbol{v}_{\mathrm{sw}}$ is the velocity of the dust relative to the solar wind, and $\boldsymbol{B}$ is the magnetic field vector carried with the

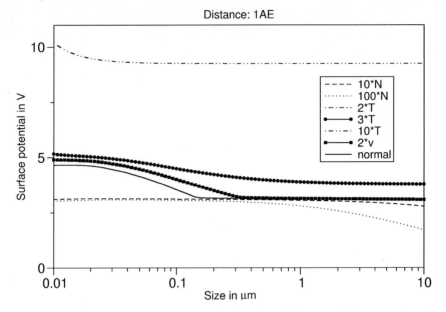

Fig. 7.8. The calculated surface potential of dust particles at 1 AU varying the solar wind parameters that were used for the calculations shown in Fig. 7.7: compared to the calculations for the solar wind parameters (*solid line*) enhanced number density by factors 10 and 100 (denoted as "10 N" and "100 N") are assumed, enhanced plasma temperature by factor of 2, 3, and 10 (2 T, 3 T, 10 T) and the calculation for a double solar wind speed coincides with a calculation for doubled temperature

solar wind. As the dust particles cross the boundary between different sectors of magnetic field with alternating polarities, the Lorentz force changes direction. For particles in bound orbit Lorentz force changes the orbital elements, semimajor axis, a, eccentricity e, and inclination i [64].

Coulomb Solar Wind Drag

Distant encounters between the solar wind particles and the dust particles cause dynamical friction described as the indirect or Coulomb solar wind drag [64]. The dynamic effects of the Coulomb drag are the same as those of the direct drag, but its strength is by about three orders of magnitude less.

7.4 Entry of Interstellar Matter

Stars are moving with typical speeds of several to several 10 km/s relative to the surrounding interstellar medium (see discussion above) and as a result interstellar medium dust enters the planetary debris disk. The entry of interstellar dust into a planetary debris disk is determined by stellar gravity force,

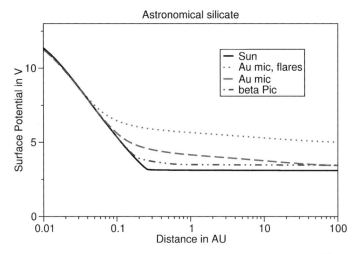

Fig. 7.9. The calculated surface potential of dust particles around of β Pictoris and AU Mic compared to the Sun for 1.5 μm particles

F_G, radiation pressure force, F_{RAD}, and Lorentz force, F_L. For $\beta < 1$ (that is F_G larger than F_{RAD}) and small $\frac{q}{m}$ the interstellar particles that approach the star as a results of the gravitational attraction are focused behind the star. When considering the slope of β as function of the particle size in Fig. 7.6, this is the case for the large particles on the right side from the maximum. Particles with $\beta > 1$ (i.e., the maximum of the same curve) approach the star against a repulsive force and do not reach the close vicinity of the star.

The distance of minimum approach is given by:

$$r(1 + \cos\ \theta) = \left(\frac{4GM_*}{v_{is^2}}\right)(1 - \beta) \qquad (7.16)$$

where θ is the angle from the interstellar upwind direction, G is the gravitational constant, M_* is the mass of the star, and r is heliocentric distance. The same conditions apply for interstellar neutral atoms.

A different picture arises for particles with large $\frac{q}{m}$ for which the Lorentz force F_L is comparable to or larger than F_g and F_{rad}. Therefore, in presence of a magnetic field for small particles (the left side from the maximum in Fig. 7.6), radiation pressure and gravity are not the dominant effects on the dust motion. If an astrosphere exists around the star, then the particles are possibly deflected from entering the disk at the boundaries of the astrosphere. This is illustrated for two interstellar particles at the boundary of the heliosphere in Fig. 7.2: they gyrate and slide along the magnetic field lines carried by the interstellar plasma flow. Particles that entered the heliosphere or the astrosphere may also be deflected by the stellar magnetic field within the astrosphere.

For the solar system, one may distinguish the following cases of interstellar dust dynamics: depletion at the heliopause for $m < 10^{-18}$ kg, magnetic field deflection inside theheliosphere for masses $m < 10^{-17}$ kg, repulsion by radiation pressure force for masses 10^{-17} kg $< m < 10^{-16}$ kg, and gravitational focusing for masses $m > 10^{-16}$ kg [52]. The importance of the different deflection mechanism, as a result of the material dependence of radiation pressure and of the rates for surface charging, varies also with the dust composition. The distributions of small interstellar dust particles are sensitive to the structure of the heliospheric transition region [21, 58].

So far, there is no clear observational evidence for interstellar dust entry into a planetary debris disk [3]. Interstellar dust particles entering the solar system were first noticed in the analysis of impact measurements near 1 AU [7] and later were confirmed with extended measurements aboard Ulysses outward to 5.4 AU [29]. The interstellar dust was identified from the data by means of orbital parameters. For earlier measurements of dust in the solar system beyond 5.4 AU, the unambiguous identification of interstellar dust fluxes is not possible [52]. The entering interstellar dust will interact with the small planetary objects and the interplanetary dust particles. The flux of the interstellar dust into the solar system is, for instance, considered as a source of dust production by impact erosion in the trans-Neptunian region [69, 88].

7.5 Collisional evolution

7.5.1 Dust Collision Model for CircumStellar Disks

Mutual collisions of dust act as a sink and as a source of particles. They determine the size distribution and influence the spatial distribution of dust. The discussion below is based on our study of the collisional effects in the solar system and in the β Pictoris system [57, 22].

It is commonly assumed [3] that the large dust particles in the disk make up the observed brightness and therefore it is required that the geometrical cross-section of the large dust particles accounts for the observed normal optical depth of the disk. The number density distribution of the large dust is derived then assuming a size distribution, for example, the collisional distribution:

$$\frac{dn}{da} = C(r)a^{-3.5} \tag{7.17}$$

where a is the dust particle radius and the parameter $C(r)$ facilitates the connection to the optical depth $\tau(r)$ at the distance r. For describing the optical depth, we use the relation [3]:

$$\tau(r) = \frac{2\tau_m}{(r/r_m)^{-2} + (r/r_m)^2} \tag{7.18}$$

where r_m=60 AU, τ_m=0.01 [42]. We point out that this model assumes that the inner region of several 10 AU distance from the star is comparatively

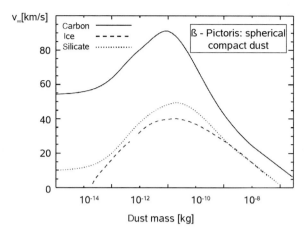

Fig. 7.10. The asymptotic velocities v_∞ versus mass for β-meteoroids created at the distance $r_0 = 10$ AU from β Pictoris. For any other initial distance r, the value of v_∞ is obtained by multiplication by the factor $(r/10 \text{ AU})^{-1/2}$ (see [22])

free of dust particles. Several other studies discuss the existence of dust in the inner region [28, 47, 81]. Assuming a dust component in the regions near the star would further increase the gas production by dust collisions that is discussed here.

Consider two subpopulations of the dust particles in the disk: the dust particles in bound orbits, with the radiation pressure to gravity ratio below the size limit for bound orbits ($\beta < 1/2$ for dust particles in circular Keplerian orbits), and the "β- meteoroids," with β above the size limit for bound orbits.

The β-meteoroids that are generated by collision need not significantly contribute to the observed brightness since they quickly leave the system. (Czechowski and Mann [23] have shown that for most of plausible assumptions about the dust properties in the β Pictoris disk this is indeed the case.)

The boundary between the bound dust particles and the β-meteoroids in the β Pictoris disk corresponds to the dust particle size $a \sim 2$ μm, the exact value depending on the dust particle material and structure (see Fig. 7.6). The very small dust particles have the β value below the size limit: they are deflected and possibly also ejected due to interaction with the magnetic field of the star [56].

The average relative velocity of colliding large dust particles is calculated assuming that the bound dust particles are in circular orbits uniformly distributed within 7° inclination from the disk central plane. The parameters of the individual collisions between the dust particles can be obtained from semi-empirical models and laboratory data [33, 78, 83] and while results may vary slightly between different models they are mainly determined by the dust number density distributions and by the relative velocities in the system.

The β-meteoroids are generated in subsequent generations: the 0th generation consists of the β-meteoroids produced in collisions between dust particles in bound orbits. The nth $(n > 0)$ generation is produced by collisions between the $(n - 1)$th generation β-meteoroids and the bound dust particles (see Fig. 7.11). The production of dust particles from the collisions between the β-meteoroids is negligible, compared to the populations listed before. The higher generations of the β-meteoroids are particularly important for the generation of impact vapor discussed below (see Fig. 7.12).

The collisions modify the size distribution of the dust and namely the power law $a^{-3.5}$ collisional distribution must be modified in the presence of the lower size cutoff (generated by the particles in hyperbolic orbits) and it is predicted that a flattening of the size distribution appears near the size limit for bound orbits [42]. Augereau et al. [6] have presented a model in which they account for the radiation pressure force acting on collision fragments that remain in bound (but elongated) orbits. They introduce an orbital ellipticity as function of the β of the fragments, which leads to a position-dependent size distribution for the bound dust particles and enhances collision rate within the component of dust in bound orbits. The collisions with the β-meteoroids influence the collision lifetime of the β Pictoris dust disk and moreover may locally generate collision avalanches [4]. While it was mentioned that the models to describe the collision process led to similar results, it should be noted that in all cases the models describing the collision process bear a large uncertainty.

Aside from the fragmentation, a fraction of the dust material is vaporized by mutual collisions. This vapor production is only significant for high relative velocities and those occur (i) between dust particles in bound orbits very close to the central star where the Keplerian velocities are high and (ii) with β-meteoroids formed close to the star and accelerated away from the star. The amount of produced vapor depends on the mass and relative velocity of

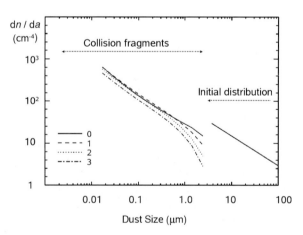

Fig. 7.11. The size distribution of dust particles at distance 100 AU from the star for the collision properties of ice. Different generations of the β-meteoroids are shown for sizes below the critical value below which ejection occurs. Ice particles with the sizes below 0.015 μm have $β < 0.5$ and were not included in the calculations

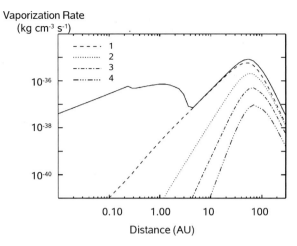

Fig. 7.12. Calculated mass vaporization rate from dust collisions in the disk of β Pictoris as a function of distance from the star assuming collision behavior of ice particles. Mutual collisions of the grains in bound orbits ($\beta < 0.5$) as well as the collisions between the first four generations of β-meteoroids and the bound grains (shown with the different lines) are included. At larger distances only the collisions with β-meteoroids contribute to vaporization. The *solid line* denotes the total gas production from collisions of dust in the disk. The corresponding size distribution is shown in Fig. 7.11

the impacting projectile and on the material composition of both particles. Figure 7.12 shows the mass vaporization rate from dust collisions in the disk of β Pictoris for vapor production from ice particles. The calculations are based on the size distribution (Fig. 7.11) discussed above. The mass vaporization rate inside 3 AU arises from mutual collisions of dust in bound Keplerian orbits. The mass vaporization rate beyond 3 AU is higher and arises from the collisions of the grains in bound orbits with the first four generations of β-meteoroids.

It is suggested that a similar process of gas production is possible in other circumstellar debris disks [22]. Generating considerable amount of gas by dust collisions requires (i) a high dust number density and sufficient acceleration of the produced collision fragments which is facilitated by large β-values which needs (ii) high temperatures of the stellar photosphere, as well as (iii) sufficiently high dust number densities in the inner disk so that the collision fragments reach a high radial velocity. For cool stars dust acceleration by stellar wind pressure should be considered as an alternative process to the radiation pressure ejection.

7.6 Second Generation Gas Components

As pointed out before, the amount of gas in the debris disks is small and in most cases is not observed. Nevertheless, some upper estimates have been made for the β Pictoris system. The fact that the gas component has no apparent influence on the motion of the dust was used to determine the upper limit of the total gas mass in the disk to 0.4 Earth masses [82]. From the absence of observable line emission from H_2, S I, Fe II, and Si II in recent Spitzer data, the upper limit of the total gas mass was placed at 17 Earth masses [19]. While these estimates are based on assumptions such as gas composition and temperature, it is generally assumed that the gas is not only much smaller than in the protoplanetary disks, but it is also not remnants of the protoplanetary disk. Most likely "second generation" gas components exist and they are observed around β Pictoris.

The circumstellar gas is observed in absorption lines superimposed to the stellar spectrum, as well as in emission lines detected around the star in the direction of the dust disk [12, 15, 18, 45, 46]. The gas comprises a stable and a time-variable component. In the time variable component, the gas with high redshift indicating inward motion toward the star is predominant. This is often interpreted as due to comet-like objects falling into the star [8, 9, 19]. The second, stable gas component is detected both in absorption lines superimposed to the stellar photospheric brightness [43] and by emission lines off-set from the star [71].

Both the absorption and the emission spectra show no evidence for gas flowing radially outward, as one would expect for gas that is repelled by radiation pressure force. The emission observations detect gas between at least 30 and 120 AU and the Doppler shift suggests that the gas is in Keplerian motion about the star [50, 71]. The results stimulated a discussion of possible braking mechanisms [26]. It was shown, however, that the observed Doppler shifts in the emission lines may arise from an observation effect, since particles in radial motion have comparatively small contribution to the observed Doppler profiles [22]. Figure 7.14 shows the calculated velocity distribution along the line of sight for species released from a parent body in Keplerian motion that is quickly ejected from the initial orbit by radiation pressure. The peak in the velocity distribution occurs at a speed that is identical to the Keplerian speed at the closest distance from the star that the LOS passes.

The apparent Keplerian motion, the predominance of non-volatile elements (see Fig. 7.13), and the spatial distribution of the gas suggest its origin from the dust disk. Processes of gas production from dust particles are collisional vaporization, sublimation, stellar wind – dust interactions, and sputtering from the dust surface. For the case of the inner solar system, it was shown that all other listed processes of gas production from the dust have production rates far below the collisional vaporization [57]. Collisional vaporization is a possible explanation for the second-generation gas observed around β Pictoris [22]. It is possible that the dust-related gas component influences the stellar wind in the system, but neither the stellar wind nor any evidence of this interaction was directly observed.

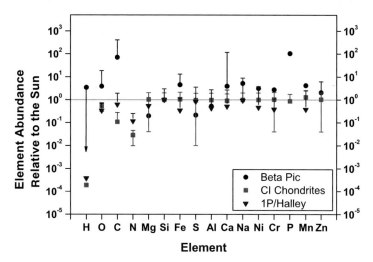

Fig. 7.13. The element abundance of gas in the β Pictoris circumstellar disk, of carbonaceous chondrites and of comet Halley dust measured in situ. The values from [74] were normalized to Si abundance and are given relative to solar elemental composition

7.7 Dust Lifetimes

It is helpful to consider which effects limit the lifetime of particles in a particular system. Mutual collisions of dust act as a sink and as a source of particles. In systems with high dust number density, the average lifetime of particles is determined by the collision rate. In systems with low dust density, the Poynting–Robertson effect limits the lifetime of dust. We may refer to the former case as collision-dominated debris disks and the later case as migration-dominated debris disks. The disk of β Pic is a collision dominated disk and most of the debris disks that are observed until today have high dust densities and are most likely collision-dominated. The solar system dust cloud is often seen as an example for a migration-dominated debris disk. Even in the solar system though, dust production by collisions plays a crucial role.

Figure 7.15 shows which effects limits the lifetime of the systems characterized by the stellar wind mass loss (given relative to the mass loss of the solar wind) and the total dust mass in the system. While systems with a large total dust mass are determined by collision lifetime, systems with a smaller dust amount are limited by the Poynting–Robertson effect. For stellar mass loss rates comparable to the early solar system, the plasma Poynting–Robertson effect exceeds the photon Poynting–Robertson effect.

Since the estimate of dust lifetimes is not always straight forward, the lifetimes that the figure is based on are given below. The time scale for a particle to move from distance $r = 50$ AU to the star as a result of the photon Poynting–Robertson effect [16] is (equation after [61]):

$$\tau_{\rm ph} = 2 \times 10^6 \, {\rm yr} \left(\frac{r}{50 \, {\rm AU}}\right)^2 \times \left(\frac{a}{\mu {\rm m}}\right) \left(\frac{\rho_d}{{\rm g \, cm}^{-3}}\right) \left(\frac{\pi a^2}{\bar{C}_{\rm ph}}\right) \left(\frac{L}{L_\odot}\right)^{-1} \quad (7.19)$$

where L and $L\odot$ are luminosities of the star and the Sun, respectively, a is the radius of the particle, and $\bar{C}_{\rm ph} = A < Q_{\rm RAD} >$ is the cross-section for radiation pressure.

The lifetime corresponding to the plasma Poynting–Robertson effect is (equation after [61]):

$$\tau_{\rm sw} = 6 \times 10^6 \, {\rm yr} \left(\frac{r}{50 \, {\rm AU}}\right)^2 \times \left(\frac{a}{\mu {\rm m}}\right) \left(\frac{\rho_d}{{\rm g \, cm}^{-3}}\right) \left(\frac{\pi a^2}{\bar{C}_{\rm sw}}\right) \left(\frac{\dot{M}}{\dot{M}_\odot}\right)^{-1} \quad (7.20)$$

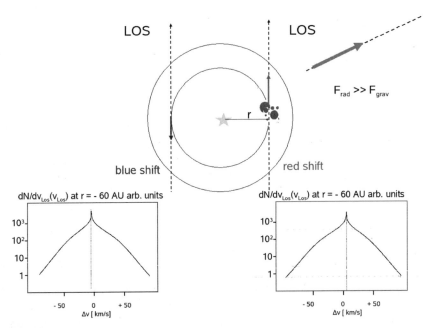

Fig. 7.14. Scheme to explain the observed Doppler profiles: The upper part illustrates that released dust or gas released from a parent body in Keplerian motion is quickly ejected from the initial orbit if radiation pressure forces are effective. The lower left and right sides of the figure denote the calculated Doppler profile for the integrated line of sight (LOS) brightness: the peak occurs at Doppler shifts that correspond to the Keplerian speed at the closest distance from the star that the LOS passes

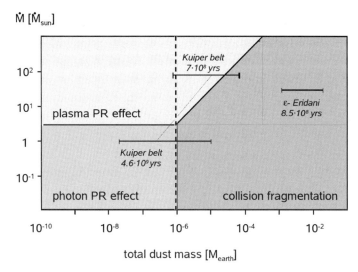

Fig. 7.15. Comparison of dust-removal processes for different values of stellar wind given by the total stellar mass loss rate compared to that of the Sun and the total amount of dust within the disk given in Earth masses. The different removal processes are discussed in the text and the dominant process is determined by the shortest lifetime (from [61])

with $\bar{C}_{sw} = A < Q_{SW} >$ the cross section for momentum transfer from solar wind. In the case of a cloud that is dominated by collision cascaded, the equation for collisional lifetime is:

$$\tau_{coll} = 2 \times 10^3 \, \text{yr} \left(\frac{r}{50 \, \text{AU}} \right)^{7/2} \times \left(\frac{a}{\mu m} \right) \left(\frac{\rho_d}{\text{g cm}^{-3}} \right) \left(\frac{\pi a^2}{\bar{S}_z} \right) \times$$
$$\times \left(\frac{M}{M_\odot} \right)^{-1/2} \left(\frac{M_d}{M_\oplus \times 10^{-3}} \right)^{-1} \tag{7.21}$$

where M_d, M, $M_\odot$, and $M_\oplus$ are the masses of the dust disk, the central star, the Sun, and the earth, respectively. The latter equation was adapted from the expression used by Najita & Williams [70] for the collision lifetime and was used as a basis for the figure comparing lifetimes.

It should be noted that aside from these major effects, the sublimation of dust occurs in the vicinity of the star, as well as sublimation of volatiles and destruction by sputtering even at larger distances. The lifetime of icy particles in the outer solar system is, for instance, limited by sputtering [88]. And Artymowicz suggested that the lifetime of water ice particles in the β Pictoris system is limited by photo sputtering. His comparison of the respective lifetimes of water ice particles around β Pictoris is shown in Fig. 7.16.

7.8 Dust Rings – Rings Caused by Dust Sublimation

Dust rings may form due to fragmentation of large objects, due to direct orbital resonance (see Ishiguro and Ueno, this issue) and due to orbital

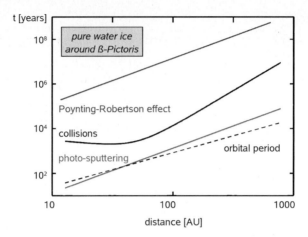

Fig. 7.16. Time scales of Poynting–Robertson drag, collsions, and ice photo-sputtering for the disk around β Pic from [2]. Further discussions are given in the text

resonances of the planetesimal parent bodies (see Wyatt, this issue) within a debris disk, as well as due to gas-dust coupling in a transitional disk stage (see [79]). Another path of dust ring formation as a result of dust sublimation was discussed in the past for the near solar dust cloud [65] and was recently suggested for migration-dominated debris disks [38, 39].

The particles initially migrate toward the Sun due to Poynting–Robertson drag (β-value below the size limit for which ejection occurs). When sublimation reduces the particle size, the radiation pressure to gravity ratio (β-value) increases and the particles are pushed outward. The decrease in particle size with distance is shown for particles sublimating near the Sun in Fig. 7.17: Solid silicate particles (thick solid lines) directly disappear due to rapid sublimation when they approach the Sun. Solid carbon particles (dashed lines) can form a dust ring. The thin horizontal line indicates the size limit of carbon particles below which they are rejected by radiation pressure.

For the larger solid carbon particles, the radiation pressure force compared to solar gravity quickly increases as their size is reduced by sublimation. Their orbital eccentricity increases and they are pushed outward. Temperature decreases at the aphelia of the dust orbits and therefore sublimation rate decreases. The particles migrate inward again. The exact slope of this size–distance curve depends on the initial conditions. This interplay of radiation pressure and sublimation can cause an enhancement of dust number density at the edge of the sublimation zone.

Dust rings were not clearly identified near the Sun. The non-detection is probably a result of the heterogeneous dust composition that leads to a broad sublimation zone and therefore a widely distributed zone in which the described orbital distribution of the particles occurs. The enhancement in dust density is distributed over a larger range of solar distances and therefore the enhancement relative to the surrounding dust density is small [51].

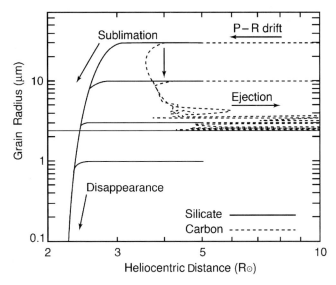

Fig. 7.17. The particle size versus distance for particles sublimating near the Sun: This figure shows the case of solid carbon particles (*dashed lines*) and silicate particles (*thick solid lines*) [54]

Recently, Kobayashi et al. [38, 39] have proposed that dust rings may form due to sublimation of icy dust particles in migration-dominated disks. These dust rings would form at larger distances from the star where the ice component disappears. While the solar system dust cloud is a migration-dominated disk, observation of our own outer solar system dust cloud is not feasible since the observed dust brightness is dominated by the brightness produced by dust near Earth orbit. While a couple of in situ measurements were carried out in the region where sublimation of ice particles should occur, the uncertainties and low statistics of the data prevent us from drawing a conclusion about the existence of sublimation-generated dust rings. Future astronomical observations of migration-dominated dust disks may reveal the dust ring formation induced by ice sublimation.

7.9 Dust Material Composition and Dust Alteration

Studies of the mineralogy of cosmic dust are numerous and their review beyond the scope of these *Lecture Notes*. This section shortly describes the basic connection between dust material composition and the origin and evolution of planetary systems. From recent observations of crystalline silicates in comets and in circumstellar disks, the relevance of the physical processes and the particle environment for the dust material alteration becomes apparent.

7.9.1 Characteristic Infrared Emission from Dust

Observed infrared spectra provide valuable information to study dust compositions in other planetary debris disks. This is so, especially in view of recent progress of infrared observations and of the wealth of new data expected for the near future (see Itoh, this issue). Deriving the mineralogy of dust from infrared spectra bears some ambiguity. Thermal emission spectra vary not only with the material and its microstructure, but also, for instance, with dust size and temperature. Moreover, the observed spectra result from the superposition of different materials and some components may not be identified from the observed spectra. In this context, Bowey and Adamson [13] point out that while the spectra of crystalline silicates are narrow and highly structured, the superposition of different crystalline silicate spectra may produce the smooth thermal emission bands around 10 and 18 μm that are commonly attributed to amorphous silicates. They show that, using their method of spectral analysis, the mass fraction of amorphous silicate in several considered spectra (of diffuse interstellar cloud, molecular cloud, and low mass young stellar objects) could be much lower than previously suggested. The analyses of infrared observations should therefore be treated with some caution. Nevertheless, they provide a valuable tool to study dust evolution.

7.9.2 Microstructure: Amorphous vs. Crystalline

Depending on the material and its microstructure, the spectral variations of infrared thermal emission brightness show distinctly different features. In this way, the microstructure, amorphous or crystalline, is sometimes derived from observations. In case of an amorphous material, the basic building blocks in the minerals – such as the SiO_4 tetrahedra in the silicate – are randomly arranged. The crystalline state forms a regular lattice structure. Energy transfer alters the microstructure: it leads to ion diffusion and breaking of chemical bonds in the solid so that the microstructure may change from amorphous to crystalline or vice versa. This process is called annealing. When heated above sublimation temperature, the silicate will sublimate and subsequently condense. Depending on the time scale t_{cool} of cooling, it will form a crystalline or amorphous phase. Figure 7.18 shows the schematics of the thermal history from amorphous to crystalline phase. This sketch was adapted from a figure by Kouchi et al. [40] for the evolution of water ice from the molecular cloud (marked with "a") to the steady state (marked with "c") of the primordial solar nebula. It can serve as a basis for discussion of other materials or other types of heating events. Assume an amorphous silicate at initial temperature ("a") is heated to a maximum temperature (marked with "b") and then cools down to a stable temperature ("c"). The microstructure that forms during condensation depends on the time span for cooling. The required temperature and time span for annealing below sublimation temperature changes in the presence of energetic particle bombardment: If energetic particle impacts damage the microstructure, the temperature, and hence the mobility of the atoms in the solid will determine at which rate the microstructure will be restored. Annealing depends on the thermal history and the energetic particle environment of the dust. It is not restricted to heating events and occurs at different stages during the life cycle of cosmic dust.

7.9.3 From Interstellar Dust to Protoplanetary Disk

One can draw the following picture of dust evolution (see Fig. 7.19): Dust particles are formed predominantly in cool stellar atmospheres and super noval and injected into the interstellar medium. These particles are processed in the interstellar medium: dust destruction, dust growth, and dust material alteration occur. The dust component of molecular clouds in which star formation occurs is incorporated into the protoplanetary accretion disks. Heating in the inner zone of the protoplanetary disk destroys the dust and dust particles recondense from the gas phase. Dust in the outer, cooler disk may survive. The radiation of the central star, impacting energetic particles, accretion processes, shock formation, and radioactive decay can cause dust alteration and destruction. Some of these processes are not confined to the inner disk. Moreover, the radial mixing may transport outward the particles that were formed or processed in the inner zones.

Detailed descriptions of the evolution of minerals and silicates, in particular, in accretion disks and the mineralogy of interstellar and circumstellar silicates are given in the chapters by Gail [30] and by Molster and Waters [63] published elsewhere in the *Lecture Notes in Physics Series*. Observations of

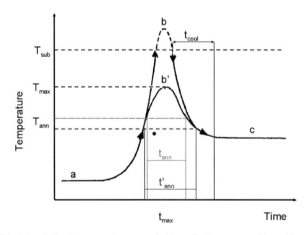

Fig. 7.18. Sketch of the temperature evolution during annealing: the microstate of material that is formed by condensation depends on the degree and the speed of cooling. If the heating is below sublimation temperature, annealing depends on the time span of the heating and may further be influenced by non-thermal processes

the newly condensed dust particles around stars in a late evolutionary stage indicate that the silicates have a crystalline structure when they are ejected into the interstellar medium. This crystalline structure is either destroyed by galactic cosmic ray bombardment, or the dust particles are destroyed and the recondensed particles form as amorphous silicates. Observations indicate that the amount of crystalline silicate is low in the diffuse interstellar medium and in molecular clouds [48]. (Carbon grains are also formed in stellar environment and may form SiC grains. However, the amount of carbon particles seems smaller and the amount of Si in interstellar medium dust that is bound in SiC is less than 5%.) Depending on the environment, ice mantles may form around the silicate grains. From those icy components by reactions in the mantle material organic refractory species are formed. These interstellar particles are incorporated into the protoplanetary accretion disk and further processed there. Heating in the inner zone will destroy the dust and silicates will re-condense to crystalline phase. Dust in the outer disk, including amorphous silicates, does not sublimate entirely.

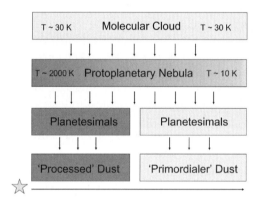

Fig. 7.19. A sketch illustrating the path of dust from the interstellar medium to a planetary debris disk: molecular cloud dust is incorporated in the proto-planetary nebula. The outer nebula has low temperatures, while the inner nebula reaches sublimation temperature. As a result planetesimals that form in the outer nebular (such as cometary nuclei) may consist of relatively primordial dust (Note that this picture does not include local heating or non-thermal processes.)

7.9.4 From Parent Bodies to Dust Debris

Dust particles are incorporated in larger parent bodies and after that the parent body fragments make up the dust that is observed in debris disks. Processing in the parent bodies can be substantial and in the solar system, meteorites, in most cases originating from asteroids are severely altered in the parent body. In spite of this, the meteorite material contain information about the processes prior to the formation of the parent bodies. Moreover, the chondrules and calcium aluminum rich inclusions (CAIs) that are found in the primitive meteorites form during processes in the inner protoplanetary disk before the parent bodies are formed. Laboratory analyses of meteoritic material provide important clues for the conditions of the protoplanetary disk of the early solar system, but its consideration is beyond the scope of this chapter.

In contrast to meteoritic material, dust particles that are incorporated in planetsimals at large distances from the star, like the cometary nuclei in the solar system, are presumably less severely altered in these icy objects. The alteration of the cometary dust is therefore regarded as indicative of the evolution in the molecular cloud and in the protoplanetary disk (as opposed to the evolution of the parent body). Since planetary debris disks are observed at distances beyond the asteroidal belt of the solar system, the dust evolution and optical properties [47, 80] in planetary debris disks are considered being similar to the cometary dust. Information about the composition of cosmic dust is in most cases limited to the optical properties. (In the case of cometary dust, space measurements provide additional information and at present, for instance, analyses of samples collected during the Stardust mission are carried out.) The major components identified from observations are ices, silicates, carbon, and organics. The evolution of ices and refractory compounds is indicative of the protoplanetary disk evolution. For instance, Yamamoto [86] pointed out that cometary volatiles may originate either as solar nebula condensates or as the sublimation residue of interstellar ices and that this should be reflected in the microstucture of the cometary water ice. Along with the evolution of ices, the evolution of volatile and refractory organic compounds is important. Recently, observers compared, in particular, the evolution of silicates in cometary dust and circumstellar dust.

7.9.5 Crystalline Silicates in Comets and Circumstellar Disks

The following discussion will focus on silicates, since they are presently best studied by astronomical observations and allow the comparison to circumstellar disk observations. For a review of silicate observations in comets and their implications for the evolution in the protoplanetary disk we refer the reader to a recent work by Wooden and colleagues [85]. Infrared spectroscopy observations with Spitzer were also recently reported for 59 main-sequence stars with debris disks, though the majority of observed spectra were featureless [17].

Since comet nuclei are formed in the colder zones of the protoplanetary disk, one would expect that they contain originally amorphous silicates. On the other hand, features due to crystalline silicate have been observed in comets as well as around β Pictoris and also in circumstellar disks of young stars. The possible explanations of crystalline silicates in the outer disks are (i) transport outward from the inner region, (ii) generation by transient heating or by non-thermal alteration of amorphous silicates in the outer zone, (iii) generation as fragments of larger parent bodies in which crystalline silicates are generated by internal processes, or (iv) crystallization after or during release from the parent body.

It is often assumed that crystalline silicates form near the star and then are transported to larger distances where they are incorporated in the planetesimals. This requires, however, a mixing process to take place before or during the formation of planetesimals. According to Gail [30] the protoplanetary accretion disks consist of three regions of different dust mixtures: (1) amorphous dust with a strong non-equilibrium composition in the outer parts of the disk, (2) crystalline dust with a strong non-equilibrium composition in a certain zone of the inner disk, and (3) crystalline dust with chemical equilibrium composition in the innermost parts of the disk. He predicts for the dust composition between 2 and 20 AU a mixture of amorphous and crystalline silicates, with the fraction of amorphous material increasing with distance from the star (see Fig. 7.20). Using a similar turbulent evolutionary model of the solar nebula Bockelee-Morvan and collaborators [11] showed that turbulent diffusion can transport silicates from the inner hot regions outward to the comet formation zone on a time scale of 10^4 years and therefore the crystalline dust could be incorporated in cometary nuclei.

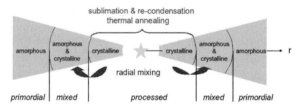

Fig. 7.20. A scenario to explain the appearance of crystalline silicates in cometary dust. Further explanations are given in the text

On the other hand, recent observational findings suggest that non-thermal processes be considered for modifying the microstructure of silicates. Silicate dust around a Herbig Be star shows an increasing mass fraction of crystalline silicate dust with distance from the star, which is not in accord with the turbulent radial mixing hypothesis [14]. From a study of solar mass T Tauri stars and intermediate-mass Herbig Ae stars with circumstellar disks, Kessler-Silacci et al. [35] infer possible variations of the silicate spectra with either the spectral type of the star or the other spectral-type-dependent factors like X-rays, UV radiation, and stellar or disk winds.

7.9.6 Non-Thermal Alteration

There are several scattered attempts to study the non-thermal alteration of dust material in debris disks. Recently, Yamamoto and Chigai [87] pointed out the opportunity that silicates may crystallize after leaving the comet nucleus and suggested the energy from chemical reaction in organic mantle material to trigger this process. Laboratory studies of the material alteration caused by ion impacts are scattered, but show the importance for material alteration for silicates and ices [77].

Irradiation of crystalline olivine with He+ ions at energies of 4 and 10 keV generated an amorphized layer of the order of several 10 nm and also changed the composition and the porosity of the silicates [23]. The material alteration caused by energetic particle impact (mainly electrons and protons) can be estimated from consideration of the particle energies. The penetration depths as function of energy are shown for the case of protons in Fig. 7.21. Solar wind particles, at the low energy end of the distribution with penetration depth smaller than the particle size, play a role in surface charging and possibly in material alteration in the surface layer. For energy ranges where penetration depth and particle size are of similar order, material alteration and sputtering occur. Particles in this energy range also cause surface erosion of larger planetary system objects. Aside from that, the penetration depths of particles with energies >100 MeV per nucleon are large compared to the dust particles and there is no significant alteration of the microstructure, although nuclear reactions may change the elemental and isotopic composition of the material.

Some clues may also be derived from the study of interplanetary dust particles in the solar system and from laboratory measurements. The processes that are caused by in-falling particles are many-fold and not clearly distinguished: surface sticking, damage of material structure, heating, ionization, sputtering, and evaporation [68]. Measurements of collected samples summarized by Jessberger et al. [34] show that the interplanetary dust has been altered by exposure to the solar wind and cosmic ray particles. The samples show an enhanced abundance of solar wind species, indicating that the solar wind particles are accumulated in the particles or in their surface layer. Also particles with higher energy seem to accumulate in the dust: The cosmic dust particles that were collected from Antarctic and Greenland ice show similar He and Ne isotope abundances than the solar energetic particles (SEP). These dust particles have sizes between 50 and 400 μm. On the other hand, SEPs with high energy pass the dust and generate the so-called solar-flare tracks, i.e., mechanical damage when they pass through the grains. Within this range of energies, impacting ions cause material alteration and particles in the energy range from MeV to several hundred MeV cause significant material alteration. Since particles in this energy range are produced by several different mechanisms, their fluxes in circumstellar systems are not known from simple extrapolation.

7.10 Summary

The vast majority of dust in planetary systems is produced by fragmentation of planetesimals and the dust is influenced by gravity, stellar radiation, and

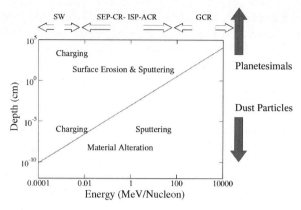

Fig. 7.21. The upper panel of the figure indicates the energy ranges of different ions in the interplanetary medium: solar wind ions (SW), solar energetic particles (SEP), particles accelerated at co-rotating interaction regions (CR), particles accelerated at interplanetary shocks (ISP), anomalous (ACR), and galactic cosmic ray particles (GCR). The calculate penetration depth of protons into SiO_2 of density $3\,g\,cm^{-3}$ for the limit of high energies is shown as a *solid line*. This is determined by interaction with the nuclei and the penetration depth at low energies is different. The size ranges of objects in planetary systems are indicated on the right side. Significant material alteration is likely to occur for energy ranges where penetration depth and particle size are of similar order

stellar wind pressure. Radiation pressure force ejects small particles that are produced by the larger parent bodies. For systems with high stellar luminosity, the fragments generated at small distances from the star reach high asymptotic speeds. Mutual collisions of the dust and small bodies produce smaller debris and, together with the condition of dust ejection, determine the size distribution of dust, which may vary within the system. Models of a steady-state collisional evolution apply to some systems or apply at least as a first-order approximation, but in general the collisional evolution may be time variable. Distributions are influenced by collision events of large parent bodies and for the systems with high β-values (i.e., high speed collision fragments) collision avalanches may occur. The entry of interstellar dust can also generate further fragments and ejecta particles.

The lifetimes of debris disks are limited by collisional fragmentation, i.e., the disk has a tendency toward self-destruction ("collision-dominated disk"). Dust lifetimes in disks with low dust density are typically limited by the Poynting–Robertson effect, or possibly by the plasma Poynting–Robertson effect for the case of cool stars ("migration-dominated disk"). While the dust in a collision-dominated disk is mainly observed in the region of the planetesimal parent bodies, the dust cloud in migration-dominated disk ranges from the region of the parent bodies inward. Local features in the dust number density are either a result of the spatial distribution of the parent bodies, or produced by forces acting directly on the dust. Namely orbital resonances with planets existing in a system will occur for both the planetesimals and the dust. Since the dust is more prone to other forces (radiation pressure, Lorentz force) than the planetesimals, it is likely that planets shape the distribution of planetesimals more clearly than they shape the dust distribution.

Dust collisions due to the impact vaporization process may also generate a component of gas within debris disks. Such a gas component was so far only studied by detailed observations for the β Pictoris system and the production by dust collisions is one among several suggested mechanisms. It is possible that the dust-related gas component influences the stellar wind in the system, but neither the stellar wind nor any evidence of this interaction was directly observed. In some cases, the strength of the stellar wind was estimated from the extension of the astrosphere that can evolve around a star as a result of stellar wind plasma interacting with interstellar medium plasma. Connected with the star and the formed astrosphere is the acceleration of different components of energetic particles: electrons, protons, and heavier highly charged ions.

The idea of the material composition of dust in the debris disks being "primordial" should be treated with some caution. Material alteration may occur during the collision process that generates the dust fragments. Energetic particles, UV- and X-ray fluxes also significantly alter the chemical appearance and microstructure of the dust material (as well as of the surface material of the small atmosphere-less bodies). Considerable processing is possibly caused by energetic particles originating from the star or from the circumstellar system, rather than by the galactic cosmic rays. These processes should be taken into account when comparing cosmic dust properties observed in debris disks to other dust populations, like dust around young stars, dust around comets, and dust in the interstellar medium.

Acknowledgement

The author wishes to thank Andrzej Czechowski, Hiroshi Kobayashi, Melanie Koehler, Richard Mewaldt, Andreas Morlok, Edmond Murad, and Tetsuo Yamamoto for helpful discussions, comments, and advice on particular topics mentioned in this chapter. Arne Manthey helped in preparing the manuscript and the figures. This research is supported by "The 21st Century COE Program of Origin and Evolution of Planetary Systems" in Ministry of Education, Culture, Sports, and Technology (MEXT).

References

1. C. W. Allen: Astrophysical Quantities, London: Athlone (3rd edition) (1976)
2. P. Artymowicz: β Pictoris: an early solar system? Annu. Rev. Earth. Planet. Sci. **25**, 175 (1997)
3. P. Artymowicz and M. Clampin: Dust around main-sequence stars: nature or nurture by , the interstellar medium? Astrophs. J. **490**, 863 (1997)

4. P. Artymowicz: β Pictoris and other solar systems, Space Sci. Revs. **92**, 69 (2000)

5. J.-C. Augereau and H. Beust: On the AU Microscopii debris disk. Density profiles, grain properties, and dust dynamics, Astron. Astrophys. **455**, 987 (2006)

6. J.C. Augereau, R.P. Nelson, A.M. Lagrange, J.C.B. Papaloizou and D. Mouillet: Dynamical modelling of large scale asymmetries in the β Pictoris dust disk, Astron. Astrophys. **370**, 447 (2001)

7. J.L. Bertaux and J.E. Blamont: Possible evidence for penetration of interstellar dust into the solar system, Nature **262**, 263–266 (1976)

8. H. Beust, A.-M. Lagrange, F. Plazy and D. Mouillet: The β Pictoris circumstellar disk. XXII. Investigating the model of multiple cometary infalls, Astron. Astrophys., **310**, 181–198 (1996)

9. H. Beust, A.-M. Lagrange, I. A. Crawford, C. Goudard, J. Spyromilio and A. Vidal-Madjar: The β Pictoris circumstellar disk. XXV. The CaII absorption lines and the Falling Evaporating Bodies model revisited using UHRF observations, Astron. Astrophys. **338**, 1015–1030 (1998)

10. H. Beust and A. Morbidelli: Falling evaporating bodies as a clue to outline the structure of the β pictoris young planetary system, Icarus **143**, 170–188 (2000)

11. D. Bockelee-Morvan, D. Gautier, F. Hersant, J.-M. Hur and F. Robert: Turbulent radial mixing in the solar nebula as the source of crystalline silicates in comets, Astron. Astrophys. **384**, 1107–1118 (2002)

12. J.C. Bouret and M. Deleuil: Investigating a chromosphere plus disk scenario for the hot gas component of β Pictoris, as revealed by FUSE. Proceedings of 12th Cambridge Workshop on Cool Stars, Stellar Systems and The Sun, 710–716 (University of Cambridge, Cambridge, 2003)

13. J. E. Bowey and A. J. Adamson: A mineralogy of extrasolar silicate dust from 10 micrometer spectra, Monthly Notices Royal Astron. Society **334**, 94–106 (2002)

14. J. Bouwman, A. de Koter, C. Dominik and L.B.F.M. Waters: The origin of crystalline silicates in the Herbig Be star HD 100546 and in comet Hale-Bopp, Astron. Astrophys. **401**, 577–592 (2003)

15. A. Brandeker, R. Liseau, G. Olofsson and M. Fridlund: The spatial structure of the β Pictoris gas disk, Astron. Astrophys. **413**, 681–691 (2004)

16. J.A. Burns, P.L. Lamy and S. Soter: Radiation forces on small particles in the solar system, Icarus, **40**, 1–48 (1979)

17. C. H. Chen, B. A. Sargent, C. Bohac, K. H. Kim, E. Leibensperger, M. Jura, J. Najita, W. J. Forrest, D. M. Watson, G. C. Sloan and L. D. Keller: Spitzer IRS spectroscopy of IRAS-discovered debris disks, Astrophys J. **166**, 351–377 (2006)

18. C. H. Chen: Dust and gas debris around main sequence stars. New horizons in astronomy: Frank N. Bash Symposium, Astronomical Society of the Pacific Conference Series, Vol. 352, p.63 (2006)

19. C. H. Chen, A. Li, C. Bohac, K. H. Kim, D. M. Watson, J. van Cleve, J. Houck, K. Stapelfeldt, M. W. Werner, G. Rieke, K. Su, M. Marengo, D. Backman, C. Beichman and G. Fazio: The Dust and Gas Around β Pictoris, Astrophys. J. **666**, 466–474 (2007)

20. V.W. Chow, D.A. Mendis and M. Rosenberg: Role of grain size and particle velocity distribution in secondary electron emission in space plasmas, Geophys. Rev. Lett. **98**, 19065 (1993)

21. A. Czechowski and I. Mann: Local interstellar cloud grains outside the heliopause, Astron. Astrophys. **410**, 165–173 (2003)

22. A. Czechowski and I. Mann: *Collisional vaporization of dust and production of gas in the β pictoris dust disk*, Astrophs. J. **660**, 1541–1555 (2007)
23. K. Demyk, Ph. Carrez, H. Leroux, P. Cordier, A. P. Jones, J. Borg, E. Quirico, P. I. Raynal and L. d'Hendecourt: *Structural and chemical alteration of crystalline olivine under low energy He+ irradiation*, Astron. Astrophys. **368**, L38–L41 (2001)
24. E.D. Feigelson, P. Broos, J.A. Gaffney III, G. Garmire, L.A. Hillenbrand, S.H. Pravdo, L. Townsley and Y. Tsuboi: *X-Ray-emitting young stars in the orion nebula*, Astrophys. J. **574**, 258–292 (2002)
25. E.D. Feigelson and T. Montmerle: *High-energy processes in young stellar objects*, Annu. Rev. Astron. Astrophys. **37**, 363–408 (1999)
26. R. Fernández, A. Brandeker and Y. Wu: *Braking the gas in the β Pictoris disk*, Astrophys. J. **643**, 509–522 (2006)
27. M. Giampapa, J. Hall, R. Radick and S. Baliunas: *A survey of chromospheric activity in the solar-type stars in the open cluster M671*, Astrophys. J. **651**, 444461 (2006)
28. A. Grigorieva, P. Artymowicz and Ph. Thébault: *Collisional dust avalanches in debris discs*, Astron. Astrophys. **461**, 537–549 (2007)
29. E. Gruen, B. Gustafson, I. Mann, M. Baguhl, G.E. Morfill, P. Staubach, A. Taylor and H.A. Zook: *Interstellar dust in the heliosphere*, Astron. Astrophys. **286**, 915–924 (1994)
30. H.-P. Gail: Formation and evolution of minerals in accretion disks and stellar outflows. In: Henning, T.K. (ed.) *Astromineralogy*, Lect. Notes Phys. **609**, 55–120 (2003)
31. L. Heroux, M. Cohen and J.E. Higgins: *Electron densities between 110 and 300 km derived from solar EUV fluxes of August 23, 1972*, J. Geophys. Res. **79**, 5237–5244 (1974)
32. J.E. Higgins: *The solar EUV flux between 230 and 1220 Å on November 9, 1971*, J. Geophys. Res. **81**, 1301–1305 (1976)
33. K. Hornung, Y.G. Malama and K.S. Kestenbohm: *Impact vaporization and ionization of cosmic dust particles*, Astrophys. Spa. Sci. **274**, 355–363 (2000)
34. E. Jessberger, T. Stephan, D. Rost, P. Arndt, M. Maetz, F.J. Stadermann, D.E. Brownlee, J.P. Bradley and G. Kurat: *Properties of Interplanetary Dust: Information from Collected Samples*, Interplanetary Dust, Springer 2001, 445–507 (2001)
35. J. Kessler-Silacci, J.-C. Augereau, C.P. Dullemond, V. Geers, F. Lahuis, N.J. Evans, II, E.F. van Dishoeck, G. A. Blake, A. C. Boogert, J. Brown, J. K. Jrgensen, C. Knez and K. M. Pontoppidan: *c2d spitzer IRS spectra of disks around T tauri stars. I. silicate emission and grain growth. Astrophys*, J. **639**, 275–291 (2006)
36. H. Kimura and I. Mann: *The electric charging of interstellar dust in the solar system and consequences for its dynamics*, Astrophs. J. **499**, 454 (1998)
37. H. Kimura, I. Mann and E.K. Jessberger: *Elemental abundances and mass densities of dust and gas in the local interstellar cloud*, Astrophys. J. **582**, 846–858 (2003)
38. H. Kobayashi, S.-I. Watanabe, H. Kimura and T. Yamamoto: *Dust ring formation due to ice sublimation in migration-dominated disks*, Icarus, **195**, 871–881 (2008)

39. H. Kobayashi, S.-I. Watanabe, H. Kimura and T. Yamamoto: *Dust ring formation due to sublimation: analytical treatment*, Icarus (submitted) (2008)

40. A. Kouchi, T. Yamamoto, T. Kozasa, T. Kuroda and J. M. Greenberg: *Conditions for condensation and preservation of amorphous ice and crystallinity of astrophysical ices*, Astron. Astrophys. **290**, 1009–1018 (1994)

41. M. Köhler and I. Mann: *Model calculations of dynamical forces and effects on dust in circumstellar debris disks*. In: Asteroids, Comets, Meteorites 2002, B. Wambein (Ed.), (ESA Publications Division, Noordwijk, 2002) 771–774

42. A.V. Krivov, I. Mann and N.A. Krivova: *The distribution of dust in circumstellar debris disks*, Astron. Astrophys. **362**, 1127–1137 (2000)

43. A.M. Lagrange, A. Vidal-Madjar, M. Deleuil, C. Emerich, H. Beust and R. Ferlet: *The β Pictoris circumstellar disk. XVII. Physical and chemical parameters of the disk*, Astron. Astrophys. **296**, 499–508 (1995)

44. R. Lallement: *Heliopause and asteropauses*, Astrophys. Spa. Sci. **277**, 205–217 (2001)

45. A. Lecavelier des Etangs, L. M. Hobbs, A. Vidal-Madjar, H. Beust, P. D. Feldman, R. Ferlet, A.-M. Lagrange, W. Moos and M. McGrath: *Possible emission lines from the gaseous β Pictoris disk*, Astron. Astrophys. **356**, 691–694 (2000)

46. A. Lecavelier des Etangs, A. Vidal-Madjar, A. Roberge, P. D. Feldman, M. Deleuil, M. André, W.P. Blair, J.-C. Bouret, J.-M. Désert, R. Ferlet, S. Friedman, G. Hébrard, M. Lemoine and H.W. Moos: *Deficiency of molecular hydrogen in the disk of β Pictoris*, Nature **412**, 706–708 (2001)

47. A. Li and J.M. Greenberg: *A comet dust model for the β Pictoris disk*, Astron. Astrophys. **331**, 291–313 (1998)

48. M.P. Li, G. Zhao and A. Li: *On the crystallinity of silicate dust in the interstellar medium*, Mon. Not. R. Astron. Soc. Doi:10.1111/j1745-3933.2007.00382.x (2007)

49. J.L. Linsky, P.L. Bornmann, K.G. Carpenter, E.K. Hege, R.F. Wing, M.S. Giampapa and S.P. Worden: *Outer atmospheres of cool stars. XII – A survey of IUE ultraviolet emission line spectra of cool dwarf stars*, Astrophys, J. **260**, 670–694 (1982)

50. R. Liseau: *Gas in dusty debris disks*, ESA Publications Division, ISBN 92-9092-849-2, 2003, 135–142 (2003)

51. I. Mann: *The solar F-corona: modellings of the optical and infrared brightness of near solar dust*, Astron. Astrophys. **261**, 329–335 (1992)

52. I. Mann and H. Kimura: *Interstellar dust properties derived from mass density, mass distribution, and flux rates in the heliosphere*, J. Geophys. Res. **105**, 10317–10328 (2000)

53. I. Mann, H. Okamoto, T. Mukai, H. Kimura and Y. Kitada: *Fractal aggregate analogues for near solar dust properties*, Astron. Astrophys. **291**, 1011–1018 (1994)

54. I. Mann, H. Kimura, D.A. Biesecker, B.T. Tsurutani, E. Grn, B. McKibben, J.C. Liou, R.M. MacQueen, T. Mukai, L. Guhartakuta and P. Lamy: *Dust near the sun*, Space Sci. Rev. **110**, 269–305 (2004)

55. I. Mann, M. Koehler, H. Kimura, A. Czechowski and T. Minato: *Dust in the solar system and in extra-solar planetary systems*, Astron. Astrophys. Rev. **13**, 159–228 (2006)

56. I. Mann, E. Murad and A. Czechowski: *Nanoparticles in the inner solar system*, Planet. Sp. Sci. **55**, 1000–1009 (2007)

57. I. Mann and A. Czechowski: *Dust destruction and ion formation in the inner solar system*, Astrophs. J. Let. **621**, L73–L76 (2005)
58. I. Mann, A. Czechowski and S. Grzedzielski: *Dust measurements at the edge of the solar system*, Adv. Space Res. **34**(1), 179–183 (2004)
59. R. A. Mewaldt, C. M. S. Cohen, G. M. Mason, A. C. Cummings, M. I. Desai, R. A. Leske, J. Raines, E. C. Stone, M. E. Wiedenbeck, T. T. von Rosenvinge and T. H. Zurbuchen: *On the differences in composition between solar energetic particles and solar wind*, Space Sci. Rev. **130**, 207–219 (2007)
60. T. Minato, M. Köhler, H. Kimura, I. Mann and T. Yamamoto: *Momentum transfer to interplanetary dust from the solar wind*, Astron. Astrophys. **424**, L13–L16 (2004)
61. T. Minato, M. Köhler, H. Kimura, I. Mann and T. Yamamoto: *Momentum transfer to fluffy dust aggregates from stellar winds*, Astron. Astrophys. **452**, 701 (2006)
62. U. Mitra-Kraev, L.K. Harra, M. Güdel, M. Audard, G. Branduardi-Raymont, H.R.M. Kay, R. Mewe, A.J.J. Raassen and L. van Driel-Gesztelyi: *Relationship between X-ray and ultraviolet emission of flares from dMe stars observed by XMM-Newton*, Astron. Astrophys. **431**, 679–686 (2005)
63. F.J. Molster and L.B. F.M. Waters: *The Mineralogy of Interstellar and Circumstellar Dust. Astromineralogy*, Lect. Notes Phys. **609**, 121–170 (2003)
64. G.E. Morfill and E. Gruen: *The motion of charged dust particles in interplanetary space. I – The zodiacal dust cloud. II – Interstellar grains*, Planet. Sp. Sci. **27**, 1269–1292 (1979)
65. T. Mukai and T. Yamamoto: *A model of the circumsolar dust cloud*. Publications Astron. Society Japan **31**, 585–596 (1979)
66. T. Mukai and G. Schwehm: *Interaction of grains with the solar energetic particles*, Astron. Astrophys. **95**, 373–382 (1981)
67. T. Mukai and T. Yamamoto: *Solar wind pressure on interplanetary dust*, Astron. Astrophys. **107**, 97–100 (1982)
68. T. Mukai, J. Blum, A.M. Nakamura, R.E. Johnson and O. Havnes: *Physical processes on interplanetary dust*, Interplanetary Dust, Springer 2001, 445–507 (2001)
69. T. Mukai, A. Higuchi, P.S. Lykawka, H. Kimura, I. Mann and S. Yamamoto: *Small bodies and dust in the outer solar system*, Adv. Space Res. **34**(1), 172–178 (2004)
70. J. Najita and J. Williams: *An 850 m survey for dust around solar-mass stars*, Astrophys. J. **635**, 625–635 (2005)
71. G. Olofsson, R. Liseau and A. Brandeker: *Widespread atomic gas emission reveals the rotation of the β Pictoris disk*, Astrophs. J. **563**, L77–L80 (2001)
72. E.N. Parker: *Dynamics of the interplanetary gas and magnetic fields*, Astrophs. J. **128**, 664 (1958)
73. D.V. Reames, C.K. Ng and A.J. Tylka: *Energy-dependent ionization states of shock-accelerated particles in the solar corona*, Geophys. Rev. Lett. **26**, 3585–3588 (1999)
74. A. Roberge, P.D. Feldman, A.J. Weiberger, M. Deleuil and J.-C. Bouret: *Stabilization of the disk around β Pictoris by extremely carbon-rich gas*, Nature **441**, 724 (2006)
75. D. Samain: *Solar continuum data on absolute intensities, center to limb variations and Laplace inversion between 1400 and 2100 Å*, Astron. Astrophys. **74**, 225–228 (1979)

76. K, Scherer, H. Fichtner, T. Borrmann, J. Beer, L. Desorgher, E. Flkiger, H.-J. Fahr, S. E. S. Ferreira, U. W. Langner, M. S. Potgieter, B. Heber, J. Masarik, N. Shaviv and J. Veizer: *Interstellar-terrestrial relations: variable cosmic environments, the dynamic heliosphere, and their imprints on terrestrial archives and climate.* Sp. Sci. Rev., 127, 327–465 (2006)

77. G. Strazzulla: *Ion irradiation and the origin of cometary materials*, Sp. Sci. Rev. **90**(1/2), 269–274 (1999)

78. S. Sugita, P.H. Schultz and S. Hasegawa: *Intensities of atomic lines and molecular bands observed in impact-indiced luminescence*, Journ. Geophys. Res. **108**, E12, 14–1 (2003)

79. T. Takeuchi and P. Artymowicz: *Dust migration and morphology in optically thin circumstellar gas disks*, Astrophs. J. **557**, 990–1006 (2001)

80. M. Tamura, M. Fugukawa and H. Kimura, et al.: *First two-micron imaging polarimetry of β pictoris*, Astrophs. J. **641**, 1172–1177 (2006)

81. P. Thébault, J.C. Augereau and H. Beust: *Dust production from collisions in extrasolar planetary systems. The inner β Pictoris disc*, Astron. Astrophys. **408**, 775–788 (2003)

82. P. Thébault and J.C. Augereau: *Upper limit on the gas density in the β Pictoris system. On the effect of gas drag on the dust dynamics*, Astron. Astrophys. **437**, 141–148 (2005)

83. A.G.G.M. Tielens, C.F. McKee, C.G. Seab and D.J. Hollenbach: *The physics of grain–grain collisions and gas–grain sputtering in interstellar shocks*, Astrophys. J. **431**, 321–340 (1994)

84. B.E. Wood, H.R. Müller, G.P. Zank, J.L. Linsky and S. Redfield: *New mass loss measurements from astrospheric Ly-α absorption*, Astrophys. J. **628**, 143–146 (2005a)

85. D. Wooden, S. Desch, D. Harker, H.-P. Gail and L. Keller: *Comet grains and implications for heating and radial mixing in the protoplanetary disk.* In: Protostars and Planets V, B. Reipurth, D. Jewitt, and K. Keil (Eds.), (University of Arizona Press, Tucson, 951, 2007) 815–833

86. T. Yamamoto: *Chemical theories on the origin of comets*, Comets in the post-Halley era., 1, 361–376 (1991)

87. T. Yamamoto and T. Chigai: *A mechanism of crystallization of cometary silicates.* Highlights of Astronomy, 13, XXVth General Assembly of the IAU – 2003, 522 (2005)

88. S. Yamamoto and T. Mukai: *Dust production by impacts of interstellar dust on Edgeworth-Kuiper Belt objects*, Astron. Astrophys. **329**, 785–791 (1998)

8

Observational Studies of Interplanetary Dust

M. Ishiguro[1] and M. Ueno[2]

[1] Department of Physics and Astronomy, Seoul National University, Seoul
151-742, Korea,
ishiguro@astro.snu.ac.kr
[2] Graduate School of Arts and Sciences, University of Tokyo, Tokyo 153-8902,
Japan,
m.ueno@exo-planet.org

Abstract We describe recent developments in observations of interplanetary dust
particles. These developments are largely due to the introduction of cooled charge
coupled device detectors and two-dimensional infrared array detectors with infrared
space telescopes. The new observational data show not only the global structure of
the interplanetary dust cloud, e.g., its symmetric plane, but also the faint structures,
such as the asteroidal dust bands and the cometary dust trails seen as a brightness
enhancement of a few percents above that of the smooth component. Spectrographic
observations provide some knowledge about the dynamics and composition of these
local components. We mention sources of interplanetary dust particles revealed by
these observations. In the last chapter, we introduce ongoing and future projects
related to the observational study of interplanetary dust.

8.1 Introduction of Zodiacal Light

The zodiacal light is a faint glow seen along the ecliptic plane. It is clearly
seen in the western sky after the evening twilight or in the eastern sky before
dawn. Figure 8.1 shows a photograph of the morning zodiacal light taken from
the top of Mauna Kea, Hawaii. In this picture, the zodiacal light appears as
a cone-shaped glow behind the telescope domes. The central axis of the cone
roughly coincides with the orientation of the ecliptic plane as seen at the
night sky. It reflects the well-established finding that the majority of dust in
the solar system is concentrated near the ecliptic plane, as most of the dust
sources (e.g., comets and asteroids) are distributed near the ecliptic plane (see
Sects. 8.2.2–8.2.4).

The zodiacal light brightness decreases with distance from the sun. In
the antisolar direction, there is a faint enhanced oval glow (Gegenschein).
Several hypotheses have been constructed to explain this excess brightness:
sunlight reflected by particles concentrated near the libration point in the
Earth–Moon system, emission from the earth's gaseous tail, sunlight scattered by the Earth's dust tail cloud, and scattered sunlight by interplanetary

Ishiguro, M., Ueno, M.: *Observational Studies of Interplanetary Dust*. Lect. Notes Phys. **758**,
231–257 (2009)
DOI 10.1007/978-3-540-76935-4_8   © Springer-Verlag Berlin Heidelberg 2009

Fig. 8.1. Photograph of the morning zodiacal light seen from the top of Mauna Kea (4200 m), Hawaii. The central axis of the cone roughly coincides with the orientation of the ecliptic plane. The shadows of domes are Keck I (*leftmost*), IRTF, CFHT, and Gemini (*rightmost*), respectively

dust cloud. After the Gegenschein was also noticed in the images taken from Pioneer 10 spacecraft at 1.86 AU from the sun [80], it became widely accepted that the Gegenschein is a part of the zodiacal light. It is attributed to enhanced scattered light in the backward direction.

At angular distances from the sun smaller than 20°, where the zodiacal light is unobservable from the ground due to atmospheric scattering light, the zodiacal light continues to increase its intensity. During total solar eclipses or from spaceborne observatories with coronagraphs, it is found that the zodiacal light is smoothly connected to the solar corona. The brightness of the solar corona consist of mainly two components, i.e., K- and F-corona. The K-corona is produced by Thomson scattering of sunlight in an atmosphere of free electrons surrounding the sun, while the F-corona is a sunlight scattered by the interplanetary dust particles. Therefore, the F-corona is the inner region of the zodiacal light. The existence of a coronal brightness hump is a controversial issue; on some of the observations that the near-infrared brightness of solar corona does not increase smoothly, but shows a deviation of the slope from continuous increase [30, 47]. The humps could be explained by a dynamical effect that produces dust ring around the sun [55]. Another explanation of the hump is that it could be produced as a result of the superposition of scattered light and thermal emission brightness when the line of sight crosses the beginning of a dust free zone [49]. However, observers of the 1991 and 1998 eclipse could not confirm the existence of humps in the near-infrared wavelength [23, 63]. It was pointed out that since the relative amount of thermal emission to scattered light brightness depends on the optical properties of dust, the existence or non-existence of the hump may indicate variations of the composition in the near-solar dust cloud [50]. The hump might also be a transient feature such as an injection of dust into near-solar space by sun-grazing comet.

The zodiacal light in the ultraviolet, visual, and near-infrared wavelengths is scattered sunlight by the interplanetary dust particles. Therefore, the color is similar to the solar spectrum, while the middle and far-infrared wavelength, it is dominated by the thermal emission of those particles (zodiacal emission). The typical temperature of interplanetary dust is around 285 K at 1 AU and the zodiacal emission is the dominant source of the diffuse sky brightness in the mid-infrared wavelength observed outside the earth's atmosphere. Since the earth is embedded in the interplanetary dust cloud, the zodiacal light and the zodiacal emission cover the entire sky even at high-ecliptic latitudes. It is important to notice that the zodiacal dust cloud is a prominent sign of the presence of planetesimal-sized objects in the solar system: Simple consideration of the Poynting–Robertson (see Sect. 8.2.1) lifetime shows that the presently observed dust cannot be identical with dust particles that existed during the formation epoch of the solar system. Although the total mass is small (equivalent to that of a small asteroid), the thermal emission and the scattered light from the interplanetary dust cloud are comparable to those from a terrestrial planet if the solar system were observed from outside. Therefore in a survey of extra-solar planetary systems, the interplanetary dust can obscure and also indicate the existence of planets.

8.2 Dust Sources and Sinks

8.2.1 Sink for Interplanetary Dust Particles

The lifetimes of the interplanetary dust particles are very short when compared with that of the solar system. The main physical processes affecting the interplanetary dust particles are expulsion by the radiation pressure, inward Poynting–Robertson drag, Yarkovsky effects, the solar wind pressure, the Lorentz force, mutual collisions, and gravitational perturbations of the planets. These are summarized in Mann et al. [52].

The solar radiation pressure, which is usually most important non-gravitational force which determine the orbital evolution of dust, is a function of the grain size, the density, and the albedo [2]. It is generally parameterized by $\beta (= F_{rad}/F_{gra})$. Here, F_{rad} and F_{gra} are the solar radiation pressure and the solar gravity, respectively. When considering a particle released from a parent body in a circular solar orbit, then dust particles with values of β larger than 0.5 will be in hyperbolic orbit. Such values obtain for silicate grains smaller than roughly <1 μm radius. The time scale for the expulsion is order of one orbital period. Thus, this process essentially produces a lower limit to the grain size. Unless the solar radiation pressure is strong enough to eject the particles, the solar radiation causes dust particles to slowly spiral inward (Poynting–Robertson drag). The falling time of the dust particle with circular orbit from a heliocentric distance r_h to the sun is proportional to r_h^2/β. The lifetime of a black 30 μm-radius grain with density 3 g cm^{-3} at 1 AU is about 10^5 yr [2], and the drift speed increases in inverse proportion as the heliocentric distance.

The solar wind pressure effectively works on the small (<1 μm) particles [56]. It includes not only radial term but also non-radial term (see the chapter by Mann). Known as the Yarkovsky effect is a force caused by photon thrust due to the temperature gradient across an object's surface, which means that this effects work on the large particles with noticeable temperature gradient [2]. If the charged particles move through the magnetic field of the sun, the Lorenz force changes the orbits. The Lorenz force works efficiently for the small particle (<1 μm) particularly in the strong magnetic field near the sun [50]. Mutual collision velocities of dust are around the order of 1 km s^{-1} and then catastrophic. The lifetime of a small dust particle is affected by the Poynting–Robertson drag when compared with the collision due to the fast orbital decay, while the lifetime of large particle is dominated by the collision.

The mass-loss rate of the interplanetary dust cloud inside 1 AU is estimated to be about 10^4 kg s^{-1} [17, 51]. Therefore, the source of the interplanetary dust particles is essential to sustain the present zodiacal cloud.

8.2.2 Cometary Dust

Comets are thought to be relatively unprocessed fossils with some of the material properties remaining from the time of the formation of the solar system, or before. When a comet approaches the sun, it develops tails of dust as well as gas. It is, however, important to notice that most of the dust particles that are observed in the vicinity of distinguished comets (e.g., C/1995 O1 Hale–Bopp and 153P/Ikeya–Zhang) are quickly ejected from the solar system. This is a result of the high eccentricities of the orbits of the parent bodies (i.e., nuclei). When we consider particles released at perihelion with zero ejection velocity, the eccentricity e_d of the dust particle is given by

$$e_{\mathrm{d}} = \frac{e_{\mathrm{p}} + \beta}{1 - \beta}, \tag{8.1}$$

where e_{p} denotes the eccentricity of the parent comet [57]. Accordingly, it is comets with low eccentricities that can efficiently supply the dust particles into the interplanetary space. Jupiter family comets (with low inclination, semi-major axis less than that of Jupiter's orbit) are potential dust sources of the interplanetary dust because not only they have low eccentricities but also they spend most of their time within the snow line (the ice condensation line, around 3 AU), where the comets are generally active.

In previous studies, the mass-loss rates from comets were mainly deduced through the observations of cometary comae. And it was concluded that comets played a minor contribution in the interplanetary dust cloud. Here, we will show that it is difficult to estimate the mass-loss rate of a comet from the dust coma observations. The observed intensity from the cometary coma is approximately proportional to the geometrical cross section,

$$\sigma(a) = \int_{a_{\mathrm{MIN}}}^{a_{\mathrm{MAX}}} \pi a^2 N(a) \mathrm{d}a, \tag{8.2}$$

where a is the radius of the dust particles. We assume a power law distribution of the size with the range between a_{MIN} and a_{MAX} for simplicity,

$$N(a) = N a^q. \tag{8.3}$$

From (8.2) and (8.3), we obtain the cross section,

$$\sigma = \frac{\pi N}{3 + q} \left(a_{\mathrm{MAX}}^{3+q} - a_{\mathrm{MIN}}^{3+q} \right). \tag{8.4}$$

Similarly, the mass-loss rate can be expressed by

$$\begin{aligned} \dot{M}(d) &= \int_{a_{\mathrm{MIN}}}^{a_{\mathrm{MAX}}} \frac{4}{3} \pi \rho a^3 N(a) \mathrm{d}a \\ &= \frac{4\pi \rho N}{3(4 + q)} \left(a_{\mathrm{MAX}}^{4+q} - a_{\mathrm{MIN}}^{4+q} \right), \end{aligned} \tag{8.5}$$

where ρ is the density of the particles. The power index q of the size distribution is generally between -3 and -4 [11, 12]. In this case, the dust mass-loss rate strongly depends on the largest particle, while the brightness depends on the smallest (0.1–1 µm-sized) grains. Accordingly, observing the large cometary particles is the key issue to derive the contribution of cometary particles to the zodiacal light.

Intriguing brightness enhancements were discovered by the Infrared Astronomical Satellite (IRAS) for directions of the line of sight crossing the orbits of eight short-period comets (2P/Encke, 7P/Pons–Winnecke, 9P/Tempel 1, 10P/Tempel 2, 22P/Kopff, 29P/Schwassmann–Wachmann 1, 65P/Gunn, and

67P/Churyumov–Gerasimenko; [73]). These features result from dust particles that are distributed on the orbits of the parent comets. They are referred to as "dust trails" and are composed of larger particles (>1 mm).

Let us recall the comet appearance (Fig. 8.2). The streams of dust and gas form a large atmosphere (coma) around the nucleus. The dust particles are initially carried outward with a steady gas flow, but may also be ejected from active areas. The streams of dust and gas of each form their own distinct tails, pointed in slightly different directions. The ion tail, made of gases that are ionized by solar radiation, is shaped by the solar wind and always points directly away from the sun. The dust particles form a curved tail: The motion of dust is mainly influenced by solar gravity and radiation pressure. Here the trajectory of the dust particles is determined by β, and β is the fundamental parameter to understand the size of particles. In other word, the solar radiation pressure is indicative of the mass of particles. The larger particles

Fig. 8.2. Photograph of C/1995 O1 Hale–Bopp obtained by Kiso 1.05 m Schmidt telescope. In this image, the ion tail points directly away from the sun, while the dust tail diffuses in the different direction. Note that Hale–Bopp do not possess the dust trail due to its large eccentricity (e=0.995) (courtesy of Kiso observatory, Univ. of Tokyo)

($>$1 mm), which are less sensitive to solar radiation pressure, can stay near the orbit of the parent comets. They are gradually dragged toward the direction opposite to the movement direction of the comet movement because the radiation pressure enlarges the orbital period of the dust particles. Finally, the large particles form narrow extended tubes along the orbit of the parent comets (dust trails). Dust trails are observed as meteor showers when they encounter the earth's atmosphere.

The Infrared Space Observatory (ISO) complemented the IRAS observations; it revealed that one of Jupiter family comets, 2P/Encke, released dust particles up to 10 cm in diameter [67]. Recently, Reach et al. [69] observed 34 comets using the 24 µm camera on the Spitzer Space Telescope, and they found dust trail structures along the orbits of 27 comets. The detection rate is $\sim$80%, and this indicates that the dust trails are generic feature of short-period comets [69]. Another possibility to detect faint dust structure-like dust trails was opened up by wide-field Charge-Coupled Devices (CCD) [27]. Figure 8.3 shows the images of short-period comet 4P/Faye [71]. The extended dust cloud is observable along the orbital plane of the parent body. The high-resolution images allow us to deduce the size distribution and mass-loss rate precisely. The estimated number density is 10^{-9}–10^{-12} m^{-3}, and the lifetime is 10–1000 years. Recent survey of the cometary dust indicates that about 1/2–1/3 of the interplanetary dust particles could be supplied from the short-period comets [28].

There is, however, a discrepancy in the optical properties between the zodiacal dust and the cometary dust. The color of zodiacal light is slightly redder than that of the sun, whereas cometary dust particles are significantly redder than that of the sun. The color of the zodiacal light in the visual

Fig. 8.3. CCD image of Jupiter family comet 4P/Faye observed from Kiso Observatory with 2KCCD camera. Note the bright elongated elliptical structure at the left. This is the dust tail composed of fresh dust particles released during a current return of the comet to the inner solar system. The faint extended and more elongated structure is the brightness generated by the dust trail, which is composed of large grains emitted during previous apparitions (courtesy of Yuki Sarugaku, Univ. of Tokyo)

and near-infrared wavelengths is similar to those of asteroids (C-type: [46]; S-type: [53]). The albedo of the interplanetary dust particles is higher than that of the cometary dust, and similar to the surface of the asteroids. These evidences might suggest that asteroid is also abundant source of the interplanetary dust [29].

8.2.3 Asteroidal Particles

It is natural to suppose that collisions among the asteroids can produce dust particles. Signatures of collisional fragments were found by the IRAS survey as the asteroidal dust bands. They were subsequently found in the scattered light of the zodiacal light (Fig. 8.4) [26]. The dust bands were initially observed as two local maxima in the IRAS scans across the ecliptic plane and two shoulders around the ecliptic latitude of $\pm 10°$. These bands of enhanced brightness are produced by fragments that have similar inclination, eccentricity, and semi-major axis, and therefore produce a torus shape cloud (Fig. 8.5). The brightness features of the dust bands are faint and appear at the location where the line of sight passes the edge of the torus. The dust bands were originally interpreted as the result of collisions in the asteroid belt associated with three prominent asteroid families (Eos, Koronis, and Themis; [4]). The discovery of the dust bands is clear evidence that there is an important asteroidal contribution to the interplanetary dust.

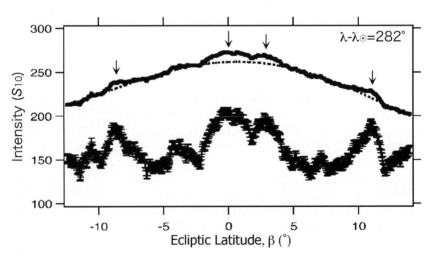

Fig. 8.4. (*Upper*) The surface brightness profile of the zodiacal light ($\lambda - \lambda_\odot = 282°$). The dotted line denotes the fit to the background component by a polynomial. (*Lower*) The difference between the brightness profile and the smoothed background. For clarity, the lower curve and error bars are scaled by a factor of 5 and shifted 150 counts. The asteroidal dust bands are seen as bumpy structures around $\pm 10°$ and near the ecliptic plane, which are indicated by arrows

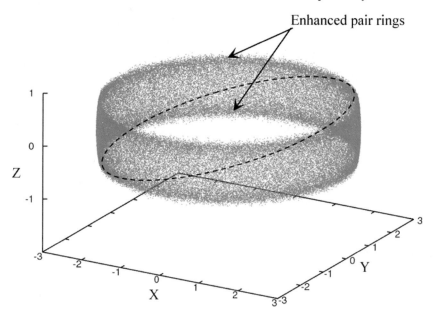

Fig. 8.5. The apparent cumulative effects on the spatial distribution of debris from an asteroid family. Note that each particle has an ellipsoidal orbit (e.g., *dashed line*). The family members have the inclination, eccentricity, and semi-major axis similar to those of each other, but have random true anomalies, longitudes of perihelion, and ascending node. These particles form the torus structure. Since the vertical velocity becomes zero at the position farthest away from the symmetric plane (nearly Jupiter orbital plane), the sojourn probability is highest on the edge of the torus. Consequently, the torus shows two paired rings (upper and lower edges of the torus) parallel to the symmetric plane

The relationship between observed asteroidal families and dust bands has been a matter of much debate. On one side, Dermott et al. [4] proposed that the dust bands consist of material produced by ongoing collisional grinding of bodies within the prominent asteroid families. This scenario is referred to as the "equilibrium model," where the contribution of the dust band material to the zodiacal cloud is much greater than that of the observed signals. The signature of the dust bands was considered as only the "tip of the iceberg" compared to the entire brightness produced by the dust bands. On the other hand, Sykes and Greenberg [74] suggested that the dust bands were produced by stochastic fragmentation of asteroids with ∼10 km diameter that occurred in the main-belt region within the last several million years ("Non-equilibrium model"). In this model, the positions of the dust bands are independent of the prominent asteroid families. The equilibrium model had been accepted widely by many researchers. The equilibrium model predicted that asteroidal families were major sources of the interplanetary dust particles (75% of the entire interplanetary dust cloud, [14]).

A difficulty in the equilibrium scenario is that the positions of the postulated fragment clouds are different from those of the observed dust bands. Especially, the position of $\pm 10°$ dust band deviates considerably from the potential parent, Eos family. The evidences of non-equilibrium model were found recently. The 10 km-sized young family Karin was discovered. It formed only 5.8 million years ago [60]. Subsequently, Veritas family, whose collisional event took place about 8.3 million years ago, was found. Indirect evidence supports the recent fragmentation events. Farley et al. [10] examined the sediment layers in the deep-sea drill cores, and found that the isotope helium-3 (^{3}He) in the 8.2 million year old layer is four times more abundant than the normal value. The authors suggest that the ^{3}He spike is due to the interplanetary dust enriched by the solar-wind ^{3}He. Thus, Poynting–Robertson drag would have caused the smallest dust particles to spiral inward immediately after the Veritas event, bathing the earth with a dust shower, and large particles are observed as dust bands. Nowadays, the non-equillibrium model is widely accepted, and the currently observed dust bands near the ecliptic plane and $\pm 10°$ are interpreted as the remnants of the Karin and Veritas events, respectively. According to the non-equillibrium model, the asteroidal contribution is 10% or less of the whole interplanetary dust cloud [62].

8.2.4 Other Sources

Kuiper belt objects (KBOs) are icy bodies beyond Neptune's orbit. The KBOs are also a potential source of dust particles either produced by mutual collisions among KBOs, or produced as ejecta particles as a result of the bombardment of KBOs with interstellar dust [9, 83]. Yamamoto and Mukai [83] estimated that the dust production rate by KBOs is 10^3–10^4 kg s^{-1}. These grains should drift toward the sun by Poynting–Robertson drag. However, for them to make a noticeable contribution to the inner solar system dust population, they must survive against (a) collisions with interplanetary/interstellar grains, which may be especially important for larger and more slowly-drifting KBO grains that are exposed to bombardment for longer intervals, and (b) gravitational perturbation by the Jovian planets that can trap grains in resonances and/or eject them from the solar system.

Dust detectors onboard Pioneer 10 and 11 conducted in situ measurements of the dust in the outer solar system and these data were reconsidered again [39]. Nevertheless, the early Pioneer measurements were severely limited in statistic and moreover results from impact measurements did not agree with optical observations about the same spacecraft [52]. An effort for the direct detection of KBO dust cloud was done using the infrared all-sky survey data by infrared telescope, and KBO dust disk is predicted to be, at most, a few percent of the brightness of the zodiacal cloud [1, 78].

Interstellar grains are expected to be a significant source of the smallest particles in the outer solar system. Ulysses detected interstellar dust in the outer solar system coming from the direction of the solar motion through the galaxy [18]. Dynamical simulation based on the in situ measurement indicates that interstellar dust brightness is less than 1% of the zodiacal emission at 12 μm at 1 AU from the sun [15].

The debate on the origins of the interplanetary dust particles is expected to continue in the future.

8.3 Smooth Zodiacal Cloud Components

8.3.1 Brightness Integral

In Sect. 2 of this chapter, we mention the dust sources. Most of the dust production is not directly observed but is apparent in the small-scale structures seen the zodiacal light and the zodiacal emission. The smooth background component gives the information about the dynamical evolution of the particles. In this section, we describe the large-scale structure of the zodiacal light, and introduce observational results which are related to the planetary perturbations.

The surface brightness of the zodiacal light $Z_{\text{opt}}(\lambda - \lambda_\odot, \beta)$ is described as a double integral over the size distribution function and along the line of sight. As a first approximation, we assume the followings:

(1) The size distribution of the dust particles in the zodiacal cloud is independent of the position.
(2) The scattering properties (e.g., albedo and the scattering function) are independent of the position too.
(3) The zodiacal cloud has a symmetry plane which contains the center of the sun and is axisymmetrical with respect to the solar axis perpendicular to the symmetry plane.
(4) The dust grains are sufficiently far from the sun so that we can regard it as point source.

Under these assumptions, the brightness of the zodiacal light $Z_{\text{opt}}(\lambda - \lambda_\odot, \beta)$ is given by

$$Z_{\text{opt}}(\lambda - \lambda_\odot, \beta) = \int_0^\infty \int_0^\infty F(r_{\text{h}})n(r_{\text{h}}; \beta')f(s)\sigma_{\text{sca}}(s)\phi(\theta; s) \, ds \, dl, \quad (8.6)$$

where r_{h}, dl, and θ denote the heliocentric distance, line element, and scattering angle, respectively, and s is the particle radius. The incident solar flux at the heliocentric distance r_{h}, $F(r_{\text{h}})$, can be replaced by $F_\odot(r_0/r_{\text{h}})^2$, where F_0 means the solar flux at r_0. $n(r_{\text{h}}; \beta')$ is the number density of the cloud particles. The geometry and the notations are summarized in Fig. 8.6.

For simplification, the mean volume scattering phase function $\Phi(\theta)$ and the mean total scattering cross-section $\bar{\sigma}_{\text{sca}}$ are defined by

$$\int_0^\infty f(s)\sigma_{\text{sca}}(s)\phi(\theta; s)ds \equiv \bar{\sigma}_{\text{sca}}\Phi(\theta) \quad (8.7)$$

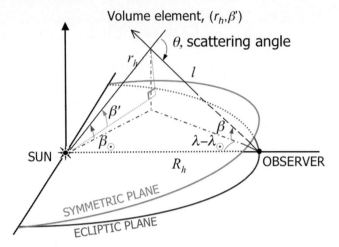

Fig. 8.6. The viewing geometry for light scattered by a dust volume element. The scattering angle θ is measured from the line of sight to the antisolar direction at the volume element

with a normalization condition

$$\int_{4\pi} \Phi(\theta)\mathrm{d}\Omega = 1. \tag{8.8}$$

The mean volume scattering phase function was empirically derived from the zodiacal light observations [6, 25, 38, 41]. In Fig. 8.7, we show the mean volume scattering phase function given by Hong [25]. It has a strong peak in the forward direction, and isotropic component at intermediate scattering angle, and a slight enhancement in the backward direction.

Similarly, the brightness of the zodiacal emission can be obtained by

$$Z_{\mathrm{IR}}(\lambda - \lambda_\odot, \beta) = \int_0^\infty n(r_\mathrm{h}; \beta')\bar{\sigma}_\mathrm{abs}B(T)\,\mathrm{d}l, \tag{8.9}$$

where $\bar{\sigma}_\mathrm{abs}$ is the mean absorption cross section of the dust particles and $B(T)$ is the Plank function.

The number density of the particles is presumed to be of a form that the separable into radial and vertical terms, i.e.,

$$n(r_\mathrm{h}; \beta') = n_0 \left(\frac{r_\mathrm{h}}{r_0}\right)^{-\nu} h(\beta'), \tag{8.10}$$

where n_0 is the reference dust number density in the symmetry plane at the heliocentric distance of r_0. The radial power law is motivated by the radial distribution expected for particles under the effect of Poynting–Robertson, which results in $\nu = 1$ for dust in circular orbit; $h(\beta')$ describes how the dust number density falls off above and below of the symmetry plane.

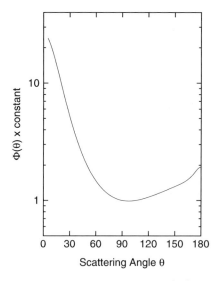

Fig. 8.7. The scattering phase function from Hong [25]. Although Hong [25] generalizes the phase function using a radial power law ν, we show the example for a $\nu=1$

8.3.2 Seasonal Variation

If the symmetric plane of the zodiacal cloud was corresponding to the ecliptic plane, the latitude of peak brightness would be in the ecliptic plane, and the brightness of the ecliptic pole would remain constant throughout the year. The discrepancy between the ecliptic plane and the symmetric plane of the zodiacal cloud causes the seasonal variation of the zodiacal light brightness.

Figure 8.8 shows the annual variation of the peak positions and the pole brightness difference. To understand the variation, we show the potential position of the symmetric plane in Fig. 8.9. Between January and May, the brightness of the north ecliptic pole (NEP) is brighter than that of the south ecliptic pole (SEP) because the line of sight crosses the region of highest dust density to the north, on the other hand, the brightness of the NEP is fainter than that of the SEP in the autumn for a similar reason. As a result, it shows the sinusoidal variation in 1-yr cycle.

The symmetry plane can be determined by investigating this variation. Helios spacecraft monitored the zodiacal light brightness in the optical wavelength from the heliocentric distance of 0.3–1.0 AU [42]. Many researchers investigated the symmetry plane in the visible and the infrared wavelength from Earth orbit (i.e., from 1 AU). We summarize their results in Table 8.1. We add the centroid of the dust bands, which are distributed around the asteroidal belt. The symmetric plane may not have a flat surface but warped structure. A warped midplane was detected in circumstellar disks (e.g., β-Pic, [54]), and is considered to be possibly formed by the gravitational perturbation of a planet.

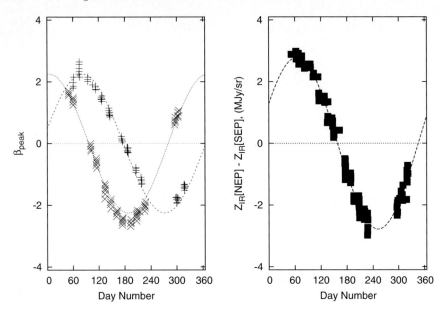

Fig. 8.8. Annual variation of the zodiacal emission obtained by IRAS in 12 μm band. (*Left*) The observed peak brightness latitude in the leading (or morning) side (*plus*) and in the trailing (or evening) side (*cross*). The original data are cited from Kwon and Hong [36]

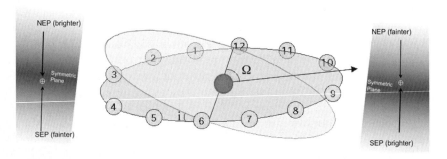

Fig. 8.9. Geometry of the earth orbit and the symmetric plane of the zodiacal cloud. The numbers denote the positions of the earth at the beginning of each month. Here we show the symmetric plane deduced by the COBE/DIRBE observations [34], i.e., $i = 2.03°$ and $\Omega = 77.7°$

8.3.3 Resonance Dust Rings

A leading/trailing brightness asymmetry in the zodiacal emission brightness around solar elongation of 90° was first found by IRAS [5]. DIRBE (Diffuse Infrared Background Experiment) onboard COBE satellite revealed the large-scale brightness enhancements around solar elongation of 90° and

Table 8.1. Plane of symmetry of the zodiacal cloud. References: (1) Leinert et al. [42], (2) Kwon et al. [37], (3) Kwon and Hong [36], (4) Kelsall et al. [34], (5) Grogan et al. [16]

Ω (°)	i (°)	Method	Position	References
87 ± 4	3.0 ± 0.3	Optical	0.3–1.0 AU	1
∼80	∼2	Optical	≈1 AU	2
38–61	1.5 ± 0.1	Infrared, Latitude of peak flux	≥1 AU	3
75 ± 2	1.8–2.4	Infrared, Pole brightness	1 AU	3
77.7 ± 0.6	2.03 ± 0.017	Infrared brightness fitting	≈1 AU	4
99.9 ± 7.8	1.16 ± 0.09	Infrared, Latitude of dust bands' midpoint	2–3	5

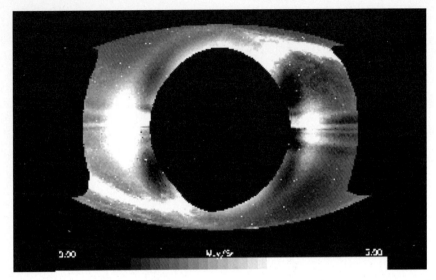

Fig. 8.10. The COBE/DIRBE image of excess zodiacal emission brightness obtained by subtraction of smooth zodiacal emission component. In the image, the sun is located in the center and the ecliptic plane runs across the sun horizontally. The NEP and the SEP are the top and bottom of the image, respectively. The two bright spots around the solar elongation of 90° near the ecliptic plane are trailing enhancement (*left*) and leading enhancement (*right*) due to the dust ring. The horizontal stripes near the ecliptic are asteroidal bands, and the "S"-shaped trail is originated from the Galaxy

the brightness in the trailing side is brighter than that in the leading side (Fig. 8.10).

Dermott et al. [5] suggested that the brightness enhancement is caused by a cloud of asteroidal dust particles that migrate toward the sun due to Poynting–Robertson drag when they are temporarily trapped in mean motion resonances with the earth. These particles corotate with the earth and form a circumsolar dust ring. The numerical simulation by Dermott et al. [5] showed not only the ring component but also the trailing blob.

Searches were also carried out for dust rings around the other planets, and a Mars resonance ring could not be detected by DIRBE [35]. On the other hand, there is weak brightness enhancement near Venus orbit which might be a resonance ring with Venus [44]. Note, that in a similar way, resonance rings or wakes in planetary debris disks could indicate the presence of planets. Such large-scale structures are indeed found in dust disks of exoplanetary systems: for instance, the brightness asymmetry observed in the HR 4796 disk [77] can be explained by the secular gravitational perturbations of a planet on an eccentric orbit 82; also clumps observed in the ε-Eri, Vega, and Fomalhaut disks [13, 24, 81] may be indicative of dust trapped in mean motion resonance with a planet in these systems [66, 83]. Aside from the formation of dust resonance rings (formed in the same way as in the solar system dust cloud), also other processes are likely to generate local structures in dust clouds of exosolar planetary disks (see Wyatt, this issue).

8.4 Wavelength Dependency and Albedo

Aside from intensity, the spectrum, the broadband color, and the polarization of the zodiacal light are measured. Since these results convey information about the dust properties, they also yield information about the dust origin. Figure 8.11 shows the spectrum of the zodiacal light from optical to mid-infrared wavelength. The color of the zodiacal light in the optical spectral regime is fundamentally similar to the solar spectrum. The brightness of the zodiacal light decreases with increasing the wavelength due to the decreasing of the incident solar flux. At the wavelength longer than $\sim 3\,\mu m$, the thermal emission from the particles arises. The zodiacal emission is mainly detectable from the space. The spectrum of the zodiacal emission measured near 1 AU around the solar elongation of 90° follows the shape of a blackbody emission curve for particles whose temperature is ~ 250–300 K.

Multiband observations revealed that the optical color of the zodiacal light is slightly redder than that of the sun. Even so, the spectral gradient of 4~5 %/(1000 Å) [43] is less reddish than the average of comet nuclei (8%/(1000 Å)), and KBOs (23%/(1000 Å)). Matsumoto et al. [53] pointed out the similarity of the spectrum of the zodiacal light at 0.5–2.5 μm to those of S-type asteroids, which is the second largest spectral group among the main-belt asteroids. It should be noted, however, that the reflectance spectrum of the S-type asteroids is highly controlled by the degree of the space weathering.

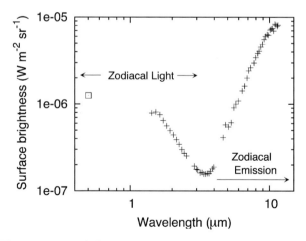

Fig. 8.11. The spectrum of the zodiacal cloud on the ecliptic plane obtained by IRTS (*cross*, [65]) and ground-based observation (*rectangle*, [45])

Space weathering is caused by the cosmic rays irradiation and bombardments of the micrometeorites [3].

When optical observations are combined with additional information about the cross-sectional area of dust (either by comparison with thermal observations, or by combination with dust in situ measurements that provide the size distribution), it is possible to derive the dust albedo. The geometric albedo A_p is widely applied in the planetary science, and is defined as the ratio of the energy scattered at an angle of 180° to that scattered according to Lambert's law by a white disk of the same geometric cross section. In practice, the definition of the geometric albedo cannot be applied to comets, because a comet is observed at a single scattering angle at a given epoch. Hanner et al. [19] generalized the geometric albedo; an albedo $A_p(\theta)$, defined as the geometric albedo times the normalized phase function at a scattering angle θ. The albedo of the interplanetary dust is $A_p(90°) \sim 0.1$ [7, 19]. The albedo of the cometary dust is smaller than that of the interplanetary dust; 0.025–0.10 in comae [20], 0.02 in the trail [72], while the albedo of asteroidal particles are generally 0.03–0.5 [75]. Therefore, it is likely that the interplanetary dust particles from bright asteroids might be detected preferentially in the optical and near-infrared wavelength due to their higher albedo. In addition, Dumont and Levasseur-Regourd [7] found that the albedo decreases as a function of the distance from the sun, that is to say, they found that grains are darker at greater distances from the sun. Renard et al. [70] found that the albedo of the interplanetary dust increases as the ecliptic latitude increases. These variations may suggest that the zodiacal dust cloud is composed of dust particles of several origins.

Even the average local polarization of interplanetary dust and the variation with scattering angle has been derived from zodiacal light observations (e.g., [25]). The function of degree of linear polarization has a negative branch around scattering angle $\theta = 160°$–180°. The negative polarization was derived from observations in the Gegenschein [40]. The degree of linear polarization increases with decreasing scattering angle, and has a maximum value around $\theta = 90°$. Similar trends are observed for both cometary dust and asteroidal surfaces, but the maximum values of the interplanetary dust is more similar to those of cometary dust.

High-dispersion spectroscopy in the optical wavelength provides unique information about the orbits of the dust particles. As we mentioned above, the brightness of zodiacal light and zodiacal emission results from the spatial distribution, the scattering phase function, and the temperature. The orbital distribution such as the distribution of the inclinations of orbits in the dust cloud are inferred from the brightness data. The sunlight that is scattered by the dust particles experiences a Doppler shift, and the measurements by optical spectroscopy can help us to understand the dust dynamics through analysis of the Doppler shift. The Fraunhofer lines of the solar spectrum are shifted as a result of the relative velocity between the sun and the dust particles and the relative velocity between the dust particles and the observer.

Consequently, the line shape is broadened as a result of the integration along the line of sight [58]. The idea of using the Doppler shift as a tool to study dust dynamics is of old standing, and the world famous guitarist, Brian May, attempted to measure the Doppler shift in early 1970s before laying aside his astronomical interests in favor of Rock music [22]. Recent measurements reveal that there are dust populations with higher eccentricities, suggesting the existence of particles of cometary origin [31]. Although the number of Doppler observations is limited, we expect the method becomes in future a powerful tool to study dynamics and therewith the origins of the interplanetary dust.

During the past decade, infrared spectroscopy became an important to study dust properties. In the 9–11 μm, some comet's spectra show an excess on the continuum thermal emission [21]. These features originate from presence of small silicate particles. These so-called infrared silicate feature have also been found in the spectra of debris disks, for instance, β-Pic and HD 14257 [48, 76]. The shape of the feature can reveal not only the composition of the silicate but also the size distribution because the appearance and the strength of the feature depends on the size of the particles and is especially present in the radiation from small particles (radii smaller than a couple of 1 μm). Reach et al. [68] reported that there is a weak excess in the zodiacal emission with an amplitude of 6%. They compared the excess with that of β-Pic, and found that the shape of the silicate feature in the zodiacal emission is different from that of β-Pic, indicating a different mineralogy. The excess in the infrared zodiacal emission was confirmed by IRTS (Infrared Telescope in Space, [64]).

8.5 Ongoing and Future Projects

8.5.1 Zodiacal Light Brightness Monitoring from the Ground and the Space

Many active observations of the zodiacal light were performed in the 1960s and 1970s using photo-multipliers attached to telescopes at high-altitude site. These efforts have provided us with an overview of the zodiacal light at the visible wavelength, but they have also revealed limits of the optical observations with the former systems, i.e., low spatial and time resolution, compared with the spaceborne observations such as with IRAS. About 20 years later, the high sensitivity and the imaging capability of CCD combined with wide-field lens optics has enabled us to obtain "snapshots" of diffuse faint objects with a portable system [32].

WIZARD (Wide-field Imager of Zodiacal light with ARray Detector) is a wide-field camera with liquid-N$_2$ cooled CCD. It covers a field-of-view of $46° \times 92°$ with good spatial resolution of $1.435'$/pixel. One of the problems of the CCD observation of the zodiacal light lies in determinating the background zero level. WIZARD is designed to determine the zero signal level by sampling the insensitive pixels. Monitoring observations of the zodiacal light by WIZARD is currently in progress at the top of Mauna Kea (Fig. 8.12).

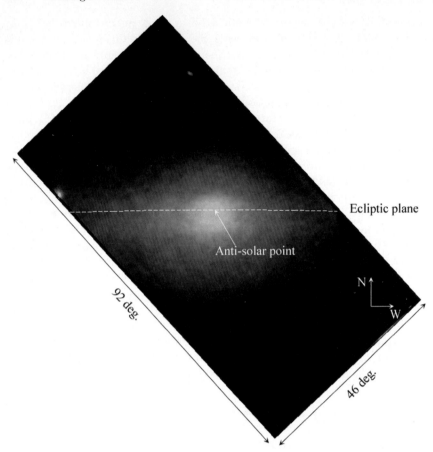

Fig. 8.12. Snapshot of the Gegenschien taken by WIZARD on October 14, 2004. The overall brightness distribution of the Gegenschein is shifted to the south around the season, which can be explained by the zodiacal cloud model that the symmetric plane deviates toward the south in the season

SMEI (Solar Mass Ejection Imager) is an optical CCD cameras designed to detect the transient Coronal Mass Ejections (CMEs). It is carried on the Coriolis satellite which was launched on January 2003. It can cover the entire sky in the course of an orbit (102 min). SEMI is composed of the three CCD cameras, and each individual camera has 3° × 60° field-of-view and ~1° resolution. SMEI provides the chance to monitor the zodiacal light brightness as well as the interplanetary phenomena associated with the CMEs. Studying the phenomena caused by CMEs is the original purpose of the mission. Nevertheless, the zodiacal light is a significant contributor to these sky maps and must be removed in the data analysis in order to detect the much fainter CMEs. The data reduction of the zodiacal light is ongoing [79].

It is expected that the global structure of the zodiacal light will be well determined based on the high time-resolution measurements with SMEI. The fine scale structures in the zodiacal light will be found in the high-statial-resolution observations with WIZARD.

8.5.2 Infrared Space Telescope Mission: AKARI and Spitzer

From the earth-bound orbits, the brightest natural light sources in the infrared wavelength are the earth and the sun. Therefore, infrared telescopes were designed to point radially outwards from the earth and nearly perpendicular to the sun. Accordingly, the zodiacal emission was/is observed only around the solar elongation around 90°. IRAS is the first infrared whole sky survey mission launched in 1983, and COBE/DIRBE is the second one launched in 1989. The achievements of IRAS were described above, and COBE/DIRBE examined the smooth components of the interplanetary dust cloud. AKARI, which was originally called ASTRO-F, is a Japanese infrared space mission to carry out an all-sky survey launched in February 2006 [59]. AKARI is the third mission to cover the whole sky in the mid- and far-infrared wavelength, and provides the unique chance to probe the origins of zodiacal cloud complex with its huge coverage of the sky and much better spatial resolution (10" × 10", in whole sky survey mode) than those of the previous missions (2' × 30', IRAS; 42' × 42', COBE/DIRBE). The regular operation has been started in May 2006 with both mid- and far-infrared detectors. The wavelength covers 6.5–11.6 µm and 13.9–25.3 µm (IRC), 60–110 µm and 110–180 µm (FIS).

Spitzer Space Telescope (formely SIRTF) is a space infrared telescope which has the specification similar to AKARI, but designed for the purpose of pointing observations rather than all-sky survey. It was launched into an earth-trailing heliocentric orbit in August 2003. Although Spitzer is not designed to cover the entire sky, the unique orbit enables to cover a wide range of solar elongation. The expected mission lifetime is 2.5–5 years.

8.5.3 Planet-C IR2 Observations During the Cruising Phase

So far, we mainly introduced the zodiacal cloud observations around a heliocentric distance of 1 AU, but also interplanetary missions will allow for experiments to probe the zodiacal light. The Planet-C/Venus Climate Orbiter (VCO) mission is scheduled to be launched in 2010, and will provide us with unique opportunity to reveal the details of the atmospheric motion on Venus, and to approach the dynamics of Venusian climate [61]. Planet-C employs four cameras to take snap shots of Venus in different wavelengths in order to observe Venusian atmosphere at various altitudes. One of the near-infrared camera, IR2 with wavelength coverage 1.5–2.5 µm, is primarily designed to observe the Venusian lower atmosphere. This camera was designed with several additional features of the optics as well as in the sensor devices to realize observations of the zodiacal light. The large baffle, which is originally designed to make a tricky observation while the viewing angle between Venus and the sun is relatively small, also advantegous for zodiacal light observations: the baffle allows us to cover a wide range of solar elongation angles from 180° to 26°. It will allow to take the first image of the Gegenschein at infrared wavelength. The advantages or uniqueness of the IR2 camera for the zodiacal light observations are the following. The wide field of view with fine spatial resolution (42″/pixel) has a high capability to subtract star light components that contribute not a small portion of sky brightness in these wavelengths. The wide coverage in solar elongation angle allows to observe the very inner part of the zodiacal emission. Determining the amount of the zodiacal light in the near-infrared wavelength will contribute not only to the planetary sciences but also to the cosmological studies: because the near-infrared excess of the extragalactic background light is an important issue of cosmological research. The cruising trajectory itself is also unique since Planet-C/VCO will track around the interplanetary dust cloud clump near the earth orbit at the beginning of the mission and will change the heliocentric distance from 1.1 AU toward 0.7 AU.

References

1. D. E. Backman, A. Dasgupta, and R. E. Stencel: *Model of a kuiper belt small grain population and resulting far-infrared emission*, Astrophys. J. **450**, L35 (1995)
2. J. A. Burns, P. L. Lamy, and S. Soter: *Radiation forces on small particles in the solar system*, Icarus **40**, 1 (1979)
3. C. R. Chapman: *S-Type asteroids, ordinary chondrites, and space weathering: The evidence from galileo's fly-bys of gaspra and ida*, Meteoritics **31**, 699 (1996)
4. S. F. Dermott, P. D. Nicholson, J. A. Burns, and J. R. Houck: *Origin of the solar system dust bands discovered by IRAS*, Nature **312**, 505 (1984)
5. S. F. Dermott, S. Jayaraman, Y. L. Xu, B. A. S. Gustafson, J. C. Liou: *A circumsolar ring of asteroidal dust in resonant lock with the Earth*, Nature **369**, 719 (1994)
6. R. Dumont and F. Sánchez: *Zodiacal light photopolarimetry. II. gradients along the ecliptic and the phase functions of interplanetary matter*, Astron. Astrophys. **38**, 405 (1975)
7. R. Dumont and A. C. Levasseur-Regourd: *Properties of interplanetary dust from infrared and optical observations. I - Temperature, global volume intensity, albedo and their heliocentric gradients*, Astron. Astrophys. **191**, 154 (1988)
8. D. D. Durda, and S. A. Stern: *Collision rates in the present-day kuiper belt and centaur regions: Applications to surface activation and modification on comets, kuiper belt objects, centaurs, and pluto-charon*, Icarus **145**, 220 (2000)
9. K. A. Farley, D. Vokrouhlický, W. F. Bottke, and D. Nesvorný: *A late miocene dust shower from the break-up of an asteroid in the main belt*, Nature **439**, 7074, 295 (2006)
10. M. Fulle: *Meteoroids from short period comets*, Astron. Astrophys. **230**, 220 (1990)
11. M. Fulle: *Motion of Cometary Dust*. In Comets II, (Eds.) M. C. Festou et al. (University of Arizona Press, Tucson, 2005) 565–576

12. J. S. Greaves, W. S. Holland, G. Moriarty-Schieven, T. Jenness, W. R. F. Dent, B. Zuckerman, C. McCarthy, R. A. Webb, H. M. Butner, W. K. Gear, and H. J. Walker, H. J: *A dust ring around epsilon Eridani: Analog to the young solar system*, Astrophys. J. **506**, L133 (1998)
13. K. Grogan, S. F. Dermott, S. Jayaraman, and Y. L. Xu: *Origin of the ten degree Solar System dust bands*, Planet. Space Sci. **45**, 1657 (1997)
14. K. Grogan, S. F. Dermott, and B. A. S. Gustafson: *An estimation of the interstellar contribution to the zodiacal thermal emission*, Astrophys. J. **472**, 812 (1996)
15. K. Grogan, S. F. Dermott, and D. D. Durda: *The size-frequency distribution of the zodiacal cloud: Evidence from the solar system dust bands*, Icarus **152**, 251 (2001).
16. E. Grün, H. A. Zook, H. Fechtig, and R. H. Giese: *Collisional balance of the meteoritic complex*, Icarus **62**, 244 (1985)
17. E. Grün, B. Gustafson, I. Mann, M. Baguhl, G. E. Morfill, P. Staubach, A. Taylor, and H. A. Zook: *Interstellar dust in the heliosphere*, Astron. Astrophys. **286**, 915 (1994)
18. M. S. Hanner, R. H. Giese, K. Weiss, and R. Zerull: *On the definition of albedo and application to irregular particles*, Astron. Astrophys. **104**, 42 (1981)
19. M. S. Hanner and R. L. Neuburn: Infrared photometry of comet Wilson (1986l) at two epochs Astrophys. J. 97, 254 (1989) 254
20. M. S. Hanner, D. K. Lynch, R. W. Russell: *The 8–13 micron spectra of comets and the composition of silicate grains*, Astrophys. J. **425**, 274 (1994)
21. T. R. Hicks, B. H. May, and N. K. Reay: *An investigation of the motion of zodiacal dust particles-1. Radial velocity measurements on Fraunhofer line profiles*, Monthly Notices Royal Astron. Soc. **166**, 439 (1974)
22. K.-W. Hodapp, R. M. MacQueen, and D. N. B. Hall: *A search during the 1991 solar eclipse for the infrared signature of circumsolar dust*, Nature **355**, 707 (1992)
23. W. S. Holland, J. S. Greaves, B. Zuckerman, R. A. Webb, C. McCarthy, I. M. Coulson, D. M. Walther, W. R. F. Dent, W. K. Gear, and I. Robson: *Submillimetre images of dusty debris around nearby stars*, Nature **392**, 6678, 788 (1998)
24. S. S. Hong: *Henyey-Greenstein representation of the mean volume scattering phase function for zodiacal dust*, Astron. Astrophys. **146**, 67 (1985)
25. M. Ishiguro, R. Nakamura, Y. Fujii, K. Morishige, H. Yano, H. Yasuda, S. Yokogawa, and T. Mukai: *First detection of visible zodiacal dust bands from ground-based observations*, Astrophys. J. **511**, 432 (1999)
26. M. Ishiguro, J. Watanabe, F. Usui, T. Tanigawa, D. Kinoshita, J. Suzuki, R. Nakamura, M. Ueno, and T. Mukai: *First detection of an optical dust trail along the orbit of 22P/Kopff*, Astrophys. J. **572**, L117 (2002)
27. M. Ishiguro, Y. Sarugaku, and M. Ueno: *Observational study of cometary gravels injection around the terrestrial orbit.* submitted to Earth, Planets and Space
28. M. Ishiguro, Y. Sarugaku, M. Ueno, N. Miura, F. Usui, M. -Y. Chun, S. M. Kwon: Dark red debris from three short-period comets: 2P/Encke, 22P/Kopff, and 65P/Gunn Icarus 189, 169 (2007)
29. S. Isobe, T. Hirayama, N. Baba, and N. Miura: *Optical polarization observations of circumsolar dust during the 1983 solar eclipse*, Nature **318**, 644 (1985)
30. S. I. Ipatov and J. C. Mather: *Migration of small bodies and dust to near-Earth space*, Adv. Space Res. **37**, 126 (2006)

31. J. F. James, T. Mukai, T. Watanabe, M. Ishiguro, and R. Nakamura: *The mor-phology and brightness of the zodiacal light and gegenschein*, Monthly Notices Royal Astron. Soc. **288**, 1022 (1997)

32. D. C. Jewitt: *From kuiper belt object to cometary nucleus: The missing Ultrared matter*, Astron. J. **123**, 1039 (2002)

33. T. Kelsall, J. L. Weiland, B. A. Franz, W. T. Reach, R. G. Arendt, E. Dwek, H. T. Freudenreich, M. G. Hauser, S. H. Moseley, N. P. Odegard, R. F. Silverberg, E. L. Wright: *The COBE diffuse infrared background experiment search for the cosmic infrared background. II. model of the interplanetary dust cloud*, Astrophys. J. **508**, 44 (1998)

34. M. J. Kuchner, W. T. Reach, and M. E. Brown: *A search for resonant structures in the zodiacal cloud with COBE DIRBE: The mars wake and jupiter's trojan clouds*, Icarus **145**, 44 (2000)

35. S. M. Kwon and S. S. Hong: *Three-dimensional infrared models of the inter-planetary dust distribution*, Earth, Planets Space **50**, 505 (1998)

36. S. M. Kwon, S. S. Hong, and J. L. Weinberg: *An observational model of the zodiacal light brightness distribution*, New Astron. **10**, 91 (2004)

37. P. L. Lamy and J. -M. Perrin: *Volume scattering function and space distribution of the interplanetary dust cloud*, Astron. Astrophys. **163**, 269 (1986)

38. M. Landgraf, J. C. Liou, H. A. Zook, and E. Grün: *Origins of solar system dust beyond jupiter*, Astron. J. **123**, 2857 (2002)

39. C. Leinert: *Zodiacal light – A measure of the interplanetary environment*, Space Sci. Rev. **18**, 281 (1975)

40. C. Leinert, H. Link, E. Pitz, and R. H. Giese: *Interpretation of a rocket pho-tometry of the inner zodiacal light*, Astron. Astrophys. **47**, 221 (1975)

41. C. Leinert, I. Richter, E. Pitz, and M. Hanner: *The plane of symmetry of interplanetary dust in the inner solar system*, Astron. Astrophys. **82**, 328 (1980)

42. C. Leinert, S. Bowyer, L. K. Haikala, M. S. Hanner, M. G. Hauser, A. C. Levasseur-Regourd, I. Mann, K. Mattila, W. T. Reach, W. Schlosser, H. J. Staude, G. N. Toller, J. L. Weiland, J. L. Weinberg, and A. N. Witt: *The 1997 reference of diffuse night sky brightness*, Astron. Astrophys. Suppl. **127**, 1 (1998)

43. C. Leinert and B. Moster: Evidence for dust accumulation just outside the orbit of Venus, Astron. Astrophys. **472**, 335 (2007) 247

44. A. C. Levasseur-Regourd and R. Dumont: *Absolute photometry of zodiacal light*, Astron. Astrophys. **84**, 277 (1985)

45. K. Lumme and E. Bowell: *Photometric properties of zodiacal light particles*, Icarus **62**, 54 (1985)

46. R. M. MacQueen: *Infrared observations of the outer solar corona*, Astrophys. J. **154**, 1059 (1968)

47. K. Malfait, C. Waelkens, J. Bouwman, A. de Koter and L.B.F.M. Waters: *The ISO spectrum of the young star HD 142527*, Astron. Astrophys. **345**, 181 (1999)

48. I. Mann: *The solar F-corona – Calculations of the optical and infrared brightness of circumsolar dust*, Astron. Astrophys. **261**, 329 (1992)

49. I. Mann, H. Kimura, D. A. Biesecker, B. T. Tsurutani, E. Grün, R. B. McK-ibben, J.-C. Liou, R. M. MacQueen, T. Mukai, M. Guhathakurta and P. Lamy: *Dust near the sun*, Space Sci. Rev. **110**, 269 (2004)

50. I. Mann and A. Chechowski: *Dust Destruction and ion formation in the inner solar system*, Astrophys. J. **621**, L73 (2005)

51. I. Mann, M. Koehler, H. Kimura, A. Czechowski, and T. Minato: *Dust in the solar system and in extra-solar planetary systems*, Astron. Astrophys. Rev. **13**, 159 (2006)
52. T. Matsumoto, M. Kawada, H. Murakami, M. Noda, S. Matsuura, M. Tanaka, and K. Narita: *IRTS observation of the near-infrared spectrum of the zodiacal light*, Publ. Astron. Soc. Japan. **48**, L47 (1996)
53. D. Mouillet, J. D. Larwood, J. C. B. Papaloizou, A. M. Lagrange: *A planet on an inclined orbit as an explanation of the warp in the Beta Pictoris disc*, Monthly Notices Royal Astron. Soc. **292**, 896 (1997)
54. T. Mukai and T. Yamamoto: *A model of the circumsolar dust cloud*, Publ. Astron. Soc. Japan. **31**, 585 (1979)
55. T. Mukai and T. Yamamoto: *Solar wind pressure on interplanetary dust*, Astron. Astrophys. **107**, 97 (1982)
56. T. Mukai: *Small grains from comets*, Astron. Astrophys. **153**, 213 (1985)
57. T. Mukai and I. Mann: *Analysis of Doppler shifts in the zodiacal light*, Astron. Astrophys. **271**, 530 (1993)
58. H. Murakami, H. Baba, P. Barthel, M. Cohen, Y. Doi, K. Enya, E. Figueredo, N. Fujishiro, M. Fujiwara, P. Garcia-Lario, T. Goto, S. Hasegawa, T. Hirao, S. S. Hong, K. Imai, M. Ishigaki, M. Ishiguro, D. Ishihara, Y. Ita, W. -S. Jeong, K. -S. Jeong, H. Kaneda, H. Kataza, M. Kawada, A. Kawamura, D. J. M. Kester, M. F. Kessler, T. Kii, D. C. Kim, W. Kim, B. -C. Koo, S. M. Kwon, H. M. Lee, S. Makiuti, H. Mutsuhara, T. Matsumoto, H. Matsuo, S. Matsuura, T. G. Muller, N. Murakami, H. Nagata, T. Nakagawa, T. Naoi, M. Narita, M. Noda, S. Oh, Y. Ohyama, Y. Okada, H. Okuda, S. Oliver, T. Onaka, T. Ootsubo, S. Oyabu, S. J. Pak, Y. -S. Park, C. P. Pearson, M. Rowan-Robinson, I. Sakon, A. Salama, R. S. Savage, S. Serjeant, H. Shibai, M. Shirahata, J. J. Sohn, T. Suzuki, T. Takagi, H. Takahashi, T. Tanabe, K. Uemizu, M. Ueno, F. Usui, T. Wada, G. J. White, I. Yamamura, C. Yamauchi, L. Wang, T. Watabe, H Watarai: Publ. Astron. Soc. Japan **59** sp2, 369 (2007)
59. D. Nesvorný, W. F. Bottke, L. Dones, and H. F. Levison: *The recent breakup of an asteroid in the main-belt region*, Nature **417**, 6890 (2002)
60. M. Nakamura, T. Imamura, M. Ueno, N. Iwagami, T. Satoh, S. Watanabe, M. Taguchi, Y. Takahashi, M. Suzuki, T. Abe, G. L. Hashimoto, T. Sakanoi, S. Okano, Y. Kasaba, J. Yoshida, M. Yamada, N. Ishii, T. Yamada, K. Uemizu, T. Fukuhara, and K. Oyama: *Planet-C: Venus climate orbiter mission of Japan*, Planetary Space Sci. **55**, 1831 (2007)
61. D. Nesvorný, D. Vokrouhlický, W. F. Bottk, and M. V. Sykes: *Physical properties of asteroid dust bands and their sources*, Icarus **181**, 107 (2006)
62. R. Ohgaito, I. Mann, J. R. Kuhn, R. M. MacQueen, and H. Kimura: *The J- and K-Band brightness of the solar F corona observed during the solar eclipse on 1998 February 26*, Astron. J. **578**, 610 (2002)
63. T. Ootsubo, T. Onaka, I. Yamamura, T. Tanabe, T. L. Roellig, L.-W. Chan and T. Matsumoto: *IRTS observation of the mid-infrared spectrum of the zodiacal emission*, Earth, Planets Space **50**, 507 (1998)
64. T. Ootsubo, T. Onaka, I. Yamamura, T. Tanabe, T. L. Roellig, K. -W. Chan, and T. Matsumoto: *IRTS Observations of the Mid-infrared spectrum of the zodiacal emission*, Adv. Space Res. **25**, 2163 (2000)
65. L. M. Ozernoy, N. N. Gorkavyi, J. C. Mather, and T. A. Taidakova: *Signatures of exosolar planets in dust debris disks*, Astrophys. J. **537**, L147 (2000)

66. W. T. Reach, M. V. Sykes, D. Lien, and J. K. Davies: *The formation of encke meteoroids and dust trail*, Icarus **148**, 80 (2000)

67. W. T. Reach, P. Morris, F. Boulanger, K. Okumura: *The mid-infrared spectrum of the zodiacal and exozodiacal light*. Icarus **164**, 384 (2003)

68. W. T. Reach, M. S. Kelly, and M. V. Sykes: *A survey of debris trails from short-period comets*, Icarus **191**, 298 (2007)

69. J. B. Renard, A. C. Levasseur-Regourd, and R. Dumont: *Properties of interplanetary dust from infrared and optical observations. II. Brightness, polarization, temperature, albedo and their dependence on the elevation above the ecliptic*, Astron. Astrophys. **304**, 602 (1995)

70. Y. Sarugaku, M. Ishiguro, J. H. Pyo, N. Miura, Y. Nakada, F. Usui, and M. Ueno: *Detection of long-extended dust trail associated with short period comet 4P/Faye in 2006 return*, Publ. Astron. Soc. Japan. **59**, 4, L25 (2007)

71. Y. Sarugaku: *Observational study of 2P/Encke dust trail: Formation and dynamical evolution of the dust trail and neckline structure*, Doctoral Dissertation, (Univ. of Tokyo, Tokyo, 2007)

72. M. V. Sykes and R. G. Walker: *Cometary dust trails. I – Survey*, Icarus **95**, 180 (1992)

73. M. V. Sykes and R. Greenberg: *The formation and origin of the IRAS zodiacal dust bands as a consequence of single collisions between asteroids*, Icarus **65**, 51 (1986)

74. E. F. Tedesco, J. G. Williams, D. L. Matson, G. J. Weeder, J. C. Gradie, L. A. Lebofsky: *A three-parameter asteroid taxonomy*, Astron. J. **97**, 580 (1989)

75. C. M. Telesco and R. F. Knacke: *Detection of silicates in the beta pictoris disk*, Astrophys. J. **372**, L29 (1991)

76. C. M. Telesco, R. S. Fisher, R. K. Piña, R. F. Knacke, S. F. Dermott, M. C. Wyatt, K. Grogan, E. K. Holmes, A. M. Ghez, L. Prato, L. W. Hartmann, R. Jayawardhana: *Deep 10 and 18 micron imaging of the HR 4796A circumstellar disk: Transient dust particles and tentative evidence for a brightness asymmetry*, Astrophys. J. **530**, 329 (2000)

77. V. L. Teplitz, S. A. Stern, J. D. Anderson, D. Rosenbaum, R. J. Scalise, P. Wentzler: *Infrared kuiper belt constraints*, Astrophys. J. **516**, 425 (1999)

78. D. F. Webb, D. R. Mizuno, A. Buffington, M. P. Cooke, C. J. Eyles, C. D. Fry, L. C. Gentile, P. P. Hick, P. E. Holladay, T. A. Howard, J. G. Hewitt, B. V. Jackson, J. C. Johnston, T. A. Kuchar, J. B. Mozer, S. Price, R. R. Radick, G. M. Simnett, S. J. Tappin: *Solar mass ejection imager (SMEI) observations of coronal mass ejections (CMEs) in the heliosphere*, J. Geophys. Res. **111**, A12101 (2006)

79. J. L. Weinberg, M. S. Hanner, H. M. Mann, P. B. Hutchison, and R. Fimmel: *Observations of zodiacal light from the pioneer 10 asteroid-jupiter probe: Preliminary results*, Space Research XIII, 1187–1192 (1973)

80. D. J. Wilner, M. J. Holman, M. J. Kuchner, and P. T. P. Ho: *Structure in the dusty debris around vega*, Astrophys. J. **569**, L115 (2002)

81. M. C. Wyatt, S. F. Dermott, C. M. Telesco, R. S. Fisher, K. Grogan, E. K. Holmes, and R. K. Piña: *How observations of circumstellar disk asymmetries can reveal hidden planets: Pericenter glow and its application to the HR 4796 Disk*, Astrophys. J. **527**, 918 (1999)

82. M. C. Wyatt and W. R. F Dent: *Collisional processes in extrasolar planetesimal discs - dust clumps in Fomalhaut's debris disc*, Monthly Notices Royal Astron. Soc. **334**, 589 (2003)
83. S. Yamamoto and T. Mukai: *Thermal radiation from dust grains in edgeworth-kuiper belt*, Earth, Planets Space **50**, 531 (1998)

9

Six Hot Topics in Planetary Astronomy

D. Jewitt

Institute for Astronomy, University of Hawaii, 2680 Woodlawn Drive, Honolulu, HI 96822, USA,
jewitt@hawaii.edu

Abstract Six hot topics in modern planetary astronomy are described: (1) light-curves and densities of small bodies, (2) colors of Kuiper belt objects and the distribution of the ultrared matter, (3) spectroscopy and the crystallinity of ice in the outer Solar system, (4) irregular satellites of the giant planets, (5) the Main Belt Comets, and (6) comets and meteor stream parents.

9.1 Introduction

The direction given to the authors of this book is to show some of the exciting recent developments in the study of the Solar system. Of course, "exciting" is a subjective term, and one which gives this author a lot of latitude. The most exciting science subjects for me are the ones I am working on, so I have written this chapter as a series of vignettes describing six topics from my own on-going research and from the research of my students and colleagues [principally Henry Hsieh (main-belt comets), Bin Yang (spectra), Jane Luu (colors and spectra), Scott Sheppard (irregular satellites and lightcurves), Pedro Lacerda (lightcurves), Nuno Peixinho (colors) and Toshi Kasuga (meteors)]. What follows is not so much a review as a window onto these six, particularly active parts of modern planetary astronomy. The reader who wants the raw science or access to the full literature on a given subject has only to go to the journals or to astro-ph: the internet makes it easy. My objective here is to focus attention mainly on newer, perhaps less-known work, the big-picture significance of which has yet to become clear. Relevant questions are listed explicitly where they crop up in each section of the text.

Research in modern planetary astronomy is concentrated on the small bodies of the Solar system rather than on, as in the past, the major planets. This is because the small bodies are relatively unstudied and much of what we find out about them is new and surprising. In fact, many of the different populations of small bodies have only recently been discovered (the Kuiper belt

Jewitt, D.: *Six Hot Topics in Planetary Astronomy.* Lect. Notes Phys. **758**, 259–295 (2009)
DOI 10.1007/978-3-540-76935-4_9   © Springer-Verlag Berlin Heidelberg 2009

and the main-belt comets are good examples) and few have much prospect
of being investigated, close-up, by spacecraft in the foreseeable future. Observations with telescopes are the main practical way to learn about these
objects.

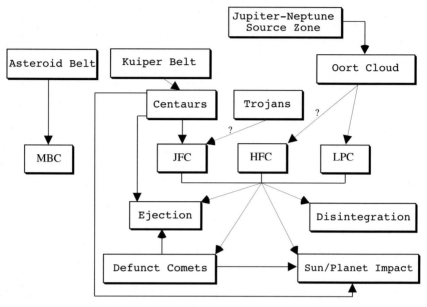

Fig. 9.1. Schematic showing the connections between some of the Solar system's
small body populations. Acronyms are MBC: Main-belt comet, JFC: Jupiter family
comet, HFC: Halley family comet, LPC: long-period comet. Arrows mark interre-
lations. For example, the Kuiper belt feeds the Centaurs which become relabeled
as JFCs when dynamically interacting with Jupiter. Most Centaurs die by being
ejected from the Solar system or by striking a planet or the Sun. The JFCs die
by one or more of four labeled processes. *Arrows* marked "?" show connections
that remain uncertain. Loss processes for the MBCs are not yet known. Figure
from [31]

The first necessary step in this chapter is to lay out the small bodies of
the Solar system in a clear way, so that we know what we are talking about.
This is done in Fig. 9.1. There, the main source regions (asteroid belt, Kuiper
belt, and Oort cloud) are shown at the top of the diagram. Objects now in the
50,000 AU scale Oort cloud were formed in the Jupiter–Neptune zone and then
scattered outwards by strong planetary perturbations. Their perihelia were
lifted by torques from passing stars and from the galactic tide. Bodies deflected
back into the planetary region from the Oort cloud are labeled long-period
comets (LPCs), distinguished by large, weakly bound, and isotropically distributed orbits. Halley family comets (HFCs) have smaller orbits that are more
often prograde than retrograde. Their source has not been established but is

likely to lie in the inner regions of Oort's cloud. (The long-period and Halley family comets are sometimes lumped together and given the mangled-English label "nearly isotropic comets," by which it is meant that the lines of apsides of the orbits of these bodies are nearly isotropically distributed.) Jupiter family comets (JFCs) have small semimajor axes, inclinations and eccentricities and dynamics controlled by strong interactions with Jupiter. Their source is thought to be somewhere in the Kuiper belt, but it is not clear which regions of the Kuiper belt actually supply the comets. Before they are trapped by Jupiter and while they are strongly scattered by the giant planets, escaped Kuiper belt objects are labeled Centaurs. (The Centaurs and JFCs typically possess modest orbital inclinations and are sometimes referred to as members of the "ecliptic comet" group for this reason.) The most recently discovered comets are the ice-rich asteroids (or main-belt comets, MBCs) probably formed in-place at ~3 AU. They do not seem to interact with the other populations and therefore constitute the third-known cometary reservoir, after the Oort cloud and the Kuiper belt. Trojan "asteroids" are likely ice-rich bodies stabilized in the 1:1 mean motion resonances of the planets (Trojans of both Jupiter and Neptune [80] have been found). The locations of their origin are unknown. Comets "die" most commonly by being ejected from the Solar system. Those not ejected disintegrate, devolatilize, or impact the planets or the Sun.

Research is ongoing into every box in Fig. 9.1 and into the arrows that symbolize the relationships between the objects in the boxes. Indeed, the key advance of the past one and a half decades is that we now clearly see both the boxes *and* the relationships that exist between them. In this sense, the six hot topics of this chapter are really one: we aim to trace the different kinds of small body populations back to their sources and so to better understand the origin of the entire Solar system.

A second schematic (Fig. 9.2) attempts to clarify some of the small-body nomenclature. It shows a two-parameter classification, reflecting the fact that both dynamical properties and physical properties are regularly used to label objects in the Solar system. The horizontal axis in Fig. 9.2 is the Tisserand parameter measured with respect to Jupiter. This is defined by

$$T_J = \frac{a_J}{a} + 2\left((1 - e^2)\, \frac{a}{a_J} \right)^{1/2} \cos(i) \qquad (9.1)$$

where a, e, and i are the semimajor axis, eccentricity, and inclination of the orbit, respectively, while $a_J = 5.2$ AU is the semimajor axis of the orbit of Jupiter. The Tisserand parameter provides a measure of the relative velocity of approach to Jupiter: Jupiter itself has $T_J = 3$, most comets have $T_J < 3$, while main-belt asteroids generally have $T_J > 3$.

The position of an object either above or below the x-axis in Fig. 9.2 shows whether the object has a measurable coma (gravitationally unbound atmosphere) or not. The presence of a coma is related, in an unclear way, to the presence of near-surface volatiles. Objects showing comae are, by the

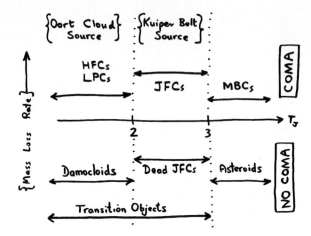

Fig. 9.2. Two parameter classification of some of the small-body populations discussed here. In the horizontal direction, objects are classified by their Tisserand parameter measured with respect to Jupiter (dynamical comets have $T_J < 3$, dynamical asteroids $T_J > 3$). In the vertical direction, objects are classified by whether or not they show evidence for mass loss, presumed to be driven by the sublimation of near-surface volatiles. Objects above the line are observationally comets because they show comae and/or tails, while objects below the line are observationally inactive and so classified as asteroids. Ideally, objects should be placed vertically in this diagram based on measurements of their mass loss rates. Given our limited knowledge, however, it is more practical at present to use a "one bit" classification in which objects are either measurably active or not

physical definition of the word, "comets." The JFCs are those comets with $2 < T_J \leq 3$. Non-outgassing objects with $2 < T_J \leq 3$ are called Transition Objects (TOs), or sometimes "dead comets" or "dormant comets." Comets with $T_J \leq 2$ fall into the LPC and HFC comet types. Non-outgassing objects with $T_J \leq 2$ are called "Damocloids:" their orbital elements suggest that most are the dead or dormant nuclei of HFCs [29]. The MBCs are like asteroids in having $T_J > 3$ but differ in showing comae.

9.2 Lightcurves and Densities

Lightcurves offer valuable opportunities to assess the shapes and rotational states of bodies that are, generally, too small in angular extent to be resolved, even with the best existing adaptive optics systems (which currently offer resolution $\sim$0.05 arcsec). It is also possible, at least for some objects, to use lightcurves to estimate the bulk density of a body.

9.2.1 Lightcurves

The first thing to acknowledge is that there are no *unique* interpretations of lightcurves. Rotational variability in the scattered light is influenced by the shape of the body, the surface distribution of materials having different albedos, the surface scattering function, the viewing geometry, etc. This non-uniqueness is unarguable, as it was when first noted as long ago as 1906 [74]. One hundred years later, the uniqueness problem is still dredged up by critics in response to new work. But, while no mathematically rigorous proof exists that a given lightcurve can be interpreted in any particular way, there is a large and growing body of exciting and illuminating work based on rotational lightcurves of small bodies. This is possible because, wherever supplementary information is available, we find that the lightcurves of small Solar system bodies, almost without exception, are dominated by rotational modulation of the projected cross section rather than by spatial variations in the albedo. That is to say, most of the available evidence shows that albedo non-uniformity is small (the exceptions tend to be pathological, like Saturn's two-faced, synchronous satellite Iapetus, and not of general relevance to objects in heliocentric orbits). Rotational modulation of the projected cross section (body shape) determines most lightcurves.

What controls the body shape? Sufficiently, strong bodies can maintain any shape against their own gravity, but evidence from the study of main-belt and near-Earth asteroids shows that large bodies are not strong. Their interiors have been fractured and weakened by past impacts (in the case of the weakly agglomerated comets, the interior strengths may have been small to start with). In the limiting case of zero strength, the shape of a body must relax to an equilibrium configuration that is a function of the body density and angular momentum. These equilibrium shapes follow a well-defined progression from spheres (no rotation) to oblate spheroids (bodies flattened along the polar direction, known as Maclaurin spheroids) to tri-axial figures (the Jacobi ellipsoids that grow longer up to a critical angular momentum content above which no single-body equilibrium shape exists). Single, strengthless bodies with specific angular momenta higher than a critical value (that depends only on the density) are unstable to rotational fission. Chandrasekhar [8] famously calculated these shapes.

Under the *assumption* of zero strength, the shape and rotation of a body can thus be used to estimate the density. The validity of the assumption is, of course, questionable and good reasons to doubt the zero-strength assumption exist. After all, small Solar system bodies are rocks, not liquids, and so they cannot literally be strengthless, especially in compression. Even if they lack overall tensile or cohesive strength, pressure-induced shear strength between components gravitationally bound in an aggregate should inhibit complete relaxation to the equilibrium state, much as grains of sand in a pile do not flow under gravity like a liquid because of frictional forces between the grains [22].

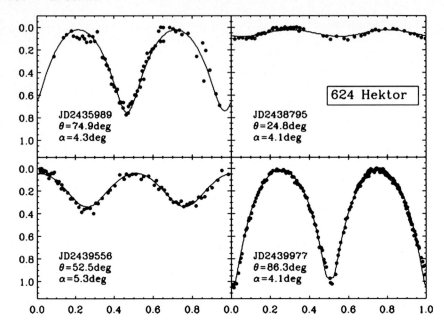

Fig. 9.3. Lightcurves of (624) Hektor at four aspect angles (the angle between the line-of-sight and the spin direction) compared with an equilibrium binary model. In each panel, the x-axis displays rotational phase (computed for period = 6.9 hr) and the y-axis shows the relative magnitude. The fits provide a remarkably good representation of the data, lending credibility to the model. Figure from [47]

Despite these legitimate reservations, the evidence suggests that equilibrium models can indeed work very well when the bodies and their lightcurve ranges (a measure of the equatorial variation of the radius) are large. As an example, I show in Fig. 9.3 the rotational lightcurves of Trojan asteroid (624) Hektor at four different epochs. The lightcurve range and shape change dramatically as the aspect angle (θ, the angle between the line-of-sight and the pole) changes, but all the variations are well modeled by an equilibrium Roche binary configuration [8], from which the density $\rho = 2480^{+80}_{-300}$ kg m^{-3} is deduced [47]. A check of this density is provided by the motion of Hektor's newly found 15 km satellite [57] and the assumption of Kepler's law. The result, $\rho \sim 2200$ kg m^{-3} (Frank Marchis, private communication, August 2006), confirms the value found from the lightcurve model. This fact, plus the remarkable quality of the fits in Fig. 9.3, suggests that the shape of Hektor cannot be far from an equilibrium (strengthless) binary. I speculate that impact jostling might explain why internal friction is unimportant: impacts energetic enough to cause bouncing or lifting of the components in an aggregate would allow the body to approach a near-equilibrium configuration by temporarily removing pressure-induced shear strength, just as strong vibrations cause a sand pile, initially at the angle of repose, to flow downhill against the inhibiting effects of inter-grain forces. Whatever the cause, the lightcurves in Fig. 9.3 show that Hektor is well described as a strengthless equilibrium figure.

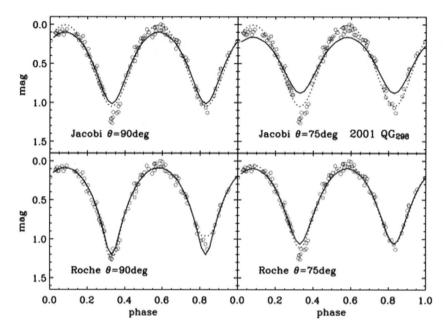

Fig. 9.4. Lightcurve of KBO 2001 QG298 compared with models. The top two panels show the best-fitting Jacobi ellipsoid models for aspect angles $\theta = 90°$ and $\theta = 75°$, respectively. The bottom two panels show best fit Roche binary models for the same aspect angles. The Roche binary model for $\theta = 90°$ (lower left panel) provides the best fit to the data, including the asymmetric lightcurve minima. No comparably good Jacobi (single-body) models were found. Data from [79], figure from [47]

Within the context of strengthless equilibrium models, we note that Jacobi ellipsoids generate lightcurves having a maximum range of ~ 0.9 magnitudes [49, 91]. At more extreme rotations, the equilibrium configuration is a double object (a contact or near-contact binary). Therefore, objects with photometric ranges >0.9 mag, like Hektor itself (Fig. 9.3), attract special attention as candidate contact binaries. Several examples exist in the literature, including some in the main-belt asteroid [49], the Kuiper belt [79], and there are others amongst the Trojans of Jupiter [47, 56].

Figure 9.4 shows the mid-sized (effective diameter ~ 240 km) Kuiper belt object 2001 QG298 [47, 79, 82]. Overplotted models confirm that the Jacobi ellipsoid models cannot fit, in particular, the deeply notched lightcurve minima. The latter are better-fitted by Roche binary models where they are interpreted as mutual eclipse phenomena in a close binary (see Fig. 9.5).

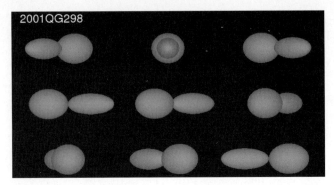

Fig. 9.5. Visualization of 2001 QG298 as a function of rotational phase based on the best-fit Roche binary model from the lower-left panel in Fig. 9.4. The binary components are elongated by mutual gravitational attraction. Figure from [47]

The density of 2001 QG298 given by a Roche binary fit to the lightcurve is $\rho = 590^{+140}_{-50} \text{kg m}^{-3}$ [47]. There is no *proof* that 2001 QG298 is a Roche binary, but the ease with which the Roche binary model fits the lightcurve data suggests that this interpretation is plausible.

Aside from the derived densities (discussed in more detail in Sect. 2.2), the contact binaries may eventually help us to discriminate between various suggestions for the formation of binaries. This is especially so in the Kuiper belt, where the fraction of binary objects is high [81] and several formation mechanisms have been proposed. Very briefly, these mechanisms include (1) binary formation in a debris ring created by a giant impact (as is thought to account for the formation of Earth's Moon and some large KBO satellites [7]) (2) permanent binding of a transient binary owing to the loss of energy by dynamical friction [17] and (3) permanent binding via three-body reactions including exchange reactions [16]. All the proposed mechanisms require Kuiper belt number densities much higher than are now found, suggesting that binaries are products of a past epoch in which the Kuiper belt mass might have been substantially (by two to three orders of magnitude) higher than now.

It is too early to reach any strong conclusion about the origin of the binary Kuiper belt objects. For example, three-body interactions are weak and so produce mainly wide binaries, of which we know many examples [81]. Persistent drag from dynamical friction would cause steady inward spiraling of binaries, perhaps ending with the production of contact or very close binary systems (c.f. [47] and Fig. 9.4). It seems likely that future determinations of the properties and statistics of the binaries, especially measurements of the contact to wide-binary ratio, will tell us a lot about the relative contributions to the binary population of different formation mechanisms.

9.2.2 Densities

Figure 9.6 (see also [31]) shows the densities of objects in various small body populations as a function of the effective diameters. Density data were obtained using a wide range of techniques, including gravitational perturbations on passing spacecraft (for the planetary satellites), mutual event data (for Pluto and Charon), the lightcurve models discussed above (for the other Kuiper belt objects), and a mixture of (mostly) indirect techniques (for the cometary nuclei).

While the range of densities at a given diameter is considerable, the tendency towards higher densities at larger sizes is self-evident in Fig. 9.6. There are no small bodies (diameters D <100 km) with high densities and no large bodies ($D > 1000$ km) with densities much less than about 1000 kg m^{-3}. The trend towards higher densities at larger sizes does not seem to be an artifact of mixing different samples having distinct sizes and densities. For instance, the planetary satellites (hollow crosses in Fig. 9.6) and the KBOs (large black

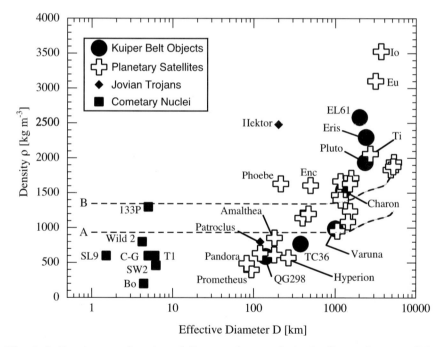

Fig. 9.6. Density as a function of diameter for mostly icy bodies in the outer Solar system. Abbreviations SL9:D/Shoemaker-Levy 9, C-G:P/Churyumov-Gerasimenko, SW2:P/Schwassmann-Wachmann 2, Bo:P/Borrelly, T1:P/Tempel 1, QG298:2001 QG298, TC36:1999 TC36, EL61:2003 EL61, Enc:Enceladus, Ti:Titan, Eu:Europa. Labeled curves are isothermal self-compression models for (A) pure water ice and (B) a 40% rock and ice mixture from [54], for comparison purposes only (see text). Figure modified from [31]

circles) both show the trend toward densification as diameter increases in the range 100–3000 km.

The effects of self-compression on solid ice and rock-ice bodies are negligible for diameters D <1000 km, and modest even at the sizes of the largest objects plotted in Fig. 9.6. This is shown by the self-compressed models plotted as dashed lines in the Figure (see [54]). Some of the observed density versus diameter variation must be compositional in origin. For example, the large dense objects Io and Europa are largely rock-dominated, while their similar sized but less dense satellite companions Ganymede and Callisto have retained larger ice fractions. On the other hand, compositional variations alone cannot account for objects with $\rho < 930$ kg m^{-3} (the density of uncompressed, pure ice [54]). Therefore, any object with a density less than 930 kg m^{-3} *must* be porous. Clear examples of such low density, *necessarily* porous objects are seen in Fig. 9.6 up to diameters of ∼500 km. From Fig. 9.6 we see that the nuclei of most comets must be porous and Saturn's small satellites Pandora and Prometheus are so underdense that they also must be porous (a probable consequence of repeated collisional disruption and reassembly [72]). Another porous body is Jupiter's satellite Amalthea ($\rho = 860 \pm 100$ kg m^{-3}, [1]), which was previously asserted to be one of the most refractory bodies in the Jupiter system but is now identified, amazingly, as a water-rich body [83] more akin to a comet. The Jovian Trojan (617) Patroclus ($\rho \sim 800$ kg m^{-3}, [58]), Kuiper belt contact binary 2001 QG298 (see above), and Saturn's tumbling moon Hyperion (see Fig. 9.7, $\rho \sim 540$ kg m^{-3}, [87]), all have low densities that require some fraction of internal void space even if they are composed of pure water ice. Of course, it is hard to see how a pure water ice object could form. Compositionally, more realistic bodies with rock/ice ratios ∼1 would have $\rho \sim 1400$ kg m^{-3} or more. Objects measured to have densities less than this value require some internal porosity. For example, KBO (20000) Varuna, with $\rho \sim 1000$ kg m^{-3}, must surely include both ice and rock and, depending on the exact rock/ice ratio and the nature of the rock component, requires a porosity ∼20% in order to explain the low density [35, 47].

The low densities could indicate microporosity (small internal voids with a scale comparable to the grain size) or macroporosity (internal void spaces with a larger scale) or some combination of the two. Macroporosity might be generated by past collisional disruption followed by chaotic reassembly of the fragments. This is a plausible explanation of the low densities ($\rho \sim 400$ kg m^{-3}) of Saturn's strongly interacting co-orbital satellites Pandora and Prometheus, each about 100 km in diameter (see Fig. 9.6). The kilometer-scale nuclei of comets could also possess internal cavity space, since their gravitational self-compression is negligible. However, I think it is unlikely that macroporosity is relevant in the deep interiors of very large, low density bodies (like Varuna [35]) where hydrostatic forces are appreciable (especially if these bodies have very low strength, as surmised from lightcurve data, above!). The central hydrostatic pressure in a spherical object having diameter, D, and density, ρ, is $P_c \sim \pi/6 \, G\rho^2 D^2$ (N m^{-2}). For example, with $D = 1000$ km and

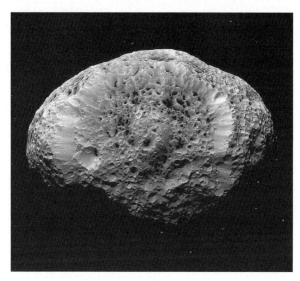

Fig. 9.7. Saturn's satellite Hyperion. This aspherical body has a mean effective diameter of 270 ± 8 km, a bulk density estimated from perturbations on a passing spacecraft as $\rho = 540 \pm 50$ kg m^{-3} and a porosity ~40% [87]. Image courtesy Cassini Imaging Team and NASA/JPL/SSI

$\rho = 1500$ kg m^{-3} the central pressure is $P_c \sim 7 \times 10^7$ N m^{-2}. This is equivalent to the hydrostatic pressure 2.5 km below the Earth's surface, deeper than any known caves.

Microporosity is a more likely candidate to explain the low measured bulk densities in Varuna-scale bodies and, if due to a loosely aggregated granular structure, would be more consistent with the low effective strengths of large bodies inferred from lightcurves. Microporosity could be produced in the early Solar system as large bodies are assembled from smaller pieces, resting together much like grains of sand on the beach. (Incidentally, although it is not directly relevant to the case at hand, it is interesting to note that terrestrial beach sand is about 40% porous at the surface and compresses to ~25% porosity at pressures of 0.5×10^8 N m^{-2}, close to the core hydrostatic pressure on Varuna.) Laboratory experiments with compositionally relevant granular rock-ice mixtures show the evolution of microporosity in the 0.8 to 8×10^8 N m^{-2} pressure range [50], suggesting its potential importance for objects in Fig. 9.6. However, the temperature and its evolution through the life of the body will play an important role in determining the strengths of ice grains in outer Solar system bodies. Therefore, it is necessary to compute coupled thermal–structural models to examine the long-term survival of porosity and this has barely been addressed [50]. Already, though, the data tell us that porosity must be significant in the outer regions of the 1000 km scale KBOs; at smaller sizes Fig. 9.6 shows that porosity can play a dominant role.

Question: To what degree do porosity variations and intrinsic compositional differences contribute to the different densities of objects of a given size in Fig. 9.6?

Question: To what extent are the porosities influenced by size-dependent thermal and ancient collisional processes?

9.3 Color Distributions

9.3.1 Distribution of Colors

One of the first results to be established from systematic physical measurements of the Kuiper belt objects was that the optical colors are very diverse, ranging from approximately "neutral" ($V - R \sim 0.35$) to "very red" ($V - R \sim 0.75$) [55]. (V and R are the apparent magnitudes in filters centered near 5500 and 6500 Å, respectively). This finding was soon extended to the near-infrared, leading to the realization that the reflection characteristics of the KBOs are determined over the wavelength range $0.45 \leq \lambda(\mu m) \leq 1.2$ by a single coloring agent [34, 59]. This is different from the case of the main-belt asteroids where, for example, distinct solid-state absorptions cause the spectral slope to vary dramatically with wavelength across this range. There is widespread suspicion (but no compelling proof) that irradiated organics are responsible for the colors of at least some KBOs: such materials display the low (few %) albedos seen on many KBOs and can be very red (e.g., see [64]). The broad color dispersion has been confirmed by numerous independent measurements over the past decade. This can be seen in Fig. 9.8, which is a compilation of published and on-line color measurements provided by Nuno Peixinho.

Explanations for the color dispersion remain controversial. In the resurfacing model [55], the color of an object is set by competition between irradiation and impact-produced resurfacing. Resurfacing excavates fresh material from beneath the surface layer susceptible to cosmic ray damage, thereby changing the surface color and (presumably) albedo. Observational evidence *against* the resurfacing hypothesis is the lack of rotational variability of the surface colors: hemispheric color asymmetries caused by partial resurfacing should be more common than the data suggest [34]. Could intrinsic differences in the compositions of the KBOs cause the color dispersion? Color differences in the main-belt asteroids are explained in this way but, in the Kuiper belt, compositional differences are less easy to understand. Colors and compositions of main-belt asteroids are clearly related to the orbital parameters (especially semimajor axis) but similar correlations are not observed in the KBOs. Furthermore, temperature differences between the inside of the Classical belt at $\sim$35 AU and the outside at $\sim$50 AU are only $\sim$10 K, seemingly too small to have a major effect on the composition.

Evidence for color-orbit correlations in the Kuiper belt is very limited. An early claim [84] that the optical colors of KBOs are distributed bimodally (i.e., that KBOs are *either* neutral *or* very red, but rarely in between) seems not to have survived independent scrutiny (Fig. 9.8). Evidence that the colors of Centaurs are bimodally distributed is more convincing ([68]; Fig. 9.8) but is unexplained. The B–R colors are related to perihelion distance [85] or to the orbital inclination [90], but only for the Classical KBOs, a relation which is also unexplained.

> *Question:* What causes the color diversity on KBOs?
>
> *Question:* Why are the Centaur colors bimodal? In particular, if the Centaurs are escapees from the Kuiper belt, why do they not show the same colors and (unimodal) color distribution as the KBOs?
>
> *Question:* Do the colors tell us something fundamental about the bulk compositions of these bodies, or do they merely reflect superficial processes acting on the optically accessible surface skin?

9.3.2 Ultrared Matter

The nearly linear reflectivity spectra of many outer Solar system bodies are usefully characterized by their gradients, expressed as S' [%/1000 Å] [33]. Spectra with $S' > 25\%/1000$ Å are defined as "ultrared" [28]. Empirically, ultrared matter is found on the surfaces of Kuiper belt objects and Centaurs

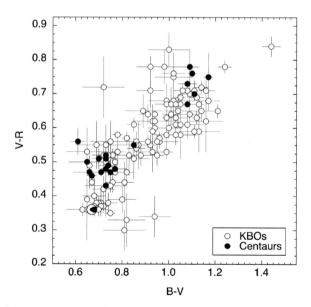

Fig. 9.8. B–V versus V–R color–color diagram showing the KBOs (*empty circles*) and Centaurs (*filled circles*). Only objects with 1σ photometric uncertainties < 0.1 mag. are plotted. The Sun is marked by a gray circle. Figure courtesy of Nuno Peixinho

but is rare or absent on the surfaces of small-bodies in other populations, including the Trojans [12], the cometary nuclei [28], dead JFCs [28], Damocloids [29], and (perhaps) the irregular satellites ([19]: however, too few of the latter have been adequately observed to be sure). This lower incidence suggests that the ultrared matter maybe thermodynamically (or otherwise) unstable in bodies which approach the Sun more closely than the Centaurs (which, by definition, have perihelia outside Jupiter's orbit).

9.4 Spectroscopy of Primitive Matter

The wavelengths of vibrational and overtone spectral features of common molecular bonds fall into the near-infrared portion of the electromagnetic spectrum. Accordingly, it is expected that near-infrared data should place the most stringent constraints on the surface compositions of primitive Solar system bodies, both in the inner and outer regions. The faintness of many of the most interesting objects demands the use of large telescopes, all of which are ground-based telescopes. Nevertheless, the utility of near-infrared spectra is limited by the faintness of the targets and by the difficulty of removing telluric signatures from the spectrum (the Earth's atmosphere contains many of the same molecular bonds as those sought in the small bodies).

Most objects studied in the near-infrared show spectra which are utterly featureless.

9.4.1 Crystallinity of Solar System Ice

Ice can form at low temperatures in the amorphous state, meaning that the geometric arrangement of the water molecules lacks periodicity. The amorphous state is distinct from the various crystalline forms in which water ice at higher temperatures is stable (e.g., the snow that falls from the sky and the ice that grows in the refrigerator is crystalline, with the molecules arranged in staggered layers having a hexagonal pattern). Amorphous ice is intrinsically unstable, and spontaneously transforms to crystalline ice on a timescale, τ_{cr} (yr), given by

$$\tau_{cr} = 3.0 \times 10^{-21} e^{\left[\frac{E_A}{kT}\right]} \qquad (9.2)$$

where E_A is the activation energy, k is the Boltzmann's constant, T is the temperature, and $E_A/k = 5370$ K [78]. The phase transition is potentially important for two reasons.

First, the transition is exothermic, with a specific energy release $\Delta E = 9 \times 10^4$ J kg^{-1}. This ΔE can heat surrounding ice, influencing the thermal regime in icy bodies, and perhaps even driving a runaway in which crystallization at one location in a body triggers crystallization over a large, thermally connected volume. Crystallization is also associated with a small change in the bulk density. Many elaborate and spectacular thermal models of comets are predicated on the assumption that the nuclei enter the middle and inner Solar system as amorphous ice bodies [70].

Second, amorphous ice possesses many nooks and crannies, giving a large surface area per unit mass (of order 10^2 m^2 kg^{-1} [2]) on which other molecules can be trapped. Empirically, a fit to experimental data [3] on the trapping efficiency (defined as $\Re = \frac{m_g}{m_i}$, where m_g is the mass of gas that can be trapped in a mass of amorphous water ice, m_i) is given by

$$\Re \sim 10^{-0.08(T-40)}. \tag{9.3}$$

Equation (9.3), which applies to CH$_4$, CO, Ar and, to a lesser extent, N$_2$, gives $\Re \sim 1$ at $T = 40$ K, falling steeply to $\Re \sim 10^{-5}$ at $T = 100$ K. At the $T \sim 40$–50 K temperatures prevalent in the Kuiper belt, it is clear that large quantities of gas, $0.1 \leq \Re \leq 1$, could be trapped within amorphous ice, in agreement with observations of comets. The trapped molecules are released as the temperature is raised above the accretion temperature, culminating with wholescale expulsion as the water molecules rearrange themselves into cubic or hexagonal lattices upon crystallization. The presence of amorphous ice can thus lead to pulses of outgassing that could be relevant to understanding the mass loss from comets.

By setting $\tau_{cr} = 4.5 \times 10^9$ yr in Equation (9.2), we find that amorphous ice formed at the beginning of the Solar system would have escaped crystallization if its temperature had always been $T < 77$ K. Because of the very strong temperature dependence in Equation (9.2), even a brief excursion above this temperature would have crystallized the ice. The temperature of an isothermal blackbody in thermal equilibrium with sunlight falls to 77 K at $R = 13$ AU, or slightly beyond the orbit of Saturn. Therefore, all else being equal, we should expect to find crystalline ice at (and inside) the orbit of Saturn, and to find amorphous ice beyond. Water ice in the inner regions is indeed crystalline, but it is also crystalline in the satellites of Uranus and Neptune and in the Kuiper belt. There is surprisingly no direct evidence for amorphous ice in the outer regions (see Fig. 9.9).

The two types of ice are observationally separable in the near-infrared. The 1.5 and 2.0 μm bands have slightly different shapes and central wavelengths, but a much better diagnostic is provided by the crystalline ice band at 1.65 μm. This band is absent in amorphous ice. If the 1.65 μm band is present then the ice must be at least partly crystalline. If it is absent then the ice might be amorphous, down to some limit set by the signal-to-noise ratio of the spectrum around the band.

However, the optically observable surfaces of bodies are bombarded by energetic particles from the Solar wind and from cosmic rays, and also by energetic photons from the Sun. These energetic particles disrupt the bonds between water molecules in ice, thereby breaking up the crystal structure and "amorphizing" the material. (It is interesting to note that silicate grains in the interstellar medium are largely amorphous for the same reason [44].) The timescale for amorphization is short, probably 10^6–10^7 yr [36]. In the

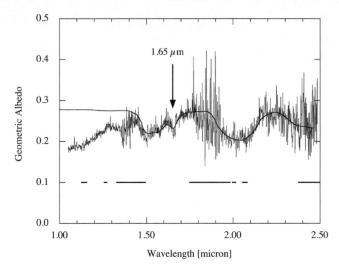

Fig. 9.9. Near-infrared spectrum of Kuiper Belt Object (50000) Quaoar showing the major ice bands at 1.5 and 2.0 μm and the narrow feature at 1.65 μm that is diagnostic of the presence of crystalline ice. The smooth line is a crystalline ice spectrum that has been plotted on top of the Quaoar spectrum for comparison: no attempt was made to fit the data but still the correspondence between Quaoar and the ice spectrum is impressive. Horizontal bands at the bottom of the figure show regions where the transparency of the Earth's atmosphere is particularly poor. From [36]

sense, the presence of crystalline ice in the outer Solar system is even more surprising and the reason for its persistence has not yet been firmly explained. One possibility is that resurfacing provides fresh material on a timescale that is short compared to the amorphization time. Resurfacing could result, for example, from impact gardening, which dredges up buried material (ice deeper than ~1 m is effectively shielded from even quite energetic cosmic rays). A more dramatic possibility is that outgassing or cryovolcanism emplaces fresh, crystalline ice on the surface. Very recent work with an ultra-high vacuum chamber in the Chemistry Laboratory of the University of Hawaii suggests a more likely explanation. We find that the amorphization efficiency is a function of temperature such that amorphization is nearly 100% efficient at $T \sim 10$ K but only ~50% efficient at $T = 50$ K. Presumably, this is because slight thermal jostling at the higher temperatures allows some water molecules to reconnect in the crystalline form even after irradiation [96]. At the surface temperature of Quaoar (Fig. 9.9), ice can remain partly crystallized forever, despite the rain of energetic particles.

While the persistence of crystalline ice is apparently now understood, what heated the ice to make it crystalline in the first place remains unknown. Several possibilities exist. In large bodies (radii > 500 km) it is possible that heating occurred upon formation by the conversion of gravitational potential energy into heat. Large bodies could also have been heated by trapped radionuclides, whether they be short-lived (half-lives $\sim 10^6$ yr) like the famous ^{26}Al and ^{60}Fe, or long-lived (half-lives $\sim 10^9$–10^{10} yr) like ^{40}K, ^{232}Th, and ^{238}U. Local surface heating by micrometeorite bombardment has the advantage that it would operate on bodies of any size, consistent with crystalline ice being common in the outer Solar system on objects of different diameters. Whatever the cause, the available evidence shows that ice on the surfaces of the large Kuiper belt objects is crystalline, which means that it has been warmed at least to twice the current surface temperatures of 40 or 50 K.

Small bodies, like the nuclei of comets, were probably not substantially heated by the above processes. Do they contain amorphous ice? Only limited direct evidence exists in the form of spectra of the dust in two long-period comets, both distinguished by showing no evidence for the 1.65 μm crystalline ice band. Other evidence comes from the distribution of the orbits of the Centaurs. These are objects recently escaped from the Kuiper belt and traveling on orbits which cross the paths of the giant planets (i.e., their defining property is that they have perihelia and semimajor axes between the orbits of Jupiter and Neptune). About 20% of the known Centaurs are also active comets. The distribution of the orbital elements of the active Centaurs is different from the Centaurs as a whole. In particular, the average perihelion distance of the active Centaurs is small compared to the average perihelion of the Centaurs as a whole (Fig. 9.10). This difference cannot be ascribed to the simple sublimation of crystalline water ice, since the latter is involatile throughout the Centaur region. Instead, activity in the Centaurs is consistent with production through the crystallization of amorphous ice, which begins at temperatures comparable to those found on the active Centaurs when at perihelion [32]. This is not iron-clad evidence for the existence of amorphous ice in the Centaurs, by any means. But it is perhaps the best evidence we possess at the moment.

Question: Can more objects be observed in order to determine whether the ice is truly crystalline in these objects? Spectra of adequate quality have been secured for only two comets. Are only the long-period comets amorphous? What about Halley-family comets?

Question: What crystallizes ice on the larger Kuiper belt objects and other bodies in the outer Solar system? Is it a global energy phenomenon as suggested (e.g., gravitational binding energy, or decay of trapped radioactive nuclei) or merely a surface effect (e.g., micrometeorite heating and crystallization of a thin surface layer)?

9.4.2 The Methanoids

Water ice is present on some large KBOs while others show instead prominent bands due to methane [52, 86]. These "methanoids" include amongst their number (134340) Pluto, as well as (136199) Eris and (136472) 2005 FY9

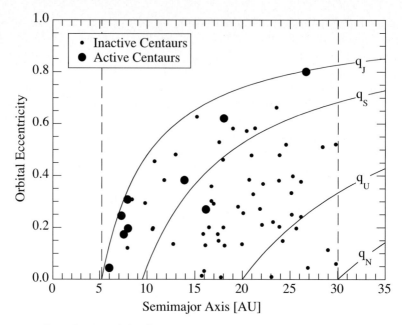

Fig. 9.10. Distribution of the Centaurs in semimajor axis versus eccentricity space. *Large circles* denote active (outgassing) Centaurs, while *small circles* show inactive Centaurs. The semi-major axes of Jupiter and Neptune, which bound the Centaur orbits, are shown with vertical *dashed lines*. Diagonal arcs show the loci of points having a fixed perihelion distances equal to the semimajor axes of the orbits of the giant planets, as marked [32]

(Fig. 9.11). Jeans (thermal) escape appears to determine which KBOs can retain CH_4 and which cannot: methane is more stable on the large, distant (cold) KBOs than on small, close (hotter) ones [77].

The source of the methane is unknown. One possibility is that the methane is produced, along with other hydrocarbons, as a by-product of energetic particle irradiation of exposed surface ices. A preexisting source of carbon would need to be present within the ice in order for CH_4 to be formed this way. In this case, one might expect all large and cold KBOs to show methane, since all are comparably irradiated by the solar wind and cosmic rays. Alternatively, perhaps methane was delivered to the KBOs at the time of their accretion in the form of clathrated ice (but this might be difficult to reconcile with the picture outlined above in which low temperature ice making up the KBOs is more likely to have been amorphous, at least at the accretion epoch). The most exciting possibility is that the methane has been created through chemical reactions in the deep interiors of the larger KBOs and has since leaked onto the surface. We know from Terrestrial experience that many serpentinization reactions (between liquid water and rocks) are exothermic and release hydrogen

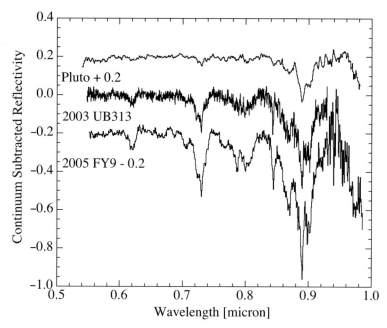

Fig. 9.11. Far-red optical spectra of the three methanoids (134340) Pluto, (136199) Eris (formerly 2003 UB313), and (136472) 2005 FY9, taken at the Keck 10 m telescope. The spectra are continuum-subtracted and vertically displaced for clarity. All the visible absorption bands in these spectra are due to solid methane

[15]. Fischer–Tropsch type reactions between the hydrogen so-produced and carbon monoxide could create methane. The main requirements for the active generation of methane would then be the existence of liquid water significantly above the triple point and intimate contact with carbon-containing rocks over a large reaction surface. Both circumstances appear likely in the larger (1000 km scale) KBOs [6, 61].

Lastly, it is good to keep in mind that while Nature always plays by the rules, it does not always play fair: it is entirely possible that more than one source contributes CH_4 to the methanoids and equally likely that the dominant source is not one that we have thought of.

Question: How can we decide between alternative production schemes for methane, and what others might exist?

Question: How could internally generated methane move from the deep interior of a KBO to the surface? Which other volatiles would move with it?

Question: Can we detect atmospheres of KBOs other than Pluto, perhaps by the occultation of background stars?

9.5 Irregular Satellites

For the most part, the satellites of the planets can be neatly separated into one of two distinct categories based on their orbits. The so-called regular satellites have small orbital inclinations and eccentricities ($e \ll 1$). By contrast, the irregular satellites (hereafter "iSats") have large inclinations (spanning the range $0 \le i \le 360°$: most irregulars are retrograde) and eccentricities ($e \sim 0.5$). Another distinction is based on the fraction of the Hill sphere occupied by the orbits of the satellites. The Hill sphere is the volume in which a planet exerts gravitational control of nearby objects in competition with the Sun. The Hill sphere radius is $r_H = a[m_p/(3M_\odot)]^{1/3}$, where $m_p/M_\odot$ is the mass of the planet in units of the Solar mass and a is the semimajor axis of the orbit of the planet. [Values of r_H are given in Table 9.1 both in AU and in apparent angle on the sky as seen from Earth. The Table also lists the (ever changing) numbers of known satellites at each planet]. A general rule is that orbits of the regular satellites are confined to the central few percent of r_H while most iSats are much more wide-ranging, with orbital semimajor axes up to ~ 0.5 r_H (Fig. 9.12 and 9.13). Although their orbits, and the effects of Solar tides, are very large, the known iSats appear to remain bound to their planets for timescales comparable to the age of the Solar system.

Table 9.1. Hill spheres of the giant planets (from [39])

]Planet	Mass ($M_\oplus$)	a (AU)	r_H (AU)	r_H (deg)	N_r	N_i
]Jupiter	310	5	0.35	5	8	55
]Saturn	95	10	0.43	2.8	21	35
]Uranus	15	20	0.47	1.4	18	9
]Neptune	17	30	0.77	1.5	6	7

NOTE: N_r (N_i) are the numbers of regular
(irregular) satellites at each planet.

These systematic differences in the orbital inclinations, eccentricities and sizes (relative to r_H) reflect different modes of formation of the regular and irregular satellites. Whereas the regular satellites are clearly the products of accretion in long-gone circumplanetary disks, the irregulars more likely formed in orbit about the Sun (but we do not know where) and were subsequently captured by the planets (we would like to know when and how).

Most of the very large (i.e., bright) satellites fall in the "regular" class and, for this reason, the regulars have captured most of our attention since Galileo discovered his four large (regular) satellites of Jupiter in 1610. Recent observational work has refocused our attention by establishing that iSats substantially out-number the known regular satellites and that the two types formed differently [39]. Irregular satellites have unambiguously emerged as a "hot topic" in planetary science.

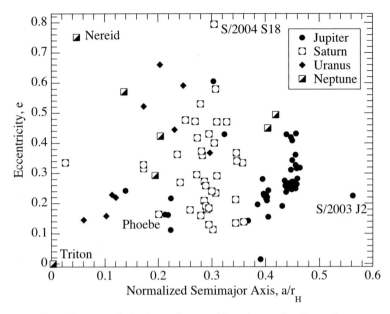

Fig. 9.12. Distribution of the irregular satellites in semimajor axis versus eccentricity space. Selected objects and dynamical groups are identified. Figure updated (to 2007 May) and adapted from [39]

There are several ideas about the origin of the iSats. Until recently, the most popular idea was that the satellites were captured from heliocentric into planetocentric orbits through the action of gas drag, in the extended atmospheres of the growing giant planets. This idea was first proposed to account for the iSats of gas giant planet Jupiter [69]. It relies on the collapse of a massive gaseous envelope to provide a transient source of drag since, if the drag persists, all satellites must ultimately spiral down into the planet. The idea might also work for the other gas giant, Saturn, but it is not so obvious that it can be applied to Uranus and Neptune, since these planets are *ice* giants. The ice giants have comparatively modest gas inventories (e.g., a few $M_\oplus$ compared with $\sim$80 $M_\oplus$ and 260 $M_\oplus$ in Saturn and Jupiter, respectively). Moreover, the timescales of formation are completely different, probably $\sim$1 Myr or less for Jupiter and Saturn but 10 or more times longer at Uranus and Neptune.

A second idea is that the satellites were captured in a phase of runaway growth, when the gas giants were pulling in gas from the adjacent protoplanetary disk. Sudden growth in mass leads to sudden expansion of the region around each planet in which the gravitational influence of the planet dominates that of the Sun [21]. In this "pull-down" model, the iSats would have been captured objects that happened to be nearby to the planets at the end phases of their runaway growth. One, apparently fatal, problem for this model

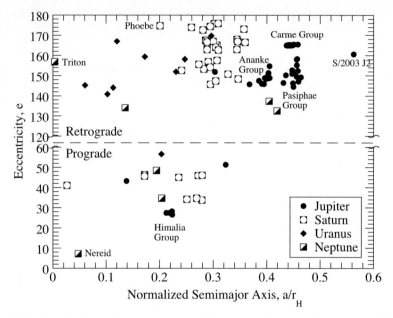

Fig. 9.13. Distribution of the irregular satellites in semimajor axis versus inclination space. No known satellites have inclinations $60 \leq i \leq 120°$ and so this region is not plotted. Orbits in this range are unstable to the Kozai resonance. Selected objects and dynamical groups are identified. Figure updated (to 2007 May) and adapted from [39]

is that the ice giant planets did not undergo runaway growth. They accreted mass by binary collisions of solid objects over a long period of time (evidently comparable to or longer than the ∼10 Myr timescale on which gas survived in the disk), with steady growth but no mass runaway.

The last idea has emerged as the most interesting, given what we now know about the young Solar system. The idea is that irregular satellites were captured from heliocentric orbits in three-body (or N-body) interactions [9]. For example, the three bodies could be two planets and a small-body initially in orbit about the Sun [65] or two asteroids could interact with each other within the Hill sphere of a planet [9]. As a result of the interaction, one of the small bodies could be ejected from the planetary region, carrying with it excess energy that would allow the other asteroid to become bound. One attraction of 3-body and N-body capture models is that the Hill spheres of the four giant planets increase in size and volume with increasing distance from the Sun (even though the masses of the giants decrease from Jupiter outwards). One consequence might be that low mass, distant Uranus and Neptune might be able to capture about as many irregulars as high mass Jupiter and Saturn, in accordance with the data [37]. However, this conjecture has not yet been placed on a quantitative basis. Indeed, 3-body and N-body capture models have received scant attention probably because, until recently, it seemed that such interactions in the Solar system must be incredibly rare. In the modern system such interactions *are* rare, but they may not always have been so, since the early Solar system was much more densely populated than it is now.

From where were the iSats captured? The evidence does not provide an answer to this question, so now we remain in a state of conjecture. The first main possibility is that the iSats were captured from initial heliocentric orbits that were close to, or at least crossing, the orbits of the giant planets. Low velocity encounters give the highest probability of capture, so local sources are in some sense preferred. The second possibility is that the iSats were captured from a remote source, perhaps the Kuiper belt. The latter possibility has been advanced in the context of the "Nice" dynamical model [18], in which the architecture of the Solar system is a consequence of an assumed crossing of the 2:1 mean-motion resonance between Jupiter and Saturn. According to initial simulations with this model, capture of the iSats of Uranus and Neptune (and perhaps Saturn) is possible but the iSats of Jupiter must have another source [65].

Question: How and when were the iSats captured? Was there a single capture mechanism or did different planets capture their satellites in different ways? How can we tell?

Question: From where were they captured? From the Kuiper belt, from orbits in the protoplanetary disk, local to the growing planets, or from elsewhere?

Question: Does ultrared matter exist on iSats? If the iSats were captured from the Kuiper belt, the presence or absence of ultrared matter might constrain the source region.

Question: How do the answers to these questions change from planet to planet?

9.6 Main-Belt Comets

Main-belt comets (MBCs) are objects with orbits in the region classically occupied by the asteroids but with physical characteristics of comets, specifically including comae and/or tails (Fig. 9.14). Three examples are known as of July 2007 [25]. Their Tisserand parameters measured with respect to Jupiter are $T_J > 3$, whereas those of comets from the Kuiper belt and Oort cloud reservoirs are $T_J < 3$. The MBCs are also completely distinct from the more familiar "transition objects" (see Sect. 9.7). The latter, in fact, are the *opposites* of the MBCs in having comet-like orbits (with $T_J < 3$) but asteroid-like physical appearances (i.e. no comae and no tails). This difference is clear in Fig. 9.15 and in the classification diagram in Fig. 9.2.

Evidence that the mass loss from MBCs is driven by the sublimation of ice is indirect. Specifically, none of the other mechanisms that we have thought of seem to fit the data. The first suggestion for 133P was that the mass loss

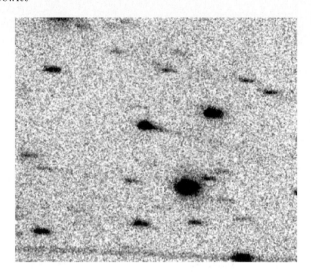

Fig. 9.14. Image of main-belt comet 133P/Elst-Pizarro taken at the University of Hawaii 2.2 m telescope on UT 2007 June 11. A dust tail is visible extending to the right of the nucleus. The region shown is approximately 70 arcsec in width and has North to the top, East to the left. The MBC has approximate apparent red magnitude 19.5

is impact debris, resulting from a small collision [88]. This explanation is now ruled out, given that the activity in 133P is periodic, having been present near perihelion in 1996, 2002 [23, 89], and now again in 2007 (Fig. 9.14). Rotational instability seems an unlikely explanation. While 133P is rotating quickly (period =3.47 hr), there is no evidence for rapid rotation in either P/Read or 176P. Moreover, there are many asteroids rotating with shorter periods, yet these are not known to be emitting dust like the MBCs. On the Moon, charge gradients in the vicinity of the terminator are known to levitate and launch dust particles from the surface [48]. The same process could eject dust from small, low escape-velocity asteroids, and comets. Two problems with this mechanism for the MBCs are (1) that dust velocities inferred from 133P and P/Read are higher than typical on the Moon and, more seriously, (2) if electrostatic ejection were important, we would have to ask why comet-like emission is not a general property of all small asteroids. There is also an issue with supply. Unlike the Lunar case, a large fraction of the small dust grains on asteroids are simply lost into space, not levitated repeatedly as the terminator sweeps by. New dust particles will be created by micrometeorite impact into the asteroid surface, but the rate of production is orders of magnitude too low to account for the escape losses to space.

The MBCs hold special significance in planetary science because they appear to be repositories of ice in a region of the Solar system that has been suggested, on independent grounds, as a potential contributor to the Earth's

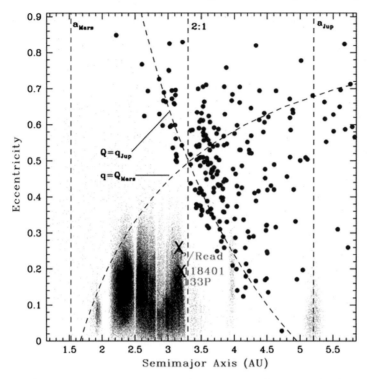

Fig. 9.15. Semimajor axis versus orbital eccentricity for asteroids (*small dots*), Jupiter family comets (*large dots*), and the three currently known main-belt comets (*marked X*). The latter are clearly associated more with the asteroid belt than with the Jupiter family comets. The semimajor axes of Mars and Jupiter are marked with *vertical dashed lines*. Two labeled arcs show the locus of orbits having perihelion inside Mars's aphelion distance and aphelion outside Jupiter's perihelion distance, respectively. Figure from [25]

oceans [63]. The reasoning behind this is as follows. The Earth probably formed too hot to have accreted *and* retained much water and so this, and other, volatiles were accreted from another source in a "late veneer" some time after the Earth had cooled down. The timing of the addition of water is uncertain. However, evidence from ^{18}O isotopes in some zircons (ancient refractory mineral grains which substantially predate the rocks in which they are found) suggests that substantial bodies of liquid water were present very early, at 4.3 Gyr [62] or even 4.404 ± 0.008 Gyr [94] ago.

Comets, being ice-rich, are one possible source of terrestrial water. Against this are measurements showing that the D/H ratios in comets are twice the D/H ratio measured in the Earth's oceans. Either the terrestrial D/H has evolved (possible), or the cometary D/H values are wrong (unlikely, see [60]) or unrepresentative (possible, because the measured comets are not the Jupiter family comets most likely to have contributed water [10]) or the comets are not the dominant source of Earth's water [63]. The mass of the oceans is about $2.5 \times 10^{-4} M_\oplus$. The mass of water trapped within the mantle is very uncertain and could be much less than or much greater than the mass on the surface. Dynamical models suggest that such a large mass is unlikely to have been trapped from the Kuiper belt and point instead to a closer source in the asteroid belt [63]. In this latter interpretation, the MBCs occupy a region that might have contributed to the oceans. It is important to note that the objects now present in the outer belt cannot be suppliers of Earth's water: there are too few and there is no clear dynamical pathway from most of the outer asteroid belt to Earth-intersecting orbits. What is imagined is that a massive, primordial asteroid belt was cleared (probably by strong perturbations from nearby Jupiter) at some earlier time, hurling ice-rich objects across the paths of the terrestrial planets.

How could ice become trapped in the main-belt asteroids? On the surface there would seem to be two possibilities. Either the ice originated there, becoming trapped in the MBCs as they formed, or the ice was delivered after formation from a more remote source. The presence of hydrated minerals in many meteorites thought to come from the outer belt requires the past presence of liquid water (e.g., [5, 40]). Perhaps the MBCs are icy asteroids in which some of the primordial ice component escaped chemical reaction with silicates and persists to the present day. I know of no evidence against this possibility. On the other hand, attempts to capture comets from the Jupiter family into orbits like the MBCs seem doomed to fail. The Tisserand parameter is approximately a constant of the motion during capture, and the fact that the MBCs and JFCs have different Tisserands indicates that simple conversion of the orbits is impossible. Additional forces, from non-gravitational accelerations due to anisotropic outgassing or from perturbations by terrestrial planets, could conceivably help transform JFC orbits into MBC orbits. I am open to this possibility and would like to see more work done to explore it. What has been published on this topic, however, gives little reason to be optimistic [51].

How can ice be stable in the main belt only $\sim$3.2 AU from the Sun? The temperature of an isothermal blackbody located at this distance is $T_{BB} = 153$ K. T_{BB} gives a good estimate of the averaged, deep temperature in kilometer-sized MBCs, while regions on the surface, for example near the subsolar point, can be expected to be hotter. The specific sublimation rate in thermal equilibrium at T_{BB} is $dm/dt \sim 3 \times 10^{-8}$ kg m^{-2} s^{-1}. An MBC surface having density $\rho = 2000$ kg m^{-3} would recede at the rate $\rho^{-1} dm/dt \sim 1.5 \times 10^{-11}$ m s^{-1}, corresponding to about 0.5 mm yr^{-1}. A 1000 m radius body could survive for only $\sim$2 Myr, if in continuous sublimation at this rate, which is very short compared to the age of the Solar system. Therefore, the ice must be stabilized against sublimation losses if it is to have survived for the age of the Solar system.

Observations show that the nuclei of comets are mantled by refractory matter. By analogy it seems reasonable to suppose that mantles also exist on the MBCs and that they stifle the gas flow from most or all of the surface, most of the time. In this way, ice might survive in the MBCs for the age of the Solar system, even at distances considerably smaller than 3 AU. Ice stability, protected by porous, refractory mantles, has been established for asteroid (1) Ceres at 2.7 AU [13] and even for Mars' satellite Phobos at 1.6 AU [14]. In order to become visibly active, the mantle of an MBC must be punctured. A likely mechanism in the main belt is collision. A meter-scale impactor would expose enough ice to drive the mass loss rates that are inferred for 133P, for example.

So, a plausible scenario for the MBCs is that they are ice-containing asteroids in which buried ice is occasionally exposed to the heat of the Sun, probably by impacts. This idea, which seems reasonable but which remains essentially untested, leads us to believe that the orbital distribution of MBCs should be determined jointly by the distribution of ice-containing objects in the main belt and by the distribution of the asteroid–asteroid collision frequency (related to the local density and other belt parameters). As for the first quantity, it is reasonable to expect that buried ice is more common in the outer belt than in the inner regions because the rotationally averaged body-temperature varies with semimajor axis as $a^{-1/2}$. Evidence for radial compositional gradients has long been recognized in the different distributions of the taxonomic classes, with S (metamorphosed) types more common at smaller R than the C (more primitive) types. The data and models of thermal stability are, however, consistent with the possibility that *all* outer belt asteroids contain ice. The ratio of MBCs (on which the ice is temporarily exposed) to outer belt asteroids would then be given roughly by the fraction of the asteroids which experience an excavating collision within the (probably short) lifetime of the exposed ice patch. Work is underway in Hawaii to begin to determine some of these quantities so that the likely incidence of buried ice can be assessed.

Lastly, note that if water could not be trapped in the hot, young Earth then neither could other, more volatile species such as the noble gases. Even the outer asteroid belt is not cold enough to trap noble gases in abundance. Sources within more distant, colder cometary reservoirs, probably the Kuiper belt ($T \sim 40$ K), seem required [67]. The full picture of the delivery of volatiles to the terrestrial planets will probably turn out to be complicated, with multiple sources.

Question: How many MBCs are there? What is their orbital element distribution and what does this tell us about the sources of these bodies?

Question: Can we obtain direct evidence (spectroscopy) for the suspected water driver of MBC activity?

Question: How are they activated?

Question: What fraction of the asteroids as a whole contain ice?

Question: What, if anything, can the MBCs tell us about the origin of the Earth's oceans and about terrestrial planet volatiles in general?

9.7 Comets and Their Debris

9.7.1 Comets Alive, Dormant, and Dead

Objects which are comet-like as judged by their orbits (Tisserand parameters $T_J < 3$), but which show no evidence for mass loss cannot be classified as comets on physical grounds. They are sometimes known as Transition Objects (TOs). The simplest interpretation is that the TOs are comets in which the lack of activity is due to the depletion of near-surface volatiles. Thermal conduction sets the relevant vertical scale for depletion to the "skin depth," of order $\ell \sim (\kappa t)^{1/2}$, where κ is the thermal diffusivity of the upper layers and t is the timescale for variation of the Solar insolation. At least three timescales and three resulting skin depths are relevant (see Table 9.2, in which I assumed $\kappa = 10^{-7} \ \mathrm{m^2 \ s^{-1}}$ as is appropriate for a powdered dielectric solid).

Table 9.2. Timescales and skin depths

Variation	Timescale, t	skin depth, ℓ (m)
Diurnal	10 hr	0.06
Orbital	10 yr	5
Dynamical	4×10^5 yr	1000

The effects of diurnal heating, in particular, can be attenuated by a very modest refractory layer ("mantle") just a few centimeters thick. Direct evidence for this comes from, for example, NASA's Deep Impact mission to comet 9P/Tempel 1 (Fig. 9.16), where remote observations have been interpreted as showing a characteristic thickness $\sim 10\,\mathrm{cm}$ [42]. Mass loss from a comet on which the mantle is much thicker than ℓ will be stifled, earning the comet the "Transition Object" label.

Whether or not cometary activity resumes depends on the long-term stability of the mantle, which itself depends on the dynamical evolution of the comet. If the mantle lacks cohesion, steady inward drift of the perihelion will lead to increasing temperatures and, eventually, to the ejection of the mantle by gas pressure forces and to the Phoenix-like rebirth of measurable mass loss [73]. (With cohesion, the mantle is potentially much more stable and the mechanism of its failure is less easily understood [46]). Since very thin mantles inhibit sublimation, the mantle formation timescales are probably very short, perhaps comparable to, or even less than, the orbital period [28, 73]. In this simple picture, it is thus likely that the mantles adjust and regrow as the orbit evolves.

Direct observations of cometary nuclei (comets 1P/Halley, Borrelly, Wild 2, and Tempel 1) confirm the existence of widespread refractory mantles (e.g., [4]) and show that mass loss is channeled through a small number of active areas which, combined, occupy 0.1–10% of the nucleus surface. However, other

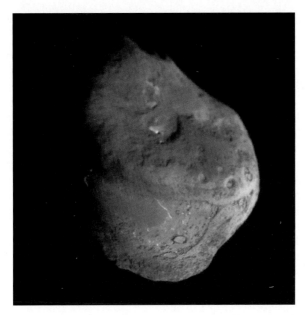

Fig. 9.16. Nucleus of comet Tempel 1. The visible surface is a refractory mantle (albedo ∼4%) which displays many intriguing landforms, few of which are understood. Image courtesy of NASA and the Deep Impact team

observations throw into doubt the role of mantles in the global control of cometary mass loss. Most important are measurements of the dust trails of comets. The dust trail masses, m_t, and the cometary mass loss rates, dm/dt, together define a trail production timescale, $\tau_t = m_t/(dm/dt)$. Separately, dynamical spreading of the trails under the action of planetary perturbations determines the dynamical age of the trail, τ_{dyn}. Where meaningful measurements of both τ_t and τ_{dyn} have been possible, the timescales are found to be very different, with $\tau_t \gg \tau_{dyn}$. In other words, cometary mass loss at the measured rates cannot supply the trail mass even if continuous over the age of the trail. This suggests that the trails are not populated by the steady, mantle-choked loss of mass from the nucleus but by some other, more impulsive phenomenon. Nucleus breakup seems to be the best explanation.

Unfortunately, we lack a quantitative understanding of why comets (other than those that are sheared apart by gravity when passing close to planets or the Sun) breakup. Suggested causes include spin-up leading to centripetal disruption [75], high internal gas pressures caused by sublimating supervolatiles (Samarasinha's most enjoyable "bomb" model [76]), impact with unseen interplanetary debris and disruption by thermally induced stresses. All of these ideas verge on the fantastic, with the exception of centripetal disruption, which is a natural outcome of torques applied to the nucleus by non-uniform outgassing. I know of no data to suggest a relationship between nucleus spin rate and breakup but this could be simply because there are too few relevant nucleus spin measurements (i.e., "absence of evidence" should not be construed as "evidence of absence," as far as the spin versus breakup connection is concerned). The lack of understanding is disconcerting given the potential importance of breakup in determining the fates of small bodies.

Question: How many TOs are there? The number of TOs relative to the number of active comets will tell us the ratio of the outgassing to the dynamical lifetimes of these bodies.

Question: What is their orbital element distribution and what does this tell us about the sources of these bodies?

Question: Do all comets evolve into TOs or do some proceed directly to disintegrate into debris streams?

Question: Are the TOs dead or dormant, or both? In other words, is the ice depleted down to the core, or just down to a few times the thermal skin depth?

Question: How do TOs die? Are their lifetimes limited by impact with the planets or the Sun, by dynamical ejection, or by a physical process such as breakup?

9.7.2 Damocloids

The Damocloids are a subset of the TO class, named after the prototype object (5335) Damocles. They are defined by having a point-source appearance and $T_J < 2$ [29]. At the time of writing (May 22, 2007), 36 objects meet this definition. The orbits of the Damocloids are statistically similar to the orbits of Halley family and long-period comets (e.g., many Damocloid orbits are retrograde), rather than with the Jupiter family. The association is further strengthened by the fact that some bodies originally classified as Damocloids have, since discovery, been found to show weak comae. Damocloids, then, are the inactive nuclei of comets recently emplaced in the planetary region of the Solar system from a source probably located in the inner Oort Cloud [11]. Curiously, although their dynamical and evolutionary histories have been quite different from those of the short-period comets, the surface properties of the two classes of comet nucleus are indistinguishable [29].

Question: How many Damocloids exist and what is the ratio of Damocloids to Halley Family Comets?

Question: What is the size distribution of the Damocloids?

Question: Is there evidence that some Damocloids might be intrinsically refractory bodies (asteroids) ejected into the Oort Cloud and then scattered back to the inner Solar system, as has been suggested for 1996 PW [92].

Question: Do any Damocloids carry Ultra-Red matter? The published sample does not, but the published sample is small.

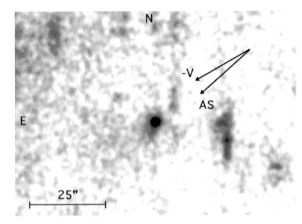

Fig. 9.17. "Asteroid" 2005 WY25 at 1.6 AU showing ultra-weak outgassing in a 1500 s, R-band image from the UH 2.2 m telescope on UT 2004 March 20. The mass loss rate inferred from the coma is very uncertain, but of the order 10 g s^{-1}. 2005 WY25 is the likely parent of the Phoenicid meteor stream, and a probable fragment of comet D/1819 W1 (Blanpain). Figure from [30]

9.7.3 Meteor Stream Parents

One fate for cometary nuclei is to disintegrate, forming a trail of solid debris particles that can be detected remotely from their thermal emission [71] and optical signatures [27] or directly, if their orbits intersect that of the Earth and produce a meteor stream e.g. Fig. 9.17. Recent work has given a boost to the study of meteor streams and the parent bodies which produce them. Significantly, some of the parent objects have now been identified with confidence [41]. One surprise is that not all the parents are comets: some streams seem to result from the breakup of bodies, like (3200) Phaethon, which are dynamically asteroids.

Asteroid (3200) Phaethon ($T_J = 4.508$) has orbital elements similar to those of the Geminid meteors. On this basis, Phaethon was long-ago proposed as a likely Geminid stream parent [93]. Recently discovered asteroid 2005 UD ($T_J = 4.504$) has very similar orbital elements and is probably related both to Phaethon and to the Geminids [66]. The albedo of Phaethon has been measured as 0.11 ± 0.02 and the diameter as 4.7 ± 0.5 km [20]. If the albedo of 2005 UD is the same, then its diameter must be only 1.3 ± 0.1 km [38]. Sensitive, high resolution imaging observations provide no evidence for on-going mass loss, either from Phaethon [24] or from 2005 UD [38], above the level of $\sim 10^{-2}$ kg s^{-1}. The age of the Geminid stream estimated from dynamical considerations is about 1000 yr [95]. In 1000 yr, mass loss at 10^{-2} kg s^{-1} would give a stream mass $M_s \sim 3 \times 10^8$ kg, whereas the mass has been independently estimated at $M_s \sim 1.6 \times 10^{13}$ kg [26]. This huge discrepancy (a factor $\sim 10^5$) indicates that meteor stream formation must be episodic or even catastrophic, not steady-state.

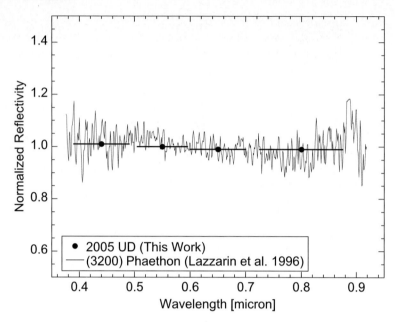

Fig. 9.18. Reflection spectrum of Asteroid 2005 UD (points) compared with the spectrum of dynamically related object (3200) Phaethon. Both objects are unusual in showing spectra slightly bluer than the Sun in reflected light. Figure from [38]

Both Phaethon and 2005 UD show slightly blue optical reflection spectra of unknown origin (Fig. 9.18). Blue reflection spectra are uncommon (only 1 out of ∼23) amongst the near-Earth objects [38, 45]. There is speculation that the blue color could reflect thermally altered minerals on these bodies at the high temperatures (perhaps 740 K) resulting from their small perihelion distances ($q \sim 0.14$ AU) [53]. Likewise, the high mean density of the Geminids ($\rho \sim 2900$ kg m^{-3}) is also unusual and has been suggested to result from compaction associated with loss of volatiles [43]. On the other hand, observational support for thermal desorption is lacking: careful spectroscopic measurements of the Na/Mg ratio show that the Geminids are not compositionally different from other meteoroids with much larger $q \sim 1$ AU [43]. A reasonable guess is that the Geminid meteors, Phaethon and 2005 UD (and probably other macroscopic bodies yet to be found) are products of the recent breakup of a precursor body but the nature of the precursor and the cause of the breakup have yet to be determined.

Question: Are all objects with small q necessarily blue as a result of thermal alteration?

Question: What kind of body was the Geminid precursor?

Question: What caused the precursor to breakup? Thermal stresses? Internal gas pressure forces? Spin-up by outgassing or radiation forces?

Question: How does the rate at which mass is input to the interplanetary medium by catastrophic disruption of meteor stream parents compare with the rates from cometary sublimation and from asteroid–asteroid collisions in the main belt?

9.8 Epilogue

A reasonable conclusion to be drawn from this chapter is that planetary astronomy is a most active and revitalized field. Key advances are being made in the determination of the contents of the Solar system, with the discovery of new populations of bodies and the unveiling of links between populations that were, until recently, unsuspected. These new observational results, combined with the rising power of computers, together motivate exciting new conjectures for the origin and evolution of the Solar system.

Acknowledgements

I thank Toshi Kasuga, Pedro Lacerda, Ingrid Mann, Nuno Peixinho, Rachel Stevenson, and Bin Yang for comments on this manuscript and NASA's Planetary Astronomy and Origins programs for support of the work described herein.

References

1. J. D. Anderson, et al.: *Amalthea's density is less than that of water*, Science **308**, 1291 (2005)
2. A. Bar-Nun, J. Dror, E. Kochavi and D. Laufer: *Amorphous water ice and its ability to trap gases*, Phys. Rev. B., **35**, 2427 (1987)
3. A. Bar-Nun, I. Kleinfeld and E. Kochavi: *Trapping of gas mixtures by amorphous water ice*, Phys. Rev. B. **38**, 7749 (1988)
4. A. T. Basilevsky and H. U. Keller: *Craters, smooth terrains, flows, and layering on the comet nuclei*, Solar System Res. **41**, 109 (2007)
5. A. J. Brearley: *The action of water.* Meteorites and the Early Solar System II, D. S. Lauretta and H. Y. McSween Jr. (Eds.), (University of Arizona Press, Tucson, 943 pp. 2006) 584–624
6. V. V. Busarev, V. A. Dorofeeva and A. B. Makalkin: *Hydrated silicates on edgeworth-kuiper objects – probable ways of formation*, Earth Moon Planets **92**, 345 (2003)
7. R. M. Canup: *A giant impact origin of pluto-charon*, Science **307**, 546 (2005)
8. S. Chandrasekhar: *Ellipsoidal Figures of Equilibrium*, Dover, (New York, 1987)
9. G. Colombo and F. A. Franklin: *On the formation of the outer satellite groups of Jupiter*, Icarus **15**, 186 (1971)

10. A. H. Delsemme: *The deuterium enrichment observed in recent comets is consistent with the cometary origin of seawater*, Planet. Space Sci. **47**, 125 (1998)

11. L. Dones, P. R. Weissman, H. F. Levison and M. J. Duncan: *Oort cloud formation and dynamics.* In Comets II, M. C. Festou, H. U. Keller and H. A. Weaver (Eds.), (University of Arizona Press, Tucson, 2004) 153–174

12. E. Dotto, et al.: *The surface composition of Jupiter Trojans*, Icarus **183**, 420 (2006)

13. F. P. Fanale and J. R. Salvail: *The water regime of asteroid (1) Ceres.* Icarus **82**, 97 (1989)

14. F. P. Fanale and J. R. Salvail: *Evolution of the water regime of PHOBOS*, Icarus **88**, 380 (1990)

15. G. L. Früh-Green, D. S. Kelley, S. M. Bernasconi, J. A. Karson, K. A. Ludwig, D. A. Butterfield, C. Boschi and G. Proskurowski: *30,000 years of hydrothermal activity at the lost city vent field*, Science **301**, 495 (2003)

16. Y. Funato, J. Makino, P. Hut, E. Kokubo and D. Kinoshita: *The formation of Kuiper-belt binaries through exchange reactions*, Nature **427**, 518 (2004)

17. P. Goldreich, Y. Lithwick and R. Sari: *Formation of Kuiper-belt binaries by dynamical friction and three-body encounters*, Nature **420**, 643 (2002)

18. R. Gomes, H. F. Levison, K. Tsiganis and A. Morbidelli: *Origin of the cataclysmic Late Heavy Bombardment period of the terrestrial planets*, Nature **435**, 466 (2005)

19. T. Grav and J. Bauer: *A deeper look at the colors of the Saturnian irregular satellites.* Icarus, **191**, pp. 267–285 (2007)

20. S. F. Green, A. J. Meadows and J. K. Davies: *Infrared observations of the extinct cometary candidate minor planet (3200) 1983TB.* MNRAS **214**, 29P (1985)

21. T. A. Heppenheimer and C. Porco: *New contributions to the problem of capture*, Icarus **30**, 385 (1977)

22. K. A. Holsapple: *Spin limits of solar system bodies: From the small fast-rotators to 2003 EL61*, Icarus **187**, 500 (2007)

23. H. H. Hsieh, D. C. Jewitt and Y. R. Fernández: *The strange case of 133P/Elstpizarro: A comet among the asteroids*, Astron. J. **127**, 2997 (2004)

24. H. H. Hsieh and D. Jewitt: *Search for activity in 3200 phaethon.* Ap. J. **624**, 1093 (2005)

25. H. H. Hsieh and D. Jewitt: *A population of comets in the main Asteroid Belt*, Science **312**, 561 (2006)

26. D. W. Hughes and N. McBride: *The mass of meteoroid streams*, MNRAS **240**, 73 (1989)

27. M. Ishiguro, et al.: *First detection of an optical dust trail along the orbit of 22P/Kopff*, Ap. J. **572**, L117 (2002)

28. D. Jewitt: *From kuiper belt object to cometary nucleus: The missing ultrared matter*, Astron. J. **123**, 1039 (2002)

29. D. Jewitt: *A first look at the damocloids*, Astron. J. **129**, 530 (2005)

30. D. Jewitt: *Comet D/1819 W1 (Blanpain): Not dead yet*, Astron. J. **131**, 2327 (2006)

31. D. Jewitt: *Kuiper Belt and Comets: An Observational Perspective*, Saas Fee Lecture Notes, N. Thomas and W. Benz (Eds.), (Springer Pub. Company, New York, 2007)

32. D. Jewitt: *The active centaurs*, Astron. J. submitted (2008)

33. D. Jewitt and K. J. Meech: *Cometary grain scattering versus wavelength, or 'What color is comet dust'?*. Ap. J. **310**, 937 (1986)

34. D. C. Jewitt and J. X. Luu: *Colors and spectra of kuiper belt objects*. Astron. J. **122**, 2099 (2001)

35. D. C. Jewitt and S. S. Sheppard: *Physical properties of trans-neptunian object (20000) Varuna*, Astron. J. **123**, 2110 (2002)

36. D. C. Jewitt and J. Luu: *Crystalline water ice on the Kuiper belt object (50000) Quaoar*, Nature **432**, 731 (2004)

37. D. Jewitt and S. Sheppard: *Irregular satellites in the context of planet formation*, Space Science Rev. **116**, 441–455 (2005)

38. D. Jewitt and H. Hsieh: *Physical observations of 2005 UD: A mini-phaethon*, Astron. J. **132**, 1624 (2006)

39. D. Jewitt and N. Haghighipour. *Irregular satellites of the planets: products of capture in the early solar system*. Ann. Rev. Astron. Astrophys. **45**, 261–295 (2007).

40. D. Jewitt, L. Chizmadia, R. Grimm and D. Prialnik: *Water in the small bodies of the solar system*. Protostars and Planets V, B. Reipurth, D. Jewitt and K. Keil (Eds.), (University of Arizona Press, Tucson, 2007) 863–878

41. P. Jenniskens: *2003 EH1 is the quadrantid shower parent comet*, Astron. J. **127**, 3018 (2004)

42. T. Kadono, et al.: *The thickness and formation age of the surface layer on comet 9P/tempel 1*, Ap. J. Lett **661**, L89 (2007)

43. T. Kasuga, T. Yamamoto, H. Kimura and J. Watanabe: *Thermal desorption of Na in meteoroids*, Astron. Ap. **453**, L17 (2006)

44. F. Kemper, W. J. Vriend and A. G. G. M. Tielens: *The absence of crystalline silicates in the diffuse interstellar medium*, Ap. J. **609**, 826 (2004)

45. D. Kinoshita, et al.: *Surface heterogeneity of 2005 UD from photometric observations*, Astron. Ap. **466**, 1153 (2007)

46. E. Kuehrt and H. U. Keller: *The formation of cometary surface crusts*, Icarus **109**, 121 (1994)

47. P. Lacerda and D. C. Jewitt: *Densities of solar system objects from their rotational light curves*, Astron. J. **133**, 1393 (2007)

48. P. Lee: *Dust levitation on asteroids*, Icarus **124**, 181 (1996)

49. G. Leone, P. Paolicchi, P. Farinella and V. Zappala: *Equilibrium models of binary asteroids*, Astron. Ap. **140**, 265 (1984)

50. J. Leliwa-Kopystynski, L. Makkonen, O. Erikoinen and K.J. Kossacki: *Kinetics of pressure-induced effects in water ice/rock granular mixtures and application to the physics of the icy satellites*, Plan. Space Sci. **42**, 545–555 (1994)

51. H. F. Levison, D. Terrell, P. A. Wiegert, L. Dones and M. J. Duncan: *On the origin of the unusual orbit of Comet 2P/Encke*, Icarus **182**, 161 (2006)

52. J. Licandro, W. M. Grundy, N. Pinilla-Alonso and P. Leisy: *Visible spectroscopy of 2003 UB313: evidence for N_2 ice on the surface of the largest TNO?*, Astron. Ap. **458**, L5 (2006)

53. J. Licandro, H. Campins, T. Mothé-Diniz, N. Pinilla-Alonso and J. de León: *The nature of comet-asteroid transition object (3200) Phaethon*, Astron. Ap. **461**, 751 (2007)

54. M. J. Lupo and J. S. Lewis: *Mass-radius relationships in icy satellites*, Icarus **40**, 157 (1979)

55. J. Luu and D. Jewitt: *Color diversity among the centaurs and kuiper belt objects*, Astron. J. **112**, 2310 (1996)

56. R. Mann, D. Jewitt and P. Lacerda: *Fraction of contact binary trojan asteroids,* Astron. J. **134**, 1133–1144 (2007)

57. F. Marchis, M. H. Wong, J. Berthier, P. Descamps, D. Hestroffer, F. Vachier, D. Le Mignant and I. de Pater: *S/2006 (624) 1,* IAUC **8732**, 1 (2006)

58. F. Marchis, et al.: *A low density of 0.8gcm-3 for the Trojan binary asteroid 617 Patroclus,* Nature **439**, 565 (2006)

59. N. McBride, S. F. Green, J. K. Davies, D. J. Tholen, S. S. Sheppard, R. J. Whiteley and J. K. Hillier: *Visible and infrared photometry of Kuiper belt objects,* Icarus **161**, 501 (2003)

60. R. Meier and T. C. Owen: *Cometary deuterium,* Space Sci. Rev. **90**, 33 (1999)

61. R. Merk and D. Prialnik: *Combined modeling of thermal evolution and accretion of trans-neptunian objects,* Icarus **183**, 283 (2006)

62. S. J. Mojzsis, T. M. Harrison and R. T. Pidgeon: *Oxygen-isotope evidence from ancient zircons for liquid water at the Earth's surface 4,300Myr ago,* Nature **409**, 178 (2001)

63. A. Morbidelli, J. Chambers, J. I. Lunine, J. M. Petit, F. Robert, G. B. Valsecchi and K. E. Cyr: *Source regions and time scales for the delivery of water to Earth,* Meteoritics Planetary Sci. **35**, 1309 (2000)

64. L. V. Moroz, G. Baratta, E. Distefano, G. Strazzulla, L. V. Starukhina, E. Dotto and M. A. Barucci: *Ion irradiation of asphaltite,* Earth Moon Planets **92**, 279 (2003)

65. D. Nesvorný, D. Vokrouhlický and A. Morbidelli: *Capture of irregular satellites during planetary encounters,* Astron. J. **133**, 1962 (2007)

66. K. Ohtsuka, T. Sekiguchi, D. Kinoshita, J.-I. Watanabe, T. Ito, H. Arakida and T. Kasuga: *Apollo asteroid 2005 UD: split nucleus of (3200) Phaethon?,* Astron. Ap. **450**, L25 (2006)

67. T. Owen, A. Bar-Nun and I. Kleinfeld: *Possible cometary origin of heavy noble gases in the atmospheres of Venus, earth, and Mars,* Nature **358**, 43 (1992)

68. N. Peixinho, A. Doressoundiram, A. Delsanti, H. Boehnhardt, M. A. Barucci and I. Belskaya: *Reopening the TNOs color controversy: Centaurs bimodality and TNOs unimodality,* Astron. Ap. **410**, L29 (2003)

69. J. B. Pollack, J. A. Burns and M. E. Tauber: *Gas drag in primordial circumplanetary envelopes – A mechanism for satellite capture,* Icarus **37**, 587 (1979)

70. D. Prialnik, J. Benkhoff and M. Podolak: *Modeling the structure and activity of comet nuclei.* In Comets II, M. C. Festou, H. U. Keller, and H. A. Weaver (Eds.), (University of Arizona Press, Tucson, 745 pp., 2004) p.359–387

71. W. Reach, M. Kelley and M. Sykes: *A survey of debris trails from short-period comets.* Icarus, **191**, 298 (2007)

72. S. Renner, B. Sicardy and R. G. French: *Prometheus and Pandora: masses and orbital positions during the Cassini tour,* Icarus **174**, 230 (2005)

73. H. Rickman, J. A. Fernandez and B. A. S. Gustafson: *Formation of stable dust mantles on short-period comet nuclei,* Astron. Ap. **237**, 524 (1990)

74. H. N. Russell: *On the light variations of asteroids and satellites,* Ap. J. **24**, 1 (1906)

75. N. H. Samarasinha, M. F. Ahearn, S. Hoban and D. A. Klinglesmith: III *ESLAB symposium on the exploration of halley's comet. Vol. 3,* Posters **250**, 487 (1986)

76. N. H. Samarasinha: *A model for the breakup of comet LINEAR (C/1999 S4).* Icarus **154**, 540 (2001)

77. E. L. Schaller and M. E. Brown: *Volatile loss and retention on kuiper belt objects*. Ap. J. Lett. **659**, L61 (2007)
78. B. Schmitt, S. Espinasse, R. J. A. Grim, J. M. Greenberg and J. Klinger: ESA SP-302, *Phys. Mech. Cometary Mater.*, **65** (1989)
79. S. S. Sheppard and D. Jewitt: *Extreme kuiper belt object 2001 QG298 and the fraction of contact binaries*, Astron. J. **127**, 3023 (2004)
80. S. S. Sheppard and C. A. Trujillo: *A thick cloud of neptune trojans and their colors*, Science **313**, 511 (2006)
81. D. C. Stephens and K. S. Noll: *Detection of six trans-neptunian binaries with NICMOS: A high fraction of binaries in the cold classical disk.* Astron. J. **131**, 1142 (2006)
82. S. Takahashi and W.-H. Ip: *A shape-and-density model of the putative binary EKBO 2001 QG298*, PASJ **56**, 1099 (2004)
83. N. Takato, S. J. Bus, H. Terada, T.-S. Pyo and N. Kobayashi: *Detection of a deep 3-micron absorption feature in the spectrum of amalthea (JV)*, Science **306**, 2224 (2004)
84. S. C. Tegler and W. Romanishin: *Two distinct populations of Kuiper-belt objects*, Nature **392**, 49 (1998)
85. S. C. Tegler and W. Romanishin: *Extremely red Kuiper-belt objects in near-circular orbits beyond 40 AU*, Nature **407**, 979 (2000)
86. S. C. Tegler, W. M. Grundy, W. Romanishin, G. J. Consolmagno, K. Mogren and F. Vilas: *Optical spectroscopy of the large kuiper belt objects 136472 (2005 FY9) and 136108 (2003 EL61)*, Astron. J. **133**, 526 (2007)
87. P. Thomas, et al.: *Hyperion's sponge-like appearance*, Nature **448**, 50–53 (2007)
88. I. Toth: *Impact-generated activity period of the asteroid 7968 Elst-Pizarro in 1996*, Astron. Ap. **360**, 375 (2000)
89. I. Toth: *Search for comet-like activity in asteroid 7968 Elst-Pizarro*, Astron. Ap. **446**, 333 (2006)
90. C. A. Trujillo and M. E. Brown: *A Correlation between inclination and color in the classical kuiper belt*, Ap. J. **566**, L125 (2002)
91. S. J. Weidenschilling: *Hektor – Nature and origin of a binary asteroid*, Icarus **44**, 807 (1980)
92. P. R. Weissman and H. F. Levison: *Origin and evolution of the unusual object 1996 PW: asteroids from the oort cloud?* Ap. J. Lett. **488**, L133 (1997)
93. F. L. Whipple: *1983 TB and the geminid meteors*, IAUC **3881**, 1 (1983)
94. S. A. Wilde, J. W. Valley, W. H. Peck and C. M. Graham: *Evidence from detrital zircons for the existence of continental crust and oceans on the Earth 4.4 Gyr ago*, Nature **409**, 175 (2001)
95. I. P. Williams and Z. Wu: *The Geminid meteor stream and asteroid 3200 Phaethon*, MNRAS **262**, 231 (1993)
96. W. Zheng, D. Jewitt and R. Kaiser: *Amorphization of crystalline water ice*, Ap. J. submitted (2008)

10

Detection of Extrasolar Planets and Circumstellar Disks

Y. Itoh

Graduate School of Science, Kobe University, 1-1 Rokkodai, Nada, Kobe, Hyogo 657-8501, Japan,
yitoh@kobe-u.ac.jp

Abstract In this chapter, various observational methods of searching for extrasolar planets and circumstellar disks are reviewed. These include Doppler-shift measurements, transit detection, astrometry, gravitational lensing, spectral energy distribution, direct detection, and coronagraphy.

10.1 Introduction

Astronomical observations, from ground-based telescopes on Earth or with spaceborn telescopes, provide direct and indirect evidence of extrasolar planets and circumstellar disks. Since each observational method has its own strengths and weaknesses, researchers need to combine various observational approaches to clarify the diversity of planetary systems. In this chapter, I describe the most frequently used methods, their current status, and their limitations in the detection of extrasolar planets and circumstellar disks. I also note some future projects. The discussion of observational methods will focus mainly on those in the optical and near-infrared wavelengths. For future observational studies, great progress is also expected from observations at all wavelengths, such as with ALMA and TPF. For learning more about these future observations, the reader is referred to some related publications (e.g., [41, 66]).

10.2 Summary of Objects

10.2.1 Extrasolar Planets

Planets orbiting a star other than the Sun were first discovered in 1992. Wolszczan & Frail [65] found a periodic delay in the pulses of the radio emission from a pulsar, indicating that two planetary-mass objects orbit a pulsar PSR $1257 + 12$. However, since these planets are located in an "extreme" environment, this discovery appeared in the spotlight only after the discovery

Itoh, Y.: *Detection of Extrasolar Planets and Circumstellar Disks.* Lect. Notes Phys. **758**, 297–323 (2009)
DOI 10.1007/978-3-540-76935-4_10   © Springer-Verlag Berlin Heidelberg 2009

of an extrasolar planet around a solar-type main-sequence star [38]. These discoveries stimulated several attempts to discover extrasolar planets using various methods. As a result of these great efforts, more than 240 extrasolar planets have been found as of mid-2007. Such extrasolar planets have masses between 0.01 M_J (3 M_{Earth}) and 20 M_J with a median value of 1.5 M_J. The semimajor axes of the planets range from 0.02 to 300 AU with the median of the semimajor axis being 0.8 AU. Most of them are less than 10 AU. Note that the semimajor axis of Neptune, the most distant planet in the Solar system, is 30.1 AU. As discussed in the section on direct detection, extrasolar planets are very faint and are located in very close proximity to the bright central star. Although direct detection of an extrasolar planet has not yet been achieved, many indirect evidences of extrasolar planets have been reported.

10.2.2 Circumstellar Disks

Extrasolar planets are thought to be born in circumstellar disks associated with young stellar objects (YSOs). A circumstellar disk evolves along with the central star itself. YSOs are usually classified into three groups according to the structure of their circumstellar material. Objects comprising the youngest group are called protostars and surrounded by an optically and geometrically thick dust envelope. The next stage of their evolution corresponds to the classical T Tauri phase. Objects in this category have a circumstellar disk of gas and dust. As the dusts aggregate into planetesimals, the central star appears as a weak-line T Tauri star.

A typical radius of the circumstellar disks around classical T Tauri stars is ∼150 AU [15, 33]. Such disks have masses between 0.001 and 1 $M_\odot$ [5]. Hayashi [23] proposed that the planets in the Solar System formed through the aggregation of grains within the circumstellar disk. The least-mass disk model which could have formed the Solar System is called the "minimum mass solar nebula." Its mass is 0.01 $M_\odot$, 10 times more massive than the total mass of the current planets in the Solar System. This mass of the model disk is roughly consistent with the mass of the circumstellar disks estimated from observations. Circumstellar disks around T Tauri stars vary widely in morphological and physical characteristics, which is believed to have given rise to a diversity of planetary systems.

Vegalike stars are main-sequence stars surrounded by dust disks. Such a dust disk was first recognized by IRAS observations of Vega [2]. Vega is a standard star, whose magnitudes are defined as 0 mag at all wavelengths. Its spectral energy distribution (SED) was ascribed to a single-temperature blackbody radiation. However, IRAS detected a large amount of excess in the mid- to far-infrared wavelengths attributable to dust in a circumstellar disk. The disk is called a "debris disk" because the dust is second generation and not primordial. Vegalike systems do not have considerable gas components [67], and the debris disks are an abundant analog to the zodiacal dust of the Solar System.

Circumstellar disks were first noticed by an indirect method, i.e., by observing the SEDs of T Tauri stars. Since their SEDs have an excess in the longer wavelengths, direct detection of the disks has been achieved primarily in the radio wavelengths. Recent instrument improvements have provided direct detection of the disks in the optical and near-infrared wavelengths by reflected light.

10.3 Indirect Detection

The presence of planets can affect the central star itself, however minor the effect is. Several methods for indirect detection have been proposed, and successful approaches include Doppler-shift measurements, transit methods, and microlensing events for the detection of extrasolar planets.

10.3.1 Doppler-Shift Measurements

Principle

By definition, a planet orbits a star. A star and a planet make a common gravitational potential within the system. Both the star and the planet orbit the potential minimum, which is not located exactly at the center of the star. The distance between the center of the star and the potential minimum is described as

$$\frac{M_P}{M_\star + M_P} a, \tag{10.1}$$

where M_P and $M_\star$ are the masses of the planet and the star respectively, and a is the semimajor axis of the planetary orbit. Imagine a system consisting of the Sun and Jupiter. Since the mass of Jupiter is $10^{-3} M_\odot$, the potential minimum is located away from the Sun by one-thousandth of the Sun–Jupiter distance. Since Jupiter orbits the Sun at a distance of 5.2 AU, the potential minimum is offset from the center of the Sun by 0.005 AU, coinciding with the photospheric surface of the Sun. The Sun and Jupiter orbit this potential minimum with a period of 12 years, which means that an observer outside the Solar System could notice the orbital motion of the Sun. The periodic change in radial velocity could be detected by high-resolution spectroscopy as a Doppler shift of the stellar spectrum. Since the orbital motion of the Sun is as small as 13 m s^{-1}, the stellar spectrum shifts up to $\pm 2 \times 10^{-5}$ nm at a wavelength of 500 nm.

Instrumentation

Doppler-shift measurements require spectral resolutions as high as $R(= \frac{\lambda}{\Delta\lambda}) \sim$ 100,000. Such high resolutions can be achieved by using an Echelle spectrograph among other means.

The observations also require precise determination of the wavelengths. One method is to use a thorium–argon (Th–Ar) lamp. The lamp is located off-axis of the instrument and the spectrum of the lamp is simultaneously taken with the object spectrum. The other method is to use an iodine cell inserted in the path of the object light. Iodine is transparent and has thousands absorption lines in the optical wavelengths. This method was originally developed for Doppler-shift measurements of the Sun [34]. With simultaneous acquisition, one can achieve a velocity accuracy as high as several meter per seconds.

One must keep the instrument in a stable temperature such that the ray of light does not vary due to thermal distortion of the instrument. A recently developed Echelle spectrograph (HARPS) at the ESO 3.6 m Telescope is situated in a vacuum, and is capable of obtaining a velocity accuracy of better than $1 \, \mathrm{m \, s^{-1}}$ [47].

Current Status

Doppler-shift measurements have been undertaken with the foregoing methods to search for extrasolar planets. Seminal pioneer studies were carried out by Walker et al. [63]. They searched for Jupiter-like planets around 21 solar-like dwarfs for 12 years using the Canada–France–Hawaii 3.6 m Telescope. However, they did not detect long-term variations in the radial velocity of the stars. The first discovery of an extrasolar planet around a dwarf star was announced by Mayor & Queloz [38]. They found periodic variations in the radial velocity of the G-type main-sequence star, 51 Peg. It was interpreted as a Jupiter-mass planet orbiting the star with a period of 4.23 days.

Since then, more than 240 extrasolar planets have been discovered with Doppler-shift measurements. It is now well known that extrasolar planets show a wide distribution of their masses and orbital elements. For example, like the planet of 51 Peg, a type of Jupiter-mass planets orbit their host stars at a few stellar radii. Known as "hot Jupiters", their expected surface temperature is as high as 1000 K. Massive planets are not expected in the inner regions of a planetary system if we assume the standard formation model of the Solar System.

With improvements in the instruments, less massive planets have been recently discovered. The least massive planet so far detected has five Earth-masses [60]. Long-term monitoring allowed the discovery of long orbital-period planets, and the planet with the longest period found so far orbits at a distance of 5.3 AU from the central star.

An evolved star also harbors a planet. Sato et al. [50] revealed a Jupiter-mass planet around a G-type giant. Since G-type giants are the successors of 2–3 $M_\odot$ main-sequence stars, this discovery proves that planets are associated not only with solar-mass stars but also with 2–3 $M_\odot$ stars. Sato et al. [51] found a planet around a G-type giant in the Hyades cluster, while 94 solar-mass dwarfs in the same cluster were found to have no Jupiter-mass planets [46]. This implies that a massive planet tends to be born around a massive star. This idea is also supported by the rareness of massive planets around less massive field stars [19].

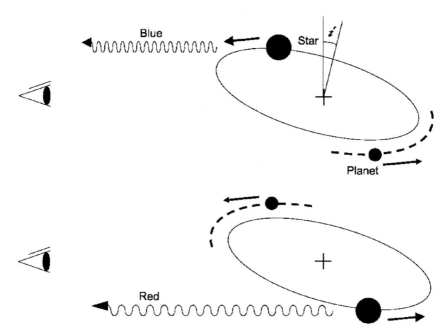

Fig. 10.1. Schematic view of the Doppler shifts of the central star induced by an accompanying planet. We observe a blueshifted spectrum of the central star when the planet moves away and the central star moves toward us. A redshifted spectrum is seen when the planet approaches toward us and the star recedes

Limitations

The Doppler-shift method is biased toward close-in giant planets because less massive planets and planets with long orbital periods have little influence on the radial velocity of the central star.

Even if we detect a periodic change in the radial velocity, the mass of a planet cannot be precisely determined. The mass is expressed as $M_P \sin i$, where i is the inclination of the planetary orbit. Hence, the derived mass is a lower-limit mass, because the Doppler-shift measurements are sensitive to orbital motion only in the radial direction.

Doppler-shift measurements cannot be applied to all types of stars. Pulsation of the star itself causes a periodic variation in the radial velocity. A large star spot on the photosphere also causes similar periodic variation. For example, pulsating occurs in M-type giants and A,F-type stars. Moreover, radial

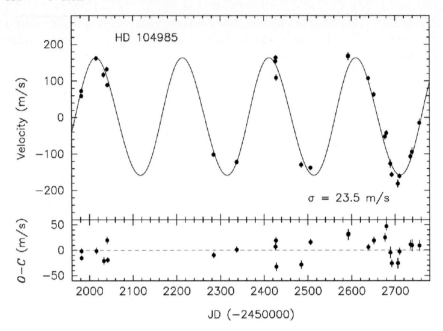

Fig. 10.2. Radial velocity of a giant star, HD 104985. The periodic variation is caused by an accompanying planet

velocity cannot be precisely measured for early-type stars because their spectra have few absorption lines and the lines are often very broad due to rapid stellar rotation ($v \sin i > 100$ km s^{-1}). The other limitation of this method is that all stars exhibit stellar oscillation (~ 0.5 m s^{-1}), which causes a small variation in the stellar radial velocity.

Future Studies

The future of Doppler-shift measurements is promising. Continuation of the measurements will lead to more discoveries of long-period planets, and less massive planets will be revealed with improved instrumentation. Current observations are limited, in most cases, by photon noise. High instrument efficiency and large-aperture telescopes are required for observations of faint targets. Several 30 m ground-based telescopes, such as the European Extremely Large Telescope (E-ELT; [22]) and the Thirty Meter Telescope (TMT; [58]), are proposed for the next decade.

10.3.2 Transit Detection

Principle

A solar eclipse occurs when the Moon intersects the light path from the Sun to an observer on Earth. An eclipse also occurs for a distant two-body system such as a close-in stellar binary. We may notice an eclipse even in an extrasolar planetary system, when the planet and its host star lie in the same line of sight. This phenomenon is called a "transit event."

Again suppose the Sun–Jupiter system. The system is located so far away from us that the area of the extrasolar planet projected onto the stellar surface is circular and equals the radius of the planet. As the radius of Jupiter is one-tenth that of the Sun, the area hidden by the planet would be one-hundredth of the projected stellar surface. As a result, we would observe a periodic 1% dip in the stellar flux with the orbital period of the planet. Note that the emission from an extrasolar planet is negligibly small, at least in the optical wavelengths.

Instrumentation

Such flux variations can be measured by photometric monitoring. As discussed below, the geometrical probability for a transit event is quite low. To detect a transit event, one needs to monitor a huge number of stars. Two methods are applied in transit detection. One involves wide-field imaging. Alonso et al. [1] discovered an extrasolar planet using a small-aperture but wide-field camera. The camera can image a $6° \times 6°$ field simultaneously. The other method is deep imaging. Urakawa et al. [61] observed near the Galactic plane using Suprime-Cam, an optical imager on the Subaru Telescope, and achieved 1% photometric accuracy for 6900 stars.

Current Status

A transit event of an extrasolar planet was first detected by Charbonneau et al. [12]. They found a 1% dip in the flux of HD 209458, a planetary system previously discovered by Doppler-shift measurements. The transit event of this planetary system was also confirmed by accurate photometry with the Hubble Space Telescope [9]. Some current searches for a transit event target planetary systems previously discovered with Doppler-shift measurements.

Many wide-field surveys are also carried out to detect unknown extrasolar planets. Several groups have developed a wide-field imager with small-aperture telescopes. Others use the long-term monitoring data obtained for other purposes, such as the OGLE (Optical Gravitational Lensing Experiment; [59]) database.

The depth of the flux dip caused by a transit event depends on the ratio of the radius of the planet and that of the star. Given a known stellar radius, the radius of the planet can be estimated. The inclination of the planetary orbit can also be derived by assuming a limb darkening model for the star. When combining these parameters with the result given by the Doppler-shift measurements, the mass and density of the planet can be determined.

During a transit, the light from the star goes through the atmosphere of the planet. The observed spectrum of this light component contains absorption lines originating in the planetary atmosphere. Vidal-Madjar et al. [62]

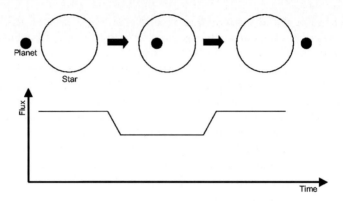

Fig. 10.3. Schematic diagram of a transit event. When a planet passes across the line of sight, the flux of the central star decreases slightly

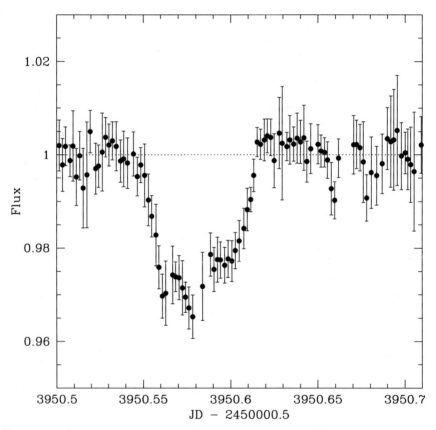

Fig. 10.4. Light curve of an extrasolar planetary system, HD 189733. The photometric monitoring was conducted with a commercial charge-coupled device (CCD) mounted on the Meade 30 cm telescope at Kobe University [28]. We see a 2% dip in the light curve

detected a hydrogen absorption feature from HD 209458 during the transit, which is believed to be caused by escaping hydrogen atoms from the planet's atmosphere.

During the opposite phase of a transit, the planet is hidden behind the star. This is called a "second transit event." A planet is rather bright in the wavelengths longer than mid-infrared region due to its thermal radiation. Since thermal radiation from the planet is blocked by the host star during the second transit event, flux from the planetary system decreases in the mid-infrared wavelengths. Deming et al. [14] and Charbonneau et al. [13] detected the second transit using the Spitzer Space Telescope. From the flux decrement of the second transit compared to an atmospheric model for the planet, one can estimate the temperature of the planetary atmosphere.

Limitations

For a transit to be observed, a planet should pass between the star and us. The geometrical probability of such a configuration is described as

$$a \cos i < R_\star + R_P, \tag{10.2}$$

where a is the semimajor axis of the planet, i is the inclination of the planet's orbit, and $R_\star$ and R_P are the radii of the star and the planet, respectively. This probability is as low as 6% for hot Jupiters, and lower for a planet with large orbital period. Thus, photometric monitoring of a large number of stars is needed.

A transit event does not provide exclusive evidence of the presence of an extrasolar planet because late-type dwarfs, brown dwarfs, and Jupiter-mass planets all have similar radii. Follow-up observations of Doppler-shift measurements are required for mass determination. A planet with small radius, such as a rocky planet, is difficult to detect using this transit method.

Future Studies

The equipment for wide-field surveys with small to mid-size telescopes is simple and inexpensive. Such observing systems will be developed in many observatories and universities.

Accurate photometry can be achieved with wide-field imagers onboard space satellites. A European satellite, CoRoT [8], was launched in 2006. Very recently, the CoRoT team announced the first detection of a transit event. A US mission, Kepler [35], will be launched in 2008.

10.3.3 Astrometry

Principle

As described in the context of the Doppler-shift measurements, a star as well as a planet orbits the nadir of the common gravitational potential, i.e., the common center of mass. For the Sun–Jupiter system, the Sun orbits a point near its surface. When observing from 5 pc away from this system, one may notice a periodic change in the position of the Sun; the amount of this variation is as small as 0.001" (1 milliarcsecond, mas).

Instrumentation

The position of a star can be measured with a simple imager. The accuracy of the position depends on the size of the instrument's point spread function (PSF). It is often stated that the position is determined within an accuracy of tenths of the PSF at best. By ground-based observations under natural seeing conditions, the size of the PSF is usually as large as 1." To sharpen the PSF, one uses adaptive optics (AO) systems, which can improve the PSF < 0.1" at near-infrared wavelengths. Interferometric technology also sharpens the PSF. Measurements with telescopes from satellites are free from atmosphere turbulence and therefore also have a sharper PSF.

Current Status

Van de Kamp [32] announced a periodic variation in the position of a nearby star, Barnard's star. The variation implied a 1.6 M_J object orbiting the central star with a period of 24 years. However, follow-up observations did not confirm this variation in stellar position.

Limitations

For precise astrometry, one needs measurements over a long-time interval, such as several decades, and/or precise measurements of the position of a star. Because long-term astrometry has to use old technology, astrometry using photographic plates is not sufficiently accurate to detect the tiny variation in the position of a star with a planet. Time intervals are too short for precise astrometry using a CCD camera.

The discrepancy between the center of the star and the potential minimum of the planetary system is large for a distant massive planet. Thus, astrometric detection is biased toward giant planets with large orbital radii, but the orbital period is long.

Future Studies

Several space missions are proposed for performing precise astrometry. The space interferometry mission (SIM; [53]) consists of two small-aperture optical telescopes, and the highest resolution of μ arcseconds (10^{-6} arcsec) will be realized. The Japanese satellite mission, JASMINE, is also proposed for highly accurate astrometry in the infrared wavelengths [21] and it will attempt to measure stellar positions with an accuracy of 10 μarcsec. The satellite is scheduled for launch around 2014.

10.3.4 Gravitational Lensing

Principle

According to the theory of relativity, light is bent by gravity. This was first confirmed by Dyson et al. [17]. During the solar eclipse in 1920, these investigators found displacements of the stellar positions located close to the Sun, which indicated that lights from the stars (the sources) were bent by the gravitational lens effect produced by the Sun (the lens). The radius (Einstein radius) and curvature of the lens depend on the mass of the lens object. A massive object has a large diameter and strong curvature, and collects much more light from the source. When the lens object passes near the light path of the source, the flux of the source increases temporally. This is known as a "microlens event" if the reimaged source object is not spatially resolved. If the lens object harbors a planet, the flux variation exhibits two peaks.

Instrumentation

Because the Einstein radius of a stellar object is very small, the likelihood of observing a microlens event is quite low. As in the case of transit searches, wide-field imaging observations are required. The OGLE project provides the most extensive data set for gravitational lens events toward the Galactic bulge and Magellanic clouds [59].

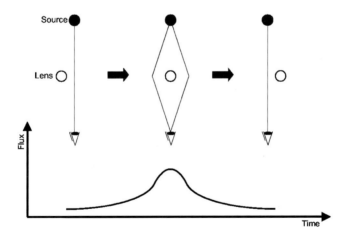

Fig. 10.5. Schematic geometry of a microlens event. The event occurs when the lens passes the line of sight of the source object. Since in most cases, the lens object is faint and is located far from us, we cannot detect the light from the lens object

Current Status

One can estimate the mass of the lens object from the duration and increase of the flux in a microlens event. Several planetary-mass companion candidates have been detected [7]. They are noticed by a sharp, short timescale (hours to days) increase of flux superposed on a long-duration time-flux increase due to a stellar object. Beaulieu et al. [4] announced the discovery of a 5.5 Earth-mass planetary companion at a separation of 2.6 AU from an M-type dwarf. A microlens event is sometimes claimed as the only way to detect an Earth-mass extrasolar planet.

Limitations

The duration time of a microlens event is not only a function of the mass of the lens, but also of the distances of the source and the lens, and the tangential velocity of the lens. Even if the distance of the source is determined, the other three parameters associated with the lens effect become inseparable. Because the lens object is too faint to be observed at times other than during a microlens event, these three parameters can never be determined. One solves this degeneracy by using a star count model of the Galaxy. Even though, the mass of an object is only determined as a probability. An inherent limitation of this method is that a planet detected by this method can never be reexamined.

Future Studies

Continuous searches for gravitational lens events will be made by small to intermediate aperture telescopes. High-stellar density regions, such as globular clusters and galaxies, will be explored with high-spatial resolution techniques (e.g., AO systems).

10.3.5 Spectral Energy Distribution

Principle

Protostars and classical T Tauri stars have continuum excess emissions at wavelengths beyond near-infrared wavelengths. The excess is attributed to an optically thick circumstellar structure, which is heated by the central star or by viscous energy in the structure itself. A protostar is deeply embedded in a geometrically thick envelope. On the other hand, since we directly detected light from the photosphere of a T Tauri star, the circumstellar structure around the T Tauri star should not be located in the line of sight. A T Tauri star is surrounded by an optically thick but geometrically thin disk.

Instrumentation

Flux measurements in the infrared and millimeter wavelengths can be made from the ground. Photometry at the $10\,\mu$m range, the $20\,\mu$m range, and several bands in the submillimeter region is also possible from the ground, although it is severely affected by rapid changes in atmospheric conditions and thermal radiation of the Earth.

Current Status

The flux ratio or difference of magnitudes between two photometric bands is called "color." The color of an object is a difference expression of its spectral distribution. One can classify YSOs in their evolutional stage by using, for example, a JHK near-infrared color–color diagram.

For classical T Tauri stars, radiation from the circumstellar disk dominates the mid-infrared flux, while the photospheric flux is still not negligible in the near-infrared wavelengths. Mid-infrared photometry has been limited due to the high-thermal background of the Earth. Recent observations using the Spitzer Space Telescope revealed YSOs, which do not show any excess in near-infrared wavelengths but do show excess in the mid-infrared wavelengths [45].

The timescale and wavelength dependence of the dissipation of circumstellar disks are important keys to understand the evolution of circumstellar disks, as well as the dust and planetesimals within the disk. Mid-infrared observations with the Spitzer Space Telescope also revealed debris disks around young stars in the Pleiades cluster [57].

Millimeter wavelength observations are also sensitive enough to detect circumstellar disks around classical T Tauri stars. In this wavelength, the disks are optically thin except for the innermost region. Beckwith et al. [5] surveyed 86 YSOs in the Taurus-Auriga molecular cloud in the 1.3 mm radio continuum. They estimated the disk masses to be between 0.001 $M_\odot$ and 1 $M_\odot$, but disks around weak-line T Tauri stars are not massive enough to be detected [16].

Limitations

The spectral energy distribution itself is an indirect signature of a circumstellar disk. If a faint low-effective temperature object is located in the close vicinity of the central star, one may confuse its infrared to radio emission as excess from a circumstellar disk. The spectral energy distribution alone cannot provide a unique solution to clarify a disk structure because several physical parameters complicate the interpretation. In addition, free–free emission is often detected in long wavelengths in the radio range (e.g., [18]), which is not direct evidence of a circumstellar disk but is indicative of an ionized jet.

Future Studies

Since a central star is still bright in the near-infrared, unambiguous detections from mid-infrared to millimeter wavelengths are critical for a circumstellar disk. The Atacama Large Millimeter Array (ALMA; [64]) is under construction in the Atacama Desert in Chile in collaboration with the United States, Europe, and Japan. Due to the low water vapor in the atmosphere, flux measurements of the continuum emission in several submillimeter bands will be achieved. It is the largest interferometric telescope and will achieve the highest spatial resolution in millimeter wavelengths.

10.3.6 Polarization

Principle

Since planets, except for the close-in planets, have maximum temperatures of a few hundreds of Kelvin, most of the flux from a planet is reflection light of the central star at short wavelengths. Because the light is reflected by solid materials, such as dust particles in the planet's atmosphere or the planet's solid surface, the light of the planet is at least partially polarized. The position angle of the polarization will rotate with the orbital period of the planet. Nevertheless, the polarized light will be diluted with the unpolarized light of the central star. Seager et al. [52] predicted that the linear polarization of an extrasolar planetary system would be as small as $10^{-4}\%$.

Instrumentation

To determine the degree and position angle of the polarization, one has to observe the object's flux in three or four position angles of the polarization plane. Each intensity is measured by rotating the optical elements in the instrument.

The accuracy of the measurement is often reduced by uncertainties in the position of the moving elements. In many cases, a Wollaston prism is used, which separates the incident light into two perpendicularly polarized beams. Using this, one simultaneously obtains the fluxes of two position angles. Hough et al. [27] developed an instrument with a Wollaston prism to conduct a polarimetric search for extrasolar planets (PlanetPol). The instrument has no moving elements, but measures intensities at different angles by rotating the instrument itself.

Current Status

Polarimetric searches for extrasolar planets are so far carried out by a very limited number of groups. Even though PlanetPol achieved polarization accuracies of $10^{-3}\%$, the expected degree of polarization due to an extrasolar planet is 10 times smaller.

Limitations

This method is also biased toward close-in giant planets. It is difficult to distinguish an extrasolar planet from a brown dwarf and a very-low mass dwarf, because both have similar radii.

Future Studies

Time-resolved polarimetry is an important observing method, but is not common. Variability of polarization is expected to be induced by not only extrasolar planets but also protoplanetary disks and circumstellar structures.

10.4 Direct Detection

Direct detection is undoubtedly desirable, but nobody has so far succeeded in obtaining an image of an extrasolar planet.

10.4.1 Direct imaging

Principle

As noted above, the brightness of a planet results mostly from the reflected light of the central star in optical and near-infrared wavelengths, which means that the contrast between a star and a planet is huge. The flux ratio of the reflected light (F_p) of a planet to the brightness (F_s) of the central star is described as

$$\frac{F_p}{F_s} = \frac{A}{2} \left(\frac{R_P}{2a} \right)^2 , \qquad (10.3)$$

where A, R_P, and a are albedo, the radius, and the semimajor axis of the planet orbit, respectively. For the Sun–Jupiter system, this flux ratio is as small as 3×10^{-9}. Since such a faint object is located in very close proximity to the bright central star, direct detection of an extrasolar planet is exceedingly difficult.

Instrumentation

To detect a faint object in the near vicinity of a bright star, high-spatial resolution observations are required. One uses AO systems, interferometric technology, speckle techniques, or makes observations from space.

AO systems consist of a wavefront sensor and a deformable mirror. For ground-based observations, light from an object is perturbed by atmospheric turbulence. The wavefront of the light is measured with a Shack–Hartmann sensor or a curvature sensor every ~100 ms. Then, the shape of the deformable mirror is reconstructed by mechanical actuators.

A speckle image consists of many exposures of short integration. Taking exposures as short as the timescale of the atmospheric turbulence, one can obtain a PSF that is not blurred by the atmospheric turbulence. By shifting the images to adjust the peak of the PSFs and then adding many exposure frames, one can obtain a sharp PSF.

Even if one obtains high-resolution images using the foregoing methods, the huge dynamic range between a star and a planet prevents us from detecting an extrasolar planet directly.

Current Status – Planets

Direct detection of an extrasolar planet has not yet been achieved. All the extrasolar planets discovered by Doppler-shift measurements or the transit methods are difficult to be detected by direct means in reflected light due to their faintness and/or their small separation from the star.

Alternative targets for direct detection are unknown extrasolar planets around nearby young main-sequence stars. Young planets in a contraction phase are expected to be hot, and thus still bright. I summarize the results of the direct imaging searches for extrasolar planets around ε Eri and Vega. These stars are surrounded by dust disks, which suggest their youthfulness. The disks are ringlike, i.e., have an inner cavity. One may imagine the situation in which an unseen planetary-mass companion clears out the dust in the inner portion of the disk.

Macintosh et al. [37] detected 10 faint objects at a distance of $17'' \sim 45''$ from ε Eri by near-infrared K-band direct imaging observations, using the Keck Telescope. While the limiting magnitude is about 21.5 mag (corresponding to 5 M_J) more than 15" from the star, the sensitivity is poor within 10" of the star. Follow-up observations of the proper motion indicated that all the objects are background objects.

Searches have also been carried out at optical wavelengths. Proffitt et al. [48] found 59 faint objects in the region between 12.5" and 58" from ϵ Eri using the Hubble Space Telescope and its optical camera, WFPC 2. Most of these are elongated and suggestive of background galaxies. Although the detection limit of their observation is as deep as 26 mag, extrasolar planets are expected to be orders of magnitude fainter than this limit in optical wavelengths.

Endeavors are also under way to directly image an extrasolar planet around Vega. Macintosh et al. [37] also searched for extrasolar planets around Vega by direct imaging observations. Their K-band limiting magnitude was $\sim$20.5 mag beyond 20" from the central star, while only $\sim$17 mag ($6 \sim 12$ M_J) at 7" away from the central star. Seven objects were found > 20" away from the central star. Based on the proper motion measurements, they are thought to be background stars. Metchev et al. [39] also attempted to directly observe extrasolar planets around Vega. Their H-band limiting magnitude was about 19 mag ($2 \sim 6 M_J$) and 14 mag ($10 \sim 20$ M_J) at 20" and 7" from the central star, respectively. They detected eight background stars.

At longer wavelengths, one may detect the thermal flux from an extrasolar planet. Hinz et al. [24] carried out M-band (5 μm) direct imaging observations of Vega. Using AO, they obtained diffraction-limited images with a detection limit of 7 M_J at 2.5" from the central star.

Planets in their early formation stages are expected to be brighter, and hence direct imaging of these objects is more likely to succeed. Neuhäuser et al. [42] announced the discovery of a faint companion around the T Tauri star GQ Lup. The companion is located at 0.7" (100 AU) from the primary. It was confirmed that both the primary and the companion have the same proper motion, indicating that they form a conjoint physical system. The companion has 13.1 mag at the K-band and the estimated mass of the object is $1 \sim 42$ M_J. However, because the mass of a protoplanet is poorly estimated with current models of evolutionary tracks, it is still unclear whether the object is a bona-fide young planet.

Another approach to achieve direct imaging lies in carrying out surveys around faint objects. Chauvin et al. [11] discovered a planetary mass object around a young brown dwarf, 2MASSWJ 1207334-393254. It is the companion of a brown dwarf and therefore is not assigned as a planet by definition.

White dwarfs are the final stage of the life of intermediate-mass ($<8M_\odot$) stars. Since the object has a very small radius, its luminosity is two to four orders of magnitude fainter than that of main-sequence dwarfs with the same spectral type. Zinnecker et al. [68] searched for extrasolar planets around seven white dwarfs in the Hyades cluster by direct imaging with the Hubble Space Telescope, but did not detect any planetary mass companion.

Objects in the mass regime of planets are not always associated with a star. Oasa et al. [43] discovered faint isolated YSOs in the Chamaeleon star-forming region. Their masses are estimated to be several Jupiter masses. Such objects are denoted as planetary-mass objects, sub-brown dwarfs, or free-floating planets. Reipurth et al. [49] proposed that they are born in a multiple system and then ejected from the system due to mutual close encounters.

Current Status – Disks

Stapelfeldt et al. [56] surveyed circumstellar structures, such as disks, envelopes, and outflows, around 153 YSOs in nearby star-forming regions. With the high-spatial resolution of the optical imaging camera of the Hubble Space Telescope, they detected circumstellar disks around 10 sources.

If one observes a YSO + disk system in edge-on geometry, the circumstellar disk blocks the light of the central star. Burrows et al. [10] imaged an edge-on disk and the jets associated with a YSO, HH 30, using the Hubble Space Telescope. To date, several edge-on disks have been revealed (see Fig. 10.6).

Limitations

The direct imaging method is sensitive to planets with large orbital radii and to large, flared circumstellar disks, but one cannot detect close-in planets using this method. An image taken with an AO system is often processed by a deconvolution technique to improve spatial resolution. However, this process sometimes creates artificial blobs.

Moreover, a point source found in the vicinity of a bright star may be a distant background star. Proper motion measurements at different epochs will confirm its companionship status, and spectroscopy will allow the determination of an effective temperature of the companion candidate.

Future Studies

Even when a high-spatial resolution instrument is available, techniques of suppressing the light from the central star are crucial in detecting an extrasolar

Fig. 10.6. Near-infrared image of an edge-on disk, Lk Hα 263 C (Itoh, unpublished). As the mid-plane of the circumstellar disk is optically thick, light from the central star is obscured. On the other hand, the light can be seen as the scattered light at the optically thin upper layer of the disk

planet. To date, application of such techniques, for example, a coronagraph (described below) has been employed.

10.4.2 Coronagraphic Imaging

Principle

It is well known that a coronagraph is a powerful instrument to detect very faint objects close to bright objects. Originally, Lyot [36] developed a coronagraph to observe the solar corona and prominences without having to wait for rarely occurring solar eclipses. The important components of a coronagraph are an occulting mask and a Lyot stop. An occulting mask is located on the first focal plane and obscures the light from a central bright source. Subsequently, the diffracted light from the pupil of the telescope and the occulting mask is blocked by a Lyot stop centered on the optical axis in the pupil plane. Finally, the image is detected at the second focal plane where the light from the central source is almost suppressed. Recent progresses in AO techniques enable us to use smaller occulting masks, thus to detect faint objects very close to the bright source.

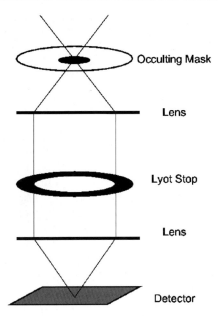

Fig. 10.7. Schematic diagram of a coronagraph. Important elements of a corona-graph are an occulting mask located on the first focal plane and a Lyot stop on the pupil plane

Instrumentation

To detect such faint objects near bright sources, a coronagraph, CIAO (Coronagraphic Imager with Adaptive Optics), was constructed for the use on the Subaru 8.2 m Telescope. CIAO is used at the near-infrared (1–5 μm) wavelengths, because an AO system effectively compensates for wavefront distortion at near-infrared wavelengths, and optical aberration, including scattered light, is smaller at near-infrared wavelengths than at optical wavelengths. In addition, great care was taken to determine the transmittance patterns of the occulting masks and the Lyot stops to increase the performance of halo suppression.

Several specifically shaped pupil masks were recently proposed for a high-dynamic contrast range. For example, Spergel [55] proposed a Gaussian-shaped pupil, which can achieve deep contrast imaging (10^{-9}) in two directions very close to a central star.

Interferometric nulling is another kind of a coronagraph. By interfering with the 180° phase-shifted light, light from the central star is strongly suppressed (e.g., [3]).

Current Status – Planets

Combining a coronagraph with an AO system, one can obtain high-spatial resolution and high signal-to-noise ratio images of faint companions and circumstellar structures.

Nakajima et al. [40] detected a faint point source 7.6" (44 AU) away from a nearby star, GJ 229. Its red color in the optical wavelengths and deep methane absorption bands in the near-infrared wavelengths [44] indicate its low-effective temperature. Indeed this is the first object definitely classified as a brown dwarf.

Itoh et al. [29] discovered a young brown dwarf companion (DH Tau B) associated with a classical T Tauri star DH Tau, using CIAO on the Subaru Telescope (Fig. 10.8). The companion has $H = 15$ mag located at 2.3" (330 AU) away from the primary DH Tau A. Comparing its position to a Hubble Space Telescope archive image, it is confirmed that DH Tau A and B share a common proper motion, suggesting that they are physically associated with each other. From the near-infrared spectra of DH Tau B, its effective temperature and surface gravity are derived to be $T_{\text{eff}} = 2700$–2800 K and $\log g = 4.0$–4.5, respectively. The location of DH Tau B on the Hertzsprung–Russell diagram (HR diagram) indicates that its mass is between 30 and 50 M_{J}.

Direct imaging of an extrasolar planet is, of course, one of the main purposes of coronagraphic surveys. Itoh et al. [30] carried out a coronagraphic imaging search for extrasolar planets around the young main-sequence stars, ε Eri and Vega. By concentrating the stellar light into the core of the PSF using the AO system, and then blocking the core with the occulting mask in the coronagraph, they achieved the highest sensitivity to date for point sources in the close vicinity of both central stars. Nonetheless, they had no reliable detection of a point source around the stars. The observations permitted determining the upper limits of the masses of potentially existing planets to $4 \sim 6$ M_{J} and $5 \sim 10$ M_{J} at a few arcseconds from ε Eri and Vega, respectively.

Current Status – Disks

Smith & Terrile [54] achieved the first remarkable success in stellar coronagraph observations, by finding a circumstellar disk around the massive (2 $M_{\odot}$) main-sequence star, β Pic. The disk generates the observed excess of the spectral energy distribution of β Pic in the mid- to far-infrared wavelengths. Recent observations indicate that a certain fraction of an early-type main-sequence star shows thermal excess. These are called "Vegalike" stars. Among them, some debris disks are spatially resolved [31].

The circumstellar disks around many T Tauri stars are now surveyed with a coronagraph. Disks are associated with both single and binary stars. A young binary system often has three disks, two circumstellar disks and one circumbinary disk. The highest angular resolution near-infrared images of the

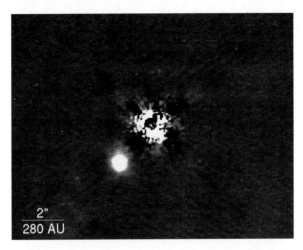

Fig. 10.8. Near-infrared coronagraphic image of DH Tau. North is up and east is to the left. The primary star, DH Tau A, is located at the center of the image but is blocked by the mask. The companion, DH Tau B, is detected ∼2.34" (330 AU) southeast of DH Tau A

GG Tau A binary system were obtained with Subaru's coronagraph, CIAO. The image clearly revealed a ringlike circumbinary disk. It is thought that the gravity of the central binary stars produced a cavity (the inner hole) in the circumbinary disk [26]. The circumbinary disk is smooth and does not have 0".1-scale structures.

In contrast, a circumbinary disk around UY Aur has a complicated structure. Hioki et al. [25] present a near-infrared coronagraphic image of UY Aur, a binary system of 0.9" separation. They detected a half-ring shaped circumbinary disk around the binary (Fig. 10.10). Its inner radius and inclination are about 520 AU and $42° \pm 3°$, respectively. The disk is not uniform but has remarkable features, including a clumpy structure along the disk, circumstellar material inside the inner cavity, and an extended armlike structure.

Disks are also associated with intermediate-mass young stars. Fukagawa et al. [20] discovered a circumstellar disk with four spiral arms at $r = 200$–450 AU around a Herbig Ae star, AB Aur, by using CIAO. The weak gravitational instability, maintained for millions of years by continuous mass supply from the envelope, might explain the presence of the spiral structure.

Limitations

The performance of coronagraph instruments is severely hampered by the imperfection of the wavefront of the incident light. To minimize this effect, coronagraphs are often used in conjunction with AO systems, although the degree of compensation of the current AO systems still limits the performance of the coronagraphs.

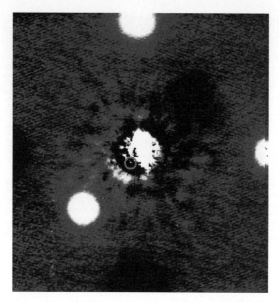

Fig. 10.9. Near-infrared coronagraphic image of ε Eri. The field of view is 20" × 21". ε Eri is located at the center of the image but is blocked by the mask. A residual halo from the PSF subtraction remains around the central star. The other three bright emissions are ghosts. The point source candidate is indicated by a circle

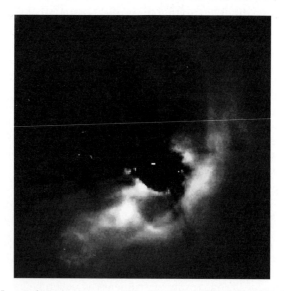

Fig. 10.10. Near-infrared coronagraphic image of UY Aur. The field of view is 13.2" × 13.2". North is up and east is to the left. The central binary is located at the center of the image but is blocked by an occulting mask. A circumbinary disk is clearly detected southwest of the star. The disk has inhomogeneous structure

Note that some coronagraph designs based on computer simulations are effective only for monochromatic light. Since an extrasolar planet is expected to be very faint, broadband imaging is necessary.

Future Studies

The Terrestrial Planet Finder (TPF; [6]) coronagraph is proposed for a future space mission. Observations are planned in the optical wavelengths. Another plan of the TPF is space interferometry with nulling techniques.

10.5 Future Prospects

The timetable for future projects described in this chapter is presented in Fig. 10.11. The reader will find that secure developments are planned for observational equipments.

It turns out that extrasolar planets and circumstellar disks exhibit a large variety of physical characteristics. Current samples of both objects are still insufficient for clarifying the diversity of planetary systems. Searches will continue for extrasolar planetary systems using various methods.

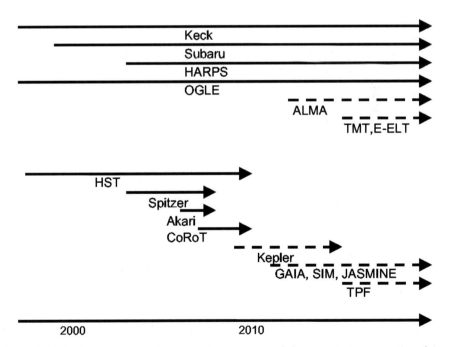

Fig. 10.11. Schematic timetable of the current and future missions mentioned in this chapter

Spectral signatures will be explored for extrasolar planets. An ultimate goal of such studies is to detect signs of life. Spectral signatures indicating habitable conditions will be investigated through theoretical studies and observational studies of the planets in the Solar System.

A planet is defined as an object with a core in which deuterium does not burn. Based on this criterion, the upper limit of the mass of a planet is 13 M_J. Some objects classified as an extrasolar planets are estimated to have mass more than 10 M_J. These may indeed be more massive objects, i.e., brown dwarfs. Determination of the mass of such objects is necessary. For objects discovered by Doppler-shift measurements, determination of the inclination angle by other methods resolves the mass. The masses of the objects discovered by direct imaging are estimated based on evolutionary tracks on the HR diagram. However, different evolutionary tracks infer different masses. Development of an evolutionary track model for low-mass objects is essential.

References

1. R. Alonso, et al.: *TrES-1: The transiting planet of a bright K0 V star*, Astrophys. J. **613**, L153 (2004)
2. H. H. Aumann, et al.: *Discovery of a shell around Alpha Lyrae*, Astrophys. J. **278**, L23 (1984)
3. N. Baba and N. Murakami: *A method to image extrasolar planets with polarized light*, Publ. Astron. Soc. Pacific **115**, 1363 (2003)
4. J. P. Beaulieu, et al.: *Discovery of a cool planet of 5.5 earth masses through gravitational microlensing*, Nature **439**, 437 (2006)
5. S. V. W. Beckwith, et al.: *A survey for circumstellar disks around young stellar objects*, Astron. J. **99**, 924 (1990)
6. C. A. Beichman: *Terrestrial planet finder: the search for life-bearing planets around other stars*, Proc. SPIE **3350**, 719 (1998)
7. D. P. Bennett, et al.: *Discovery of a planet orbiting a binary star system from gravitational microlensing*, Nature **402**, 57 (1999)
8. P. Bordè, et al.: *Exoplanet detection capability of the COROT space mission*, Astron. Astrophys. **405**, 1137 (2003)
9. T. M. Brown, et al.: *Hubble space telescope time-series photometry of the transiting planet of HD 209458*, Astrophys. J. **552**, 699 (2001)
10. C. J. Burrows, et al.: *Hubble space telescope observations of the disk and jet of HH 30*, Astrophys. J. **473**, 437 (1996)
11. G. Chauvin, et al.: *A giant planet candidate near a young brown dwarf. Direct VLT/NACO observations using IR wavefront sensing*, Astron. Astrophys. **425**, L29 (2004)
12. D. Charbonneau, et al.: *Detection of planetary transits across a sun-like star*, Astrophys. J. **529**, L45 (2000)
13. D. Charbonneau, et al.: *Detection of thermal emission from an extrasolar planet*, Astrophys. J. **626**, 523 (2005)
14. D. Deming, et al.: *Infrared radiation from an extrasolar planet*, Nature **434**, 740 (2005)

15. A. Dutrey, et al.: *Dust and gas distribution around T Tauri stars in Taurus-Auriga. I. Interferometric 2.7 mm continuum and ^{1}3CO J=1-0 observations*, Astron. Astrophys. **309**, 493 (1996)

16. G. Duvert, et al.: *A search for extended disks around weak-lined T Tauri stars*, Astron. Astrophys. **355**, 165 (2000)

17. F. W. Dyson, A. S. Eddington, and C. R. Davidson: *Astrographic catalogue, greenwich, analysis of the proper motions of the reference stars*, Mon. Not. Royal Astron. Soc. **62**, 291 (1920)

18. J. Eislöffel, R. Mundt, T. P. Ray, and L. F. Rodríguez: *Collimation and Propagation of Stellar Jets, Protostars and Planets IV*, (Univ. Arizona Press Tucson 2000) pp 815–840

19. M. Endl et al.: *Exploring the frequency of close-in jovian planets around M Dwarfs*, Astrophys. J. **649**, 436 (2006)

20. M. Fukagawa, et al.: *Spiral structure in the circumstellar disk around AB Aurigae*, Astrophys. J. **605**, L53 (2004)

21. N. Gouda et al.: *Japanese astrometry satellite mission for infrared exploration*, Astrophys. Space Sci. **280**, 89 (2002)

22. T. G. Hawarden et al.: *Critical science for the largest telescopes: science drivers for a 100 m ground-based optical-IR telescope*, Proc. SPIE **4840**, 299 (2003)

23. C. Hayashi: *Structure of the solar nebula, growth and decay of magnetic fields and effects of magnetic and turbulent viscosities on the nebula*, Prog. Theoretical Phys. **70**, 35 (1981)

24. P. M. Hinz, et al.: *Thermal infrared constraint to a planetary companion of vega with the MMT adaptive optics system*, Astrophys. J. **653**, 1486 (2006)

25. T. Hioki, et al.: *Near-infrared coronagraphic observations of the T Tauri binary system UY Aur*, Astrophys. J. **134**, 880 (2007)

26. M. J. Holman and P. A. Wiegert: *Long-term stability of planets in binary systems*, Astron. J. **117**, 621 (1999)

27. J. H. Hough, et al.: *PlanetPol: A very high sensitivity polarimeter*, Publ. Astron. Soc. Pacific **118**, 1302 (2006)

28. S. Ishiguma: Master thesis of Kobe university (in Japanese)

29. Y. Itoh, et al.: *A young brown dwarf companion to DH tauri*, Astrophys. J. **620**, 984 (2005)

30. Y. Itoh, et al.: *Coronagraphic search for extrasolar planets around ϵ Eri and vega*, Astrophys. J. **652**, 1729 (2006)

31. P. Kalas et al.: *A planetary system as the origin of structure in Fomalhaut's dust belt*, Nature **435**, 1067 (2005)

32. P. van de Kamp: *Astrometric study of Barnard's star from plates taken with the 24-inch Sproul refractor*, Astron. J. **68**, 515 (1963)

33. Y. Kitamura et al.: *Investigation of the physical properties of protoplanetary disks around T Tauri stars by a 1 arcsecond imaging survey: evolution and diversity of the disks in their accretion stage*, Astrophys. J. **581**, 357 (2002)

34. A. Koch and H. Wöhl: *The use of molecular iodine absorption lines as wavelength references for solar Doppler shift measurements*, Astron. and Astrophys. **134**, 134 (1984)

35. D. G. Koch et al.: *Kepler: a space mission to detect earth-class exoplanets*, Proc. SPIE **3356**, 599 (1998)

36. B. Lyot: *The study of the solar corona and prominences without eclipses (George Darwin Lecture, 1939)*, Mon. Not. Royal Astron. Soc. **99**, 580 (1939)

37. B. A. Macintosh et al.: *Deep keck adaptive optics searches for extrasolar planets in the dust of ε eridani and vega*, Astrophys. J. **594**, 538 (2003)
38. M. Mayor and D. Queloz: *A jupiter-mass companion to a solar-type star*, Nature **378**, 355 (1995)
39. S. Metchev et al.: *Adaptive optics observations of vega: eight detected sources and upper limits to planetary-mass companions*, Astrophys. J. **582**, 1102 (2003)
40. T. Nakajima et al.: *Discovery of a cool brown dwarf*, Nature **378**, 463 (1995)
41. Nelson, R.: *Extrasolar Planets: A Review of Current Observations and Theory, Solar and Extra-Solar Planetary Systems*. Lect. Notes Phys. **577**, 35–53 Splinger, Heidelberg (2001)
42. R. Neuhäuser et al.: *Evidence for a co-moving sub-stellar companion of GQ Lup*, Astron. Astrophys. **435**, L13 (2005)
43. Y. Oasa et al.: *A deep near-infrared survey of the chamaeleon I dark cloud core*, Astrophys. J. **526**, 336 (1999)
44. B. R. Oppenheimer et al.: *Infrared spectrum of the cool brown dwarf GL:229B*, Science **270**, 1478 (1995)
45. D. L. Padgett et al.: *The SPITZER c2d survey of weak-line T Tauri stars. I. initial results*, Astrophys. J. **645**, 1283 (2006)
46. D. B. Paulson et al.: *Searching for planets in the hyades. V. Limits on planet detection in the presence of stellar activity*, Astron. J. **127**, 3579 (2004)
47. F. Pepe et al.: *The HARPS search for southern extra-solar planets. VIII. μ Arae, a system with four planets*, Astron. Astrophys. **462**, 769 (2007)
48. C. R. Proffitt et al.: *Limits on the optical brightness of the ε eridani dust ring*, Astrophys. J. **612**, 481 (2004)
49. B. Reipurth and C. Clarke: *The formation of brown dwarfs as ejected stellar embryos*, Astron. J. **122**, 432 (2001)
50. B. Sato et al.: *A planetary companion to the G-type giant star HD 104985*, Astrophys. J. **597**, L157 (2003)
51. B. Sato et al.: *A planetary companion to the hyades giant ε tauri*, Astrophys. J. textbf661, 527 (2007)
52. S. Seager et al.: *Photometric light curves and polarization of close-in extrasolar giant planets*, Astrophys. J. **540**, 504 (2000)
53. M. Shao: *SIM: the space interferometry mission*, Proc. SPIE **3350**, 536 (1998)
54. B. A. Smith and R. J. Terrile: *A circumstellar disk around beta pictoris*, Science **226**, 1421 (1984)
55. D. N. Spergel: *A new pupil for detecting extrasolar planets*, astro-ph 0101142 (2001)
56. K. R. Stapelfeldt et al.: *A hubble space telescope imaging survey of T Tauri stars*, IAUS. **221**, 276 (2003)
57. J. R. Stauffer et al.: *Spitzer space telescope observations of G dwarfs in the pleiades: circumstellar debris disks at 100 Myr age*, Astron. J. **130**, 1834 (2005)
58. L. M. Stepp and S. E. Strom: *The thirty-meter telescope project design and development phase*, Proc. SPIE **5382**, 67 (2004)
59. D. Udalski et al.: *The optical gravitational lensing experiment: The discovery of three further microlensing events in the direction of the galactic bulge*, Astrophys. J. **426**, L69 (1994)
60. S. Udry et al.: *The HARPS search for southern extra-solar planets. XI. Super-Earths (5 and 8 M_E) in a 3-planet system*, Astron. Astrophys. **469**, L43 (2007)

61. S. Urakawa et al.: *An extrasolar planet transit search with subaru suprime-cam*, Publ. Astron. Soc. Japan **58**, 869 (2006)
62. A. Vidal-Madjar et al.: *An extended upper atmosphere around the extrasolar planet HD209458b*, Nature **422**, 143 (2003)
63. G. A. H. Walker et al.: *A search for Jupiter-mass companions to nearby stars*, Icarus **116**, 359 (1995)
64. M. Walmsley: *ALMA: the atacama large millimeter array*, Mem. Soc. Astron. It. **71**, 889 (2000)
65. A. Wolszczan and D. A. Frail: *A planetary system around the millisecond pulsar PSR1257+12*, Nature **355**, 145 (1992)
66. N. Woolf and R. Angel: *Astronomical searches for earth-like planets and signs of Life*, Ann. Rev. Astron. Astrophys. **36**, 507 (1998)
67. T. Yamashita et al.: *Upper limits to the CO J = 1-0 emission around Vega-like stars – Gas depletion of the circumstellar ring around Epsilon Eridani*, Astrophys. J. **402**, L65 (1993)
68. H. Zinnecker et al.: *Search for giant extrasolar planets around white dwarfs: direct imaging with NICMOS/HST and NACO/VLT*, IAUC. **200**, 19 (2006)

Index